Behaviour and Ecology of Riparian Mammals

Many mammals, such as otters, mink and water shrews live in close association with rivers and streams, feeding in them or using them as a place of safety or means of escape from predators. The distinct adaptations that riparian mammals have evolved in order to live in these environments also handicap them for living elsewhere. They are therefore threatened by alterations to their environment. In recent years, our rivers have become highly polluted, and with bankside modifications for agriculture and forestry, enhanced or decreased water flow, and use for recreation, they become less and less suitable for these highly specialized animals. This book looks at the habitat utilization, adaptation, feeding ecology and conservation status of a range of riparian mammals, and will give insights into the problems facing these fascinating animals, and how they might be overcome.

NIGEL DUNSTONE is a lecturer in Zoology at the University of Durham. He is the author of *The Mink* (1993), and has co-edited *The Exploitation of Mammal Populations* (1996) with V. J. Taylor.

MARTYN L. GORMAN is Senior Lecturer in Zoology at the University of Aberdeen. He is the author of *Island Ecology* (1979) and *The Natural History of Moles* (1990) with D. Stone, and has also co-edited a previous book, *Mammals as Predators* (1993), with Nigel Dunstone.

T0275600

Symposia of the Zoological Society of London

The series *Symposia of the Zoological Society of London* was originally established in 1960 and published the invited contributions to international meetings held by the Society to explore a wide variety of zoological topics.

The series has been published by Cambridge University Press since 1997, and it is now evolving to focus particularly on conservation biology, while continuing to include volumes on other zoological topics. With the addition of specially commissioned contributions to augment those arising from the Society's meetings, an integrated, comprehensive and authoritative treatment of the subject is ensured.

Symposia of the Zoological Society of London 71

Behaviour and Ecology of Riparian Mammals

Edited by

Nigel Dunstone and Martyn L. Gorman

CAMBRIDGE
UNIVERSITY PRESS

CAMBRIDGE UNIVERSITY PRESS
Cambridge, New York, Melbourne, Madrid, Cape Town, Singapore, São Paulo

Cambridge University Press
The Edinburgh Building, Cambridge CB2 8RU, UK

Published in the United States of America by Cambridge University Press, New York

www.cambridge.org
Information on this title: www.cambridge.org/9780521631013

© Cambridge University Press 1998

First published 1998
This digitally printed version 2007

A catalogue record for this publication is available from the British Library

Library of Congress Cataloguing in Publication data
Behaviour and ecology of Riparian mammals / edited by Nigel Dunstone and Martyn Gorman.
     p.   c.m. — (Symposia of the Zoological Society of London; 71)
     Based on papers from a conference held in London in 1995.
     ISBN 0 521 63101 7
     1. Riparian animals – Behaviour – Congresses. 2. Riparian animals – Ecology – Congresses. 3.
Mammals – Behaviour – Congresses.
4. Mammals – Ecology – Congresses. I. Dunstone, N. (Nigel) II. Gorman, M. L. (Martyn L.) III.
Zoological Society of London. IV. Series: Symposia of the Zoological Society of London; no. 71.
QL1.Z733 no. 71
[QL141]
590 s–dc21   97-32129 CIP
[599.168]

ISBN 978-0-521-63101-3 hardback
ISBN 978-0-521-03807-2 paperback

Contents

Contributors

J. Aeschimann, Institut de Zoologie et d'Ecologie animale, Université de Lausanne, CH-1015 Lausanne, Switzerland.

P.J. Bacon, Institute of Terrestrial Ecology, Hill of Brathens, Glassel, Banchory, A31 4BY, Scotland.

G.R. Barreto, Wildlife Conservation Research Unit, Department of Zoology, University of Oxford, Oxford OX1 3PS, UK.

H.C. Biggs, Scientific Services, Kruger National Park, Private Bag X402, Skukuza 1350, South Africa.

C. Bodmer, Institut de Zoologie et d'Ecologie animale, Université de Lausanne, CH-1015 Lausanne, Switzerland.

D.N. Carss, Institute of Terrestrial Ecology, Hill of Brathens, Glassel, Banchory, A31 4BY, Scotland.

B.C. Choudhury Wildlife Institute of India, Post Box 18, Dehra Dun. 248 001, India.

S. Churchfield, Division of Life Sciences, King's College London, Campden Hill Road, London W8 7AH, UK.

J.W.H. Conroy, Institute of Terrestrial Ecology, Hill of Brathens, Glassel, Banchory, A31 4BY, Scotland.

M. Delibes, Doñana Biological Station (CSIC), PO Box 1056, Sevilla 41080, Spain.

N. Dunstone, Department of Biological Sciences, University of Durham, South Road, Durham City, DH1 3LE, UK.

P.R. Evans, Department of Biological Sciences, University of Durham, South Road, Durham City, DH1 3LE, UK.

M.L. Gorman, Department of Zoology, University of Aberdeen, Aberdeen AB9 2TN, Scotland.

E.A. Herrera, Departamento de Estudios Ambientales, Universidad Simón Bolívar, Apartado 89.000, Caracas 1080-A, Venezuela.

S.A. Hussain, Wildlife Institute of India, Post Box 18, Dehra Dun. 248 001, India.

D.J. Jefferies, The Vincent Wildlife Trust, 10 Lovat Lane, London EC3R 8DT, UK.

C. Jones, Department of Zoology, University of Aberdeen, Aberdeen AB9 2TN, Scotland.

H. Kruuk, Institute of Terrestrial Ecology, Hill of Brathens, Glassel, Banchory, A31 4BY, Scotland.

D.W. Macdonald, Wildlife Conservation Research Unit, Department of Zoology, University of Oxford, Oxford OX1 3PS, UK.

T. Maran, Wildlife Conservation Research Unit, Department of Zoology, University of Oxford, Oxford OX1 3PS and EMCC, Tallinn Zoo, 1 Paldiski Road 145, Tallinn EE0035, Estonia.

G. McLaren, Department of Zoology, University of Aberdeen, Aberdeen AB9 2TN, Scotland.

K.C. Nelson, Institute of Terrestrial Ecology, Hill of Brathens, Glassel, Banchory, A31 4BY, Scotland.

S. Palazón, Departament de Biologia Animal (Vertebrats), Facultat de Biologia, Universitat de Barcelona, Avda Diagonal 645, 08028, Spain.

P.A. Racey, Department of Zoology, University of Aberdeen, Aberdeen, AB9 2TN, Scotland.

D.T. Rowe-Rowe, Natal Parks Board, PO Box 662, Pietermaritzburg 3200, South Africa. *Present address:* 97, Frances Staniland Road, Pietermaritzburg 3201, South Africa.

V.V. Rozhnov, A.N. Severtsov Institute of Evolutionary Morphology and Animal Ecology, 117017 Moscow, Russia.

J. Ruiz-Olmo, Servei de Proteccio i Gesto de la Fauna, Direcció General del Medi Natural, Avda Corts Catalanes 612, 08037 Barcelona, Spain.

C. Schenk, Frankfurt Zoological Society, Alfred-Brehm-Platz 16, 60316 Frankfurt, Germany.

V. **Sidorovich**, Wildlife Conservation Research Unit, Department of Zoology, University of Oxford, Oxford OX1 3PS *and* Institute of Zoology, Academy of Sciences of Belarus, F. Skoriny Str. 27, Minsk, 20072 Belarus.

M.J. **Somers**, Department of Zoology, University of Stellenbosch, Private Bag X1, 7602 Matieland, South Africa.

M. **Spreng**, Institut de Zoologie et d'Ecologie animale, Université de Lausanne, CH-1015 Lausanne, Switzerland.

C. **Strachan**, The Vincent Wildlife Trust, 10 Lovat Lane, London EC3R 8DT, UK.

R. **Strachan**, Wildlife Conservation Research Unit, Department of Zoology, University of Oxford, Oxford OX1 3PS, UK.

E. **Staib**, Frankfurt Zoological Society, Alfred-Brehm-Platz 16, 60316 Frankfurt, Germany.

T.J. **Thom**, Department of Biological Sciences, University of Durham, South Road, Durham City, DH1 3LE, UK.

C.J. **Thomas**, Department of Biological Sciences, University of Durham, South Road, Durham City, DH1 3LE, UK.

P.C. **Viljoen**, Scientific Services, Kruger National Park, Private Bag X402, Skukuza1350, South Africa. *Present address*: Tanzania Wildlife Conservation Monitoring, PO Box 14935, Arusha, Tanzania.

P. **Vogel**, Institut de Zoologie et d'Ecologie animale, Université de Lausanne, CH-1015 Lausanne, Switzerland.

T.M. **Williams**, Department of Biology, Earth and Marine Science Building, A316, University of California, Santa Cruz, CA 95063, USA.

Preface

Many species of mammals live in close association with rivers and streams, often feeding in them and frequently using them as a place of safety and as a means of escape from predators. Modifications of species for swimming and diving may involve such profound anatomical and physiological specialization that they are at a relative disadvantage to terrestrial species when on land. Indeed, the degree of aquatic specialization exhibited may be a corollary of the frequency with which an animal needs, on occasion, to return to a terrestrial life. The mammal exploiting a semi-aquatic lifestyle often suffers other limitations in addition to those consequent upon its habitat-specific adaptations. For example, in recent years many of our rivers have become highly polluted, draining as they do a highly modified landscape where bankside modification for agriculture/forestry, enhanced water flow, or recreation, has further reduced their suitability for the highly specialized animals that inhabit them.

In November 1995 a conference on the subject of the Behaviour and Ecology of Riparian Mammals was held at the Meeting Rooms of the Zoological Society in London. The idea behind the symposium was to bring together scientists from a range of disciplines but all researching the behaviour and ecology of a group of mammals united by their adherence to a semi-aquatic lifestyle. It was hoped that such a comparative approach would lead to a clearer understanding and would be of value to those charged with the task of undertaking their conservation and management. Four main themes of research were identified: adaptation, habitat utilization, feeding ecology and conservation status.

In his opening chapter Dunstone reviews the adaptations of a diverse range of semi-aquatic mammals, from their anatomical, sensory and respiratory modifications to the energetics of their foraging strategies. Recent investigations into adaptations for an aquatic lifestyle have focused largely on the comparative approach to the energetics of swimming behaviour and the demands that the environment imposes upon thermoregulatory capacity. Williams provides evidence from a comparative study of swimming energetics that attempting to balance the physiological responses of obligate terrestrial and fully aquatic marine mammals is energetically costly for semi-aquatic mammals. Vogel *et al.* examine in detail the diving and foraging behaviour of the water shrew, *Neomys fodiens*, and show that foraging economics can also affect the dive endurance of these poorly adapted mammals.

The habitat utilization and preferences of a range of riparian species, including the water shrew (Churchfield), bats (Racey) and otters *Lutra lutra* (Kruuk *et al.*), are addressed. Despite their small size, these animals forage year-round in thermally challenging streams. In the UK some species of microchiropteran bats can be heavily dependent on emerging aquatic insects. Comparative studies of otters in riverine and coastal habitats suggest that they spend most time in areas with a high biomass of prey, and that bankside vegetation is a poor predictor of otter abundance. Many species of riparian mammal are critically dependent on an abundance of aquatic prey. Ruiz-Almo demonstrates that the decrease in abundance of *Lutra lutra* at increasing altitudes in north-east Spain is related to the lowered availability of prey rather than to other habitat features. An investigation by Carss *et al.* of prey selection by the European otter in streams and in lochs, where the fish communities differ, and a study on the diet of the smooth-coated otter (*Lutra perspicillata*) in India by Hussain & Choudhury found that in both cases prey were taken in proportion to their abundance.

However, co-existence of a number of riparian species might be expected to lead to resource partitioning. This was examined by Rowe-Rowe & Somers for three species of African otters (*Aonyx capensis, A. congica* and *Lutra maculicollis*) and the water mongoose (*Atilax paludinosus*), and for an assemblage of semi-aquatic carnivores (*Mustela lutreola, M. vison, Lutra lutra* and *Mustela putorius*) in Belarus by Sidorovich *et al.* The relative sizes, degree of adaptation, foraging strategy employed and differing degree of dependence on aquatic foraging reduced, to a certain extent, competitive interaction between these sympatric species.

Analysis of the population biology of the European otter by Gorman *et al.*, of hippopotami (*Hippopotamus amphibius*) by Viljoen & Biggs, and of capybaras (*Hydrochoeris hydrochaeris*) by Herrera highlight the various problems experienced by these species. The European mink (*Mustela lutreola*) is one of the most endangered species of mammal and its populations are threatened by habitat destruction and persecution throughout its range (Palazón & Ruiz-Olmo). Competition with other carnivores, particularly the introduced American mink (*Mustela vison*) is an important factor in the decline of this species (Maran *et al.*). Human activities have detrimentally affected the status of the giant otter (*Pteronura brasiliensis*) throughout its range (Schenck & Staib). The purity of the aquatic environment and the effect of this on prey populations that directly affect otter populations is demonstrated by Ruiz-Olmo *et al.* For the water vole (*Arvicola terrestris*), on the other hand, it is habitat modification and the presence of an introduced predator, the American mink, that have been of most significance in the decline of the species.   N.D.   M.L.G.

1

Adaptations to the semi-aquatic habit and habitat

N. Dunstone

Introduction

It has been asked by opponents of such views as I hold, how for instance, could a land carnivorous animal have been converted into one with aquatic habits; for how could the animal in its transitional state have subsisted?

Charles Darwin (1859), *The Origin of Species*

Mammals exhibit a fascinating array of adaptations that suit them to their chosen habit and habitats. Many species show an association with water but the aquatic environment is a challenging one and presents many problems for those mammals that have chosen to utilize it. Hence, many species show compromise in the extent of their adaptations to amphibious life. Eisenberg (1981) uses the term semi-aquatic to include those species that must spend part of each 24 hour period out of the water. I use the term 'semi-aquatic' mammals to <u>exclude</u> those species where the association with water is to a large extent obligatory; such a distinction is somewhat arbitrary, since even the most highly evolved semi-aquatic mammals (e.g. pinnipeds) retain some dependence on a terrestrial substrate for part of their life, and many species of otter (Lutrinae) seem to be inseparably tied to waterways despite being incompletely adapted. Semi-aquatic mammals are phylogenetically diverse; representatives are found in several mammalian orders, including Monotremata, Marsupialia, Insectivora, Artiodactyla, Carnivora and, most commonly, Rodentia. Species from some 24 families of mammals have an association with aquatic habitats, although they vary in the extent to which they treat water as a medium in which to forage or to escape from predators or simply to traverse.

In this chapter I discuss the limitations that the semi-aquatic habit imposes, and then review the literature on adaptations to the aquatic environment, particularly as it pertains to those mammals living along riverine ecosystems. A range of mammalian species that vary in their dependence on the aquatic medium are considered. As certain topics will be dealt with in greater detail elsewhere in this volume I will confine my review to include selected topics on thermoregulation, sensory biology (in particular visual and tactile perception), respiratory physiology and feeding behaviour including foraging strategies that serve to optimize underwater hunting behaviour.

Previous authors have emphasized the importance of the family Mustelidae in illustrating the evolutionary sequence from terrestrial through semi-aquatic adaptations (Estes, 1989; Fish, 1993*a*). Charles Darwin (1859) in *The Origin of Species* recognized the equivocal position of the mink, (*Mustela vison*), in the defence of his evolutionary theory, which required that animals could evolve from one form to another in gradual steps, while each intermediate grade must be well adapted to its place in nature. I shall frequently use the mink as a useful comparison with other semi-aquatic species since it appears to be incompletely adapted to aquatic foraging and has been widely studied in many aspects of its aquatic hunting behaviour (Dunstone, 1993).

The aquatic environment as a hunting environment

The physical properties of the aquatic medium dictate much of the form and function observed in aquatic mammals. Water offers more resistance to movement than air does, because its density and viscosity are greater than those of air (800 and 30 times greater, respectively). Water yields when it is pushed against, which leads to a loss of energy as an animal attempts to effect propulsion. As a result movement through water would be expected to impose severe limitations on the speed and energetic performance of swimming mammals. Despite all this there are advantages; aquatic locomotion can be the most economical means of transport. The density of water approaches that of body tissue, allowing for almost neutral buoyancy, hence little energy need be expended in supporting the body in the water column during locomotion.

An animal's design is acknowledged to have a major impact on its ecological performance, whilst the environment itself may be one of the major influences on an animal's design. The constraints of foraging, in what to a terrestrial animal must be considered an alien environment, have led to considerable evolutionary convergence (Howell, 1930) and to the selection of morphological, physiological and behavioural adaptations for swimming and diving efficiency. It is not unexpected that the degree to which particular species show such adaptations correlates positively with the extent to which the aquatic medium is used for foraging, predator evasion or general transport, and negatively with the necessity for efficiency e.g. in locomotion on land. Not surprisingly there is a sequence of transitional stages displayed by mammal species within the range from fully terrestrial to fully aquatic (Howell, 1930; Fish, 1992).

Energetics and metabolism

Studies of small mammals have generally assumed that environmental factors are of great, if not overriding, importance in the evolution of body shape. For example, in terrestrial mammals a more spherical body shape is assumed to minimize surface area and thus heat loss (Brown & Lasiewski, 1972). The elongated shape of mustelids, a design for the capture of prey in confined spaces, must constitute a very successful evolutionary strategy as it has been retained by nearly all members of the family, despite their diverse range of habits, over considerable evolutionary time. But it has had profound consequences for much of their behaviour and lifestyle, particularly for those such as mink and otters that have become aquatic.

Thermoregulation

Although some of the species of semi-aquatic mammal considered in subsequent chapters of this volume come from tropical regions, e.g. the giant river otter, (*Pteronura brasiliensis*), many hunt in cold northern waters, and must therefore compensate for the drain of heat from their bodies by increased insulation, metabolic heat production and a decreased surface-area to volume ratio. For several small semi-aquatic mammals, that is, those of less than 1 kg body weight, foraging time is limited by cold water temperature (Kruuk *et al.*, 1994).

Semi-aquatic mammals have a number of short-term behavioural options for reducing the energetic costs of foraging in cold water. These include shuttling between a terrestrial and aquatic environment, retreating to a burrow and postural adjustments. Such behaviour alleviates the continual loss of heat to a medium colder than the animal's own deep body temperature. Small semi-aquatic mammals living in the northern temperate zone also have to contend with the need to conserve heat when on land. No completely aquatic mammal is very small, which reflects the problem of heat loss. Small semi-aquatic mammals (e.g. water shrew (*Neomys fodiens*), desman (*Galemys pyrenaicus*), an aquatic tenrec (*Limnogale mergulus*): see Stephenson, 1994) all tend to feed on arthropods and occasionally small fish – foods that require the minimum of processing to provide a high calorific return. Intermediate-sized semi-aquatic mammals (otters) are almost always piscivorous whilst large semi-aquatic mammals (hippopotamus (*Hippopotamus amphibius*), capybara (*Hydrochoeris hydrochaeris*)) eat plants. Their larger size and concomitantly reduced metabolic rate allow them to take advantage of plant foods despite the loss of assimilation efficiency.

With little or no body fat to provide insulation, semi-aquatic mammals rely on dense waterproof fur, rather than blubber as in many marine mammals, as an insulator. Blubber actually has a lower insulative value than fur (Costa & Kooyman, 1982), but it has several advantages. The fur of many semi-aquatic mammals has a silvery appearance underwater due to the presence of trapped fur. This provides semi-aquatic mammals with a considerable degree of buoyancy, additional to that provided by the lungs. The air layer trapped in the fur compresses with depth, reducing its effectiveness – although this may not be a huge problem for semi-aquatic mammals since most are shallow divers. The alternative insulator, blubber, would cause the semi-aquatic mammal to be very cumbersome on land. Blubber may also serve as an energy reserve and help in controlling blood flow to the surface and heat loss from it. The problem of inefficient insulation is, of course, exacerbated by the generally small body size (and hence relatively large surface area) of the semi-aquatic mammals. In order to achieve good thermal properties the fur has to be conditioned, and this involves grooming between individual dives or sequences of dives, which can be an important component of the time budget. The dense fur of many northern species of mammal is made up of two types of hair, guard hairs and under-fur. The under-fur produces a dense, matted, felt-like layer, which forms an efficient insulating layer by trapping air next to the skin, providing protection against the low northern temperatures. The same fur that insulates against the cold also provides, for semi-aquatic mammals, a high degree of water repellency. A violent shake of the body when the animal leaves the water is all that is required to fluff up the coat, allowing it to regain its insulative property. Aquatic mammals have a moult pattern markedly different from that of their terrestrial relatives, which tends to be a more gradual process (Ling, 1970).

The fur of the water shrew is even more highly specialized, but it can lose its water-repellent properties even as a result of stress in captivity. Under normal conditions the fur has a hydrophobic property considered unique amongst aquatic mammals. If the fur remains dry whilst under water then the water shrew can maintain its body temperature, but if it becomes wet the body temperature can drop at a rate of 1.1 °C per minute (Köhler, 1991). This important property is maintained by the animal squeezing water out of the fur and renewing the electrostatic energy of the awn (a type of underfur) hair, usually by entering a constricting tunnel on emergence from the water (Köhler, 1991).

Many aquatic mammals are carnivores and their diet may be the primary cause of an elevated metabolic rate. Heat production may be increased periodically by bouts of activity or by energy generated from food digestion. For mustelids weighing 1 kg or more, the basal metabolic rate is 20% higher than for terrestrial mammals of equivalent size (Iversen, 1972). It is because they need the energy resources to sustain such a metabolic rate that aquatic forms are generally larger than their terrestrial relatives. There may have been a shortening of the body length of some semi-aquatic mammals (e.g. muskrat (*Ondatra zibethicus*)) but this may have taken place for hydrodynamic reasons. However, dietary constraints may also apply, since many of the larger semi-aquatic mammals are herbivorous (e.g. capybara and hippopotamus), with larger gut volumes than those of carnivores leading to greater bulk than can be accommodated in a streamlined shape.

A trend has been identified for aquatic vertebrates usually to maintain resting metabolic rate and core body temperature at levels higher than expected from their body mass, possibly to compensate for the high rates of heat loss experienced in water. This relationship may not be uniform throughout the Mammalia since neither the water opossum (*Chironectes minimus*) of South America (Thompson, 1988) nor an aquatic tenrec (*Limnogale mergulus*) from Madagascar (Stephenson, 1994) has an elevated rate when compared to terrestrial forms.

Locomotion

In general, aquatic mammals display a wide variety of locomotory adaptations due to their diverse evolutionary histories and performance requirements. The biomechanical demands of swimming through water and terrestrial locomotion are vastly dissimilar. To cope with these demands, divergent trends have evolved across a wide range of amphibious and aquatic mammals in the degree of specialization of their appendages. An excellent review of this subject area is given by Fish (1993*a*). Cetaceans and sirenians totally abandoned the terrestrial environment and swim by undulation of the body and tail. Pinnipeds maintain their link with the land, although to an extremely limited extent, and swim using paired limbs that are modified as flippers.

The overall body plan of marine mammals tends to be spindle-shaped, which allows for maximum hydrodynamic efficiency. In comparison, semi-aquatic mammals appear slow and inefficient in the water and are restricted to shallow freshwater bodies and near-shore marine environments. Hydrodynamic factors differ considerably between marine and semi-aquatic mammals. For pinnipeds and cetaceans propulsion is lift-based. This is generated by the forelimbs in otariids, whilst undulatory movements of the hind limbs and body are used by phocids.

The evolution of semi-aquatic mammals from a terrestrial towards a fully aquatic existence necessitated morphological and behavioural changes with regard to locomotion. The swimming associated with semi-aquatic mammals originated from quadrupedal paddling. Modification of the limbs for paddling is readily accomplished from terrestrial locomotion. The aquatic locomotion of semi-aquatic mammals tends to be drag-based and is usually generated from alternate dorsoventral movements of the hind limbs (Fish, 1993a).

Many semi-aquatic mammals are very inefficient swimmers: the small surface area of the paws does not provide adequate and/or efficient thrust generation, nor is the alternating use of the paired limbs as effective as the synchronous thrusts produced by better adapted species such as the otter.

The webbed appendages characteristic of many aquatic species enhance the surface area available for thrust generation during swimming. This is usually more pronounced in the hind limbs. The mink appears to be an exception since its paws have remained almost web-less, a condition more typical of high-speed terrestrial runners. The limbs are short, with partial webbing between the toes (but little more than that found in the more terrestrial polecat, and considerably less than that of the otter). The surface area of the mink's foot is relatively small, which suggests that they are adapted for locomotion on land rather than underwater (Dunstone, 1993).

The switch from a terrestrial gait to bipedal paddling in swimming mammals could not have occurred without effective control of stability. Whereas large aquatic mammals achieve and maintain buoyancy through increased body fat and large lungs, it is the non-wettable fur of semi-aquatic mammals, through trapped air, that provides most buoyancy. Without this buoyancy aid such mammals would assume a more vertical position in the water and this would increase drag, as has been shown by Fish (1993b) for the water opossum.

The costs of locomotion can be estimated by determining rates of oxygen uptake and carbon dioxide production. It is estimated for the mink that a five-to tenfold reduction in costs is achieved when it swims underwater rather than on the surface (Williams, 1983). This is due to a reduction in the drag of the poorly streamlined mink body as it ploughs through the water. Whilst swimming underwater the mink commonly makes use of solid structures as a base to push off from, giving additional propulsion (Poole & Dunstone, 1976).

The increase in hydrostatic pressure associated with diving to depth may also be important since this will compress the air in the lungs, leading to increased uptake of lung oxygen stores by the blood. The reduced lung volume will also reduce buoyancy, further lowering the cost of locomotion, since the mink will have to expend less energy in order to stay submerged at a particular depth.

Sensory adaptations to hunting underwater

The terrestrial predatory behaviour of mustelids has been described (see Dunstone, 1993); vision and olfaction are important for the detection and pursuit of prey. The considerable contribution of aquatic prey to the diet necessitates different sensory adaptations. When one considers the special case of an amphibious mammal hunting underwater, vision becomes to some extent obligatory, even though the opacity, spectral shift and low contrast of water will impair vision at anything but close range. Whereas olfaction and audition may be of use when hunting on land, these senses are of limited use underwater. Tactile information can provide a useful alternative in very close encounters or may act as a substitute when murky water precludes the use of vision.

Adaptations for vision underwater

The optical difficulty of focusing in both air and water is considered to be one of the primary environmental influences on the adaptive radiation of the vertebrate eye (Walls, 1942). Vision underwater is confounded by the similar refractive indices of the cornea and water. As a result the cornea can no longer contribute to the focusing power of the eye. Since, in terrestrial animals, the cornea is the principal refracting surface of the eye, underwater such an eye would focus the image behind the retina, resulting in blurred vision. Submergence thus causes longsightedness (hypermetropia). To maintain its acuity in water, the eye of an amphibious mammal must have greater focusing or dioptric power. In optical terms this means a higher lens curvature. Aquatic mammals have also evolved a variety of mechanisms to overcome this problem.

Whales, dolphins and porpoises have adopted the 'fish-eye lens' solution by evolving a spherical lens of high curvature and hence high dioptric power (Walls, 1942). However, this causes problems when these animals, albeit rarely, need to see in air. Seals and sea lions also use this method but manage to achieve reasonably acute vision in air by using a narrow pupil that functions like a pin-hole camera. As a result their aerial vision deteriorates markedly with decreasing light levels.

Another approach has been adopted by a number of semi-aquatic species, including birds such as the dipper (*Cinclus mexicanus*; Goodge, 1960) and the cormorant (*Phalacrocorax* sp.), and the otter, involving the use of a well developed sphincter iridis muscle to compress the outer edge of the lens in order to produce an area of high curvature and hence powerful focusing ability (Walls, 1942).

Ballard *et al.* (1989) have compared the development of the intraocular muscles in the eyes of the Canadian river otter (*Lutra canadensis*) and the Canadian beaver (*Castor canadensis*). The beaver was chosen for comparison with the otter because its herbivorous habits suggest that it is not as dependent on acute underwater vision. These authors determined that the otter may be capable of some 54 diopters of accommodation; by comparison young primates, which are acknowledged to have the greatest accommodative ability of any terrestrial mammals, could only produce 10 diopters.

Using this method of accommodation the otter achieves equivalent levels of visual acuity in air and water, at least in bright light (Balliet & Schusterman, 1971). Histological studies on the mink eye suggest that it may also employ this method. Certainly the development of the iris sphincter muscle in the mink is greater than in the ferret, although less than in the otter (Dunstone, 1993).

Vibrissae

Vibrissae are specialized tactile hairs that have a rich nerve supply in comparison to other hairs on the body. They are an important source of environmental information for many mammals, especially those with crepuscular or nocturnal habits. Because of this nerve supply vibrissae are extremely sensitive, not only to touch but also to the direction and extent of bending. It is likely that they provide information on velocity as the animal moves through its habitat. Fine-tuning of movement at close quarters is also afforded, and may be particularly important in the capture and manipulation of prey.

Green (1977) has shown that otters whose vibrissae have been trimmed show a reduced efficiency at catching fish prey. The otter utilizes its vibrissae to detect underwater prey movement, probably through turbulence. Like the otter, the mink's muzzle is surrounded by stiff whiskers. The mink's reliance upon these structures for the detection of prey in murky water and at night has been investigated in a series of laboratory experiments with mink searching underwater in the dark for submerged plastic objects. Preliminary experiments have shown mink to take longer to locate objects when their vibrissae had been shaved off. Their ability to distinguish between the objects of different shapes was also impaired when vibrissae were absent (unpublished results).

It has been suggested that vibrissae may act as a sonar net in the detection of vibrations from aquatic invertebrates. Richard (1981, 1982) has examined the tactile sensitivity of the Pyrenean desman (*Galemys pyrenaicus*) using a series of discrimination trials of objects that differ in surface relief; in preliminary trials differences in fine detail of under 0.07 mm could be discriminated.

Other senses

The electrical sense is occasionally found in the lower aquatic vertebrates, particularly fishes, where active or passive systems may be employed, and is also characteristic of the larvae of urodele amphibians, but not anurans. Electrosensitivity has been searched for in a number of semi-aquatic mammals but, so far, demonstrated only in the platypus. Scheich *et al.* (1986) used both behavioural and physiological tests to localize the sensory organs in the highly specialized bill. High frequency sensitivity to alternating current could allow the detection of muscular activity from prey, i.e. small crustaceans. Such prey, e.g. shrimp, have been shown to generate electrical discharges by the tail flick. These were detectable at up to 50 cm and would aid detection by the platypus when probing in the mud and stones. In addition the platypus's ability to avoid vertical barriers carrying a d.c. charge placed in its path was demonstrated. Overall, the experiments showed that the platypus can make use of a.c. and d.c. fields in prey detection and object avoidance.

Schlegel & Richard (1992) investigated this ability in the desman (*Galemys pyrenaicus*), a nocturnal animal with very reduced eyes that occupies a similar feeding niche to the platypus. However, they failed to demonstrate sensitivity to any stimuli of the intensity likely to be encountered in the natural environment.

Respiratory biology

Most semi-aquatic mammals dive with their lungs full of air, and often only with some considerable difficulty due to the buoyancy that must be overcome. During underwater hunting, air-breathing mammals must voluntarily suspend ventilation for an extended period. The capacity to store oxygen in the lungs and to bind it chemically in the blood haemoglobin and muscle myoglobin may need to be increased; the tolerance of particular tissues to toxic chemicals may need to be modified. These chemicals include lactic acid, which builds up during exercise under conditions of low oxygen. The ability to restrict peripheral blood supply to the skin by vasoconstriction allows sensitive tissues, i.e. heart and brain, to maintain an adequate supply of oxygen. It is doubtful whether this mechanism occurs in shallow-diving species.

Although lung volume increases with body size across many mammal species, there is no evidence to show that the lung volume of amphibious mammals is proportionately larger than that of their terrestrial relatives. Kruuk (1993) has demonstrated that there is a clear relationship between log mean body weight and log mean dive time across semi-aquatic and aquatic species of mammal.

The physiology of diving has been well studied (see Harrison *et al.*, 1968; Kooyman, 1973; Harrison, 1974, for reviews). Early studies of the physiology of diving mammals involved forced submersion of the subjects as the main technique of study. Under these circumstances it was generally observed, not unexpectedly, that the heart rate dropped dramatically when the animal was immersed. This is referred to as a diving bradycardia. Diving bradycardia has been interpreted as an oxygen-conserving response. However, reduction in heart rate alone would not be sufficient to allow an animal to remain submerged for extended periods. It is of course more appropriate, and humane, to monitor the behaviour of freely-diving, unrestrained animals. For many aquatic species an adequate supply of oxygen should be available in the storage capacity of the lungs, blood and muscles for the entire dive. This is particularly likely to be the case for shallow-diving animals that remain submerged for only a short time. Laboratory estimation of oxygen storage capacities and oxygen consumption rates by respirometry have shown that most animals dive for a shorter period than would exhaust their stored oxygen reserves and cause them to switch to anaerobic metabolism (Estes, 1989). This suggests that anaerobic metabolism may be reserved for emergency situations.

Studies conducted by Stephenson *et al.* (1988) monitored the heart rate and oxygen consumption of mink by using small implanted radio-transmitters during normal foraging dives in a large laboratory respirometer. When swimming in a familiar tank the animals did not show any reduction of heart rate during normal shallow (0.3 m) or deep (1.9 m) dives. However, when the animals encountered a novel situation, or were diving in an unfamiliar environment, they showed a reduction in heart rate that may therefore be associated with a fear response. The possibility cannot be excluded that the animals are consciously able to initiate the development of a bradycardia if the situation – for example, sustained pursuit of a fish or escape from a predator – requires it. Other constraints, such as those imposed by foraging strategy, also act to keep dives short (Dunstone & O'Connor, 1979*a*,*b*). But how do these animals decide when to terminate a dive and how can they extend dive times physiologically when circumstances require them to do so?

Aquatic hunting

Dive duration

An extension of optimal foraging theory to model diving behaviour in mammals was presented by Dunstone & O'Connor (1979*a*,*b*). More sophisticated models have since been published by Houston & McNamara (1985) and Houston & Carbone (1992).

Diving mammals can be regarded as central place foragers (Orians & Pearson, 1979) with the surface regarded as the central place. Some air-breathing mammals hunting for prey underwater must bring each item to the surface – so-called single-prey loaders (Houston & McNamara, 1985); others can eat underwater or carry more than one prey item to the surface and are referred to as multiple-prey loaders.

In the case of the diving forager, the upper limit to dive duration is reached when the oxygen supply is exhausted. A given dive duration may also require a subsequent pause at the surface before the next dive can be made, and the ratio of pause duration to dive duration may increase with dive duration as the animal depletes its oxygen reserves during a foraging session. This must be taken as part of the cost of a dive since during this recovery period the animal cannot forage. In economic terms it will be optimal for a diver to surface at some time before the lethal limit in order to replenish its supplies. Butler (1982) reviewed evidence across a large number of species of voluntarily diving animals and showed that they surface well before their theoretical oxygen stores are exhausted.

The time budget of underwater foraging semi-aquatic mammals consists of bouts of surface activity in which oxygen stores are replenished, bouts of feeding or other activity on the bottom and the travel time between these two 'habitats' (Kramer, 1988). The model assumes that the resource is only har-vested on the bottom, where we would expect the animal to attempt to maximize its time. The precise shape of the curve of oxygen gain with surface time will be affected by species-specific physiological and morphological char-acteristics of the respiratory and circulatory system. Upon arrival at the surface the mammal may have to expire before inspiring to replace the oxygen in its lungs, blood and, in some cases, muscle. As these oxygen stores are replenished over the course of several breaths, the partial pressure differentials decline and the rate of oxygen gain is reduced because of lower diffusion rates. The optimal oxygen store may then be less than the maximum (Kramer, 1988).

This leads to some interesting predictions concerning the relationship between breathing behaviour and the distance to sub-surface feeding sites (in most cases this is equivalent to depth). As depth increases, surface times should increase to allow for increased loading of the tissues with oxygen. The amount of oxygen remaining at the end of the dive should be unaffected by distance. Since larger oxygen stores permit longer periods away from the surface, dive times should also increase with increase in depth. However, because the oxygen for longer dives is acquired at a lower average rate, the percentage time at the surface should increase with increasing depth (Kramer, 1988).

Empirical evidence from studies of amphibious hunters provides general support for the hypothesis that dive times should increase with depth (Davies, 1988; Dunstone, 1993; Nolet *et al.*, 1993). Interdive interval (recovery time) also increases with depth in otters (Nolet *et al.*, 1993) but this may be confounded by the animals attempting to maximize their time on the bottom across a series of consecutive diving bouts. This behaviour is particularly likely to be shown by single-prey loaders such as otters, which return to the surface to eat each prey item captured, and even more by poorly adapted semi-aquatic mammals such as mink, which leave the water to eat every prey item captured (Dunstone, 1993).

The requirement to return to the surface at regular intervals is an important constraint on the fishing behaviour of amphibious mammals, in which the limited duration of the oxygen reserves can lead to a negative correlation between the search and pursuit phases of a given hunt: i.e. as the search for a fish is prolonged, the predator's oxygen reserves are further depleted, so that the maximum duration of any pursuit then initiated is steadily reduced (Dunstone, 1978). Such a consideration should necessarily alter the optimal duration of the search phase. We would expect a trade-off between the benefits of continued search and the cost of losing the prey because of oxygen exhaustion during pursuit.

These arguments can be extended to allow for the limited oxygen supply available to the amphibious mammal hunting underwater: any persistence of a search or pursuit manoeuvre demanded on optimization criteria is physiologically feasible only if adequate oxygen reserves are available, otherwise the time-budget of the hunt must be altered to allow for replenishment of the air supply. The disadvantage to any time-limited predator that abandons a search is that the prey may redistribute themselves while the predator is at the surface breathing. They may escape into areas that have already been searched by the predator and that, on searching economics criteria, it should not revisit. Additionally, the predator may be at some risk of disorientation while at the surface and have considerable difficulty relocating a particular foraging patch on subsequent dives (Dunstone, 1978, 1993). One method the mink employs for optimizing pursuit time is to locate aquatic prey from out of the water before diving. In the laboratory, prey location from out of the water was frequently enhanced by the mink peering intently into the tank whilst retaining a tentative grip on land with their hind feet (Poole & Dunstone, 1976).

Water depth
Both otters and mink are shallow-water divers. The reasons why the mink should choose to dive in shallow water are probably related to its poor adaptations, but it is interesting to speculate why otters should also prefer

shallow water, since they are more accomplished divers. Nolet *et al.* (1993) examined three hypotheses that could explain this observation.

1   Optimal breathing hypothesis. Kramer's (1988) model of optimal recovery time argues that oxygen stores are replenished according to a curve of diminishing returns. Assuming that the amount of oxygen remaining at the end of a dive is unaffected by the diving depth, the hypothesis predicts that both time underwater and recovery time should increase with depth. Recovery time should increase more rapidly than underwater time, hence diving efficiency should decrease with depth.

2   Efficient searching hypothesis. This predicts that an unsuccessful dive should be as long as the aerobic breath-holding limit, and underwater time should be irrespective of water depth. Since dives to greater depths will incur greater travel costs, the time available to search the bottom will diminish with water depth, i.e. searching efficiency will decrease with depth.

3   Heat loss hypothesis. Cooling rate is a function of water temperature, and hence the time spent diving is reduced in cold water. Dives to increasing depth cause the insulating air-layer in the pelt to be compressed, reducing its effectiveness.

Both the optimal breathing hypothesis and the efficient searching hypothesis were developed for the case of multiple-prey loaders such as the duck-billed platypus, i.e. species that take more than one prey item per dive and that necessarily must seek to maximize their time at the resource. The river otter is a single-prey loader and its diving behaviour could not be explained by a higher dive efficiency at shallow depths. Searching efficiency was higher because of the decreased travel times from the surface during shallow dives, and hunting bouts were generally longer. Recently, Houston & McNamara (1994) have suggested that the optimal breathing hypothesis may be applicable to single-prey loaders as well. The heat loss hypothesis is supported by a study of the oxygen consumption of the Eurasian otter in relation to body temperature by Kruuk *et al.* (1994), where it was found that the costs of diving in water at 2 °C are some 2.7 times higher than they at 20 °C.

The present results point clearly to the inability of this type of predator to hunt efficiently in open water offering low prey density. On the other hand, structured environments should allow efficient hunting through the greatly increased encounter rates. Streams and small pools are thus ideal habitats.

Habitat structure

In a series of experiments on underwater searching ability, Dunstone (1978) showed that mink seemed able to retain a spatial memory of the topography of the hunting environment, using information provided by the position of refuges for the prey and other habitat features. Successive searches of a tank did not tend to include revisits to areas previously unsuccessfully explored. The provision of refuges for prey caused a redirection of the mink's search effort to these structures where fish could frequently be found. The mink responded to a capture in or near a refuge by increased investigation of that area on subsequent dives. The mink thus quickly learns the places where it is likely to encounter prey rather than searching open water. Presumably this also occurs in nature, with the mink selectively searching areas of its habitat where it has previously encountered prey. When there were no refuges available for the fish to hide in and the fish were restricted to open water, the mink's hunting increased with the increasing number of fish present (Dunstone, 1978).

Even animals considerably better adapted for underwater hunting than the mink, e.g. otter, seem to restrict their aquatic hunting to favoured patches (Kruuk *et al.*, 1990). These authors could not, however, correlate patch use with the availability of landing sites or enhanced prey densities; access to beds of marine algae – where fish lurk – appeared to be the feature of the habitat structure most likely to be related to the availability of prey. Ostfeld (1991) suggest that it is the high mobility of the otter's predominantly demersal fish prey that allows high replenishment rates and hence predictable diving success rates. Other aquatic predators, which are more reliant on sessile or less mobile prey, may experience prey depletion, either temporarily or permanently. Such predators do not then have the ability to choose diving sites guaranteed to produce prey. Ostfeld also points out that the fish taken by otters are rather similar in their behaviour and allow the predator to employ a small range of foraging strategies compared to species with a greater dietary breadth requiring greater behavioural flexibility.

References

Ballard, K. A., Sivak, J. G. & How-land, H. C. (1989). Intraocular muscles of the Canadian river otter and Canadian beaver and their optical function. *Can. J. Zool.* **67**: 469–474.

Balliet, R. F. & Schusterman, R. J. (1971). Underwater and aerial visual acuity in the Asian

"clawless" otter (*Amblonyx cineria cineria*). *Nature, Lond.* **234**: 305–306.

Brown, J. H. & Lasiewski, R. C. (1972). Metabolism of weasels: the cost of being long and thin. *Ecology* **53**: 939–943.

Butler, P. J. (1982). Respiratory and cardiovascular control

during diving in birds and mammals. *J. exp. Biol.* **100**: 195–221.

Costa, D. P. & Kooyman, G. L. (1982). Oxygen consumption, thermoregulation, and the ef-fect of fur oiling and washing on the sea otter, *Enhydra lutris*. *Can. J. Zool.* **60**: 2761–2767.

Darwin, C. (1859). *On the origin of species by means of natural selection, or the preservation of favoured races in the struggle for life.* John Murray, London.

Davies, S. W. (1988). *An Investigation of the Effects of Various Environmental Parameters on the Underwater Foraging Behaviour of the American Mink,* Mustela vison *Schreber.* PhD thesis: University of Durham.

Dunstone, N. (1978). The fishing strategy of the mink (*Mustela vison*); time-budgeting of hunting effort? *Behaviour* 67: 157–177.

Dunstone, N. (1993). *The mink.* T. & A. D. Poyser, London.

Dunstone, N. & O'Connor, R. J. (1979a). Optimal foraging in an amphibious mammal. I. The aqualung effect. *Anim. Behav.* 27: 1182–1194.

Dunstone, N. & O'Connor, R. J. (1979b). Optimal foraging in an amphibious mammal. II. A study using principal component analysis. *Anim. Behav.* 27: 1195–1201.

Eisenberg, J. F. (1981). *The mammalian radiations: an analysis of trends in evolution, adaptation and behavior.* Chicago University Press, Chicago.

Estes, J. A. (1989). Adaptations for aquatic living by carnivores. In *Carnivore behavior, ecology and evolution*: 242–282. (Ed. Gittleman, J. R.). Chapman & Hall, London and Cornell University Press, New York.

Fish, F. E. (1992). Aquatic locomotion. In *Mammalian energetics: interdisciplinary views of metabolism and reproduction*: 34–63. (Eds Tomasi, T. E. &

Horton, R. H.). Comstock Publishing Associates, Ithaca & London.

Fish, F. E. (1993a). Influence of hydrodynamic design and propulsive mode on mammalian swimming energetics. *Aust. J. Zool.* 42: 79–101.

Fish, F. E. (1993b). Comparison of swimming kinematics between terrestrial and semiaquatic opossums. *J. Mammal.* 74: 275–284.

Goodge, W. R. (1960). Adaptations for amphibious vision in the dipper (*Cinclus mexicanus*). *J. Morph.* 107: 79–91.

Green, J. (1977). Sensory perception in hunting otters *Lutra lutra* L. *Otters* 1977: 13–16.

Harrison, R. J. (Ed.) (1974). *Functional anatomy of marine mammals* 2. Academic Press, London and New York.

Harrison, R. J., Hubbard, R. C., Peterson, R. S., Rice, C. E. & Schusterman, R. J. (Eds) (1968). *The behavior and physiology of pinnipeds.* Appleton-Century-Crofts, New York.

Howell, A. B. (1930). *Aquatic mammals.* C. C. Thomas, Springfield, IL.

Houston, A. I. & Carbone, C. (1992). The optimal allocation of time during the diving cycle. *Behav. Ecol.* 3: 255–265.

Houston, A. I. & McNamara, J. M. (1985). A general theory of central place foraging for single-prey loaders. *Theoret. Popul. Biol.* 28: 233–262.

Houston, A. I. & McNamara, J. M. (1994). Models of diving and data from otters: comments on Nolet *et al.* (1993). *J. Anim. Ecol.* 63: 1004–1006.

Iversen, J. A. (1972). Basal energy metabolism of mustelids. *J. comp. Physiol.* 81: 341–344.

Köhler, D. (1991). Notes on the diving behaviour of the water shrew, *Neomys fodiens* (Mammalia, Soricidae) *Zool. Anz.* 227: 218–228.

Kooyman, G. L. (1973). Respiratory adaptations in marine mammals. *Am. Zool.* 13: 457–468.

Kramer, D. L. (1988). The behavioral ecology of air breathing by aquatic animals. *Can. J. Zool.* 66: 89–94.

Kruuk, H. (1993). The diving behaviour of the platypus (*Ornithorhynchus anatinus*) in waters with different trophic status. *J. appl. Ecol.* 30: 592–598.

Kruuk, H., Wansink, D. & Moorhouse, A. (1990). Feeding patches and diving success of otters, *Lutra lutra*, in Shetland. *Oikos* 57: 68–72.

Kruuk, H., Balharry, E. & Taylor, P. T. (1994). Oxygen consumption of the Eurasian otter *Lutra lutra* in relation to water temperature. *Physiol. Zool.* 67: 1174–1185.

Ling, J. K. (1970). Pelage and moulting patterns in wild animals with special reference to aquatic forms. *Q. Rev. Biol.* 45: 16–54.

Nolet, B. A., Wansink, D. E. H. & Kruuk, H. (1993). Diving of otters (*Lutra lutra*) in a marine habitat: use of depths by a single-prey loader. *J. Anim. Ecol.* 62: 22–32.

Orians, G. H. & Pearson, N. E. (1979). On the theory of central-place foraging. In *Analysis*

of ecological systems: 155–177.
(Eds Horn, D. J., Mitchell, R.
D. & Stairs, G. R.). Ohio State
University Press, Columbus,
OH.

Ostfeld, R. S. (1991). Measuring
diving success of otters. *Oikos*
60: 258–260.

Poole, T. B. & Dunstone, N.
(1976). Underwater predatory
behaviour of the American
mink (*Mustela vison*). *J. Zool.,
Lond.* **178**: 395–412.

Richard, P. B. (1981). La détec-
tion des objets en milleux
aquatique et aérien par le desman des Pyrénées (*Galemys
pyrenaicus*). *Behav. Processes* **6**:
145–159.

Richard, P. B. (1982). La sen-
sibilité tactile de contact chez le
desman *Galemys pyrenaicus.*

Biol. Behav. **7**: 325–336.

Scheich, H., Langner, G.,
Tidemann, C., Coles, R. B. &
Guppy, A. (1986). Elec-
troreception and electroloca-
tion in platypus. *Nature, Lond.*
319: 401–403.

Schlegel, P. A. & Richard, P. B.
(1992). Behavioral evidence
against possible subaquatic
electrosensitivity in the
Pyrenean desman *Galemys
pyrenaicus* (Talpidae, Mammalia). *Mammalia* **56**: 527–
532.

Stephenson, P. J. (1994). Resting
metabolic rate and body temperature in the aquatic tenrec
Limnogale mergulus
(Insectivora: Tenrecidae). *Acta
theriol.* **39**: 89–92.

Stephenson, R., Butler, P. J.,
Dunstone, N. & Woakes, A. J.
(1988). Heart rate and gas exchange in freely diving American mink (*Mustela vison*). *J.
exp. Biol.* **134**: 435–442.

Thompson, S. D. (1988). Thermoregulation in the water
opossum (*Chironectes
minimus*): an exception that
'proves' a rule. *Physiol. Zool.*
61: 450–460.

Walls, G. L. (1942). *The vertebrate
eye and its adaptive radiation.*
Cranbrook Press, Bloomfield
Hills, NY.

Williams, T. M. (1983). Locomotion in the North American
mink – a semi-aquatic mammal. I. Swimming energetics
and body drag. *J. exp. Biol.* **103**:
155–168.

2

Physiological challenges in semi-aquatic mammals: swimming against the energetic tide

T.M. Williams

Introduction

Foraging in both terrestrial and aquatic habitats exposes secondarily aquatic mammals to a range of environmental conditions not usually encountered by mammalian specialists. This is due in part to the very different physical properties of air and water. Depending on temperature and salinity, the density of water approximates 800 times that of air and its viscosity is nearly 60 times that of air. The heat capacity of water is approximately 3500 times higher than air and its heat conductivity 24 times greater. Consequences of these physical properties are different locomotor and thermoregulatory demands for runners and swimmers (Dejours, 1987). Remarkably, semi-aquatic mammals have maintained lifestyles that require locomotor and thermoregulatory proficiency in these two disparate environments.

Morphological specialization is one obvious way in which mammals have responded to the challenges presented by terrestrial and aquatic environments. Several notable trends include body streamlining and enlargement of the propulsive surfaces in animals specialized for aquatic locomotion (Fish, 1993). Comparison of terrestrial, semi-aquatic and marine mammals (Fig. 2.1) shows that both features change in a predictable manner. Fineness ratio, an index of body streamlining (Hertel, 1966; Webb, 1975), approaches the optimum value of 4.5 as we move from secondarily aquatic mammals to obligate marine mammals (Fish, 1993; see Fig. 2.1). Many obligate aquatic mammals also show a reduction in limb length and expansion of the propulsive surface area. In contrast, modification of the appendages of semi-aquatic mammals may be limited by the dual role they must serve.

Many aquatic mammals also display specialized thermoregulatory morphologies. Heat dissipation and maintenance of a stable core body temperature are especially challenging for semi-aquatic mammals that must move between the thermally disparate media of air and water. Semi-aquatic mammals such as beavers, muskrats and otters maintain exceptionally dense pelage for insulation in water (Kenyon, 1969; Estes, 1989). Obligate marine mammals includ-

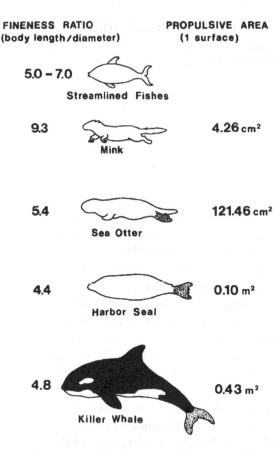

FINENESS RATIO
(body length/diameter)

PROPULSIVE AREA
(1 surface)

5.0 – 7.0   Streamlined Fishes

9.3   Mink   4.26 cm²

5.4   Sea Otter   121.46 cm²

4.4   Harbor Seal   0.10 m²

4.8   Killer Whale   0.43 m²

Figure 2.1. Fineness ratio and propulsive surface area for animals differing in aquatic specialization. Degree of specialization increases in the mammals from top to bottom of the illustration. Fineness ratio was determined from morphological measurements for animals resting in air (mink, sea otter and seal) or floating in water (whale). Propulsive area was determined by tracing the appendages used in a single stroke cycle: both front paws (mink), both hind flippers (sea otter), single surface of the hind flipper (seal) or fluke (killer whale).

ing pinnipeds and cetaceans rely on a thick blubber layer to conserve heat (Williams *et al.*, 1992; T.M. Williams unpublished data). To circumvent the insulating layer during periods of high heat production or environmental temperatures, aquatic mammals use 'thermal windows' – i.e. the sparsely haired appendages and tails of semi-aquatic mammals (Fish, 1979; Fanning & Dawson, 1980) and the large surface areas of the dorsal fin and flukes of marine mammals (Scholander & Schevill, 1955; Pabst *et al.*, 1995). These areas often contain specialized vascular arrangements that promote heat transfer or heat conservation as needed.

In contrast to morphological studies, there is little information about the trends in physiological specialization for aquatic living by mammals. In view of this, the present study evaluated the physiological responses of semi-aquatic mammals and determined the energetic consequences of locomotor and thermal specialization. A comparative approach was used in which the energetics of this group was compared to values for obligate terrestrial and marine mammals. By examining animals that vary in aquatic specialization we were able to assess physiological trends leading to energetic efficiency by mammals in the aquatic environment.

Methods

Regional heterothermy

Patterns of heterothermy were determined from measurements of skin temperature for five to seven anatomical sites in four species of mammals. Sites included central and peripheral areas appropriate for each species (Fig. 2.2(a)–(d)). Subjects included adult domestic dogs (*Canis familiaris*, $n = 5$) measured at $T_{air} = 23.8\,°C$, mink (*Mustela vison*, $n = 3$) at $T_{water} = 24.6\,°C$ (from Williams, 1986), California sea lions (*Zalophus californianus*, $n = 2$) at $T_{water} = 20.9\,°C$, and bottlenose dolphins (*Tursiops truncatus*, $n = 5$) at $T_{water} = 29.4\,°C$. Measurements were made using a thermocouple (Physiotemp, Inc.) or thermistor (Thermonetics, Inc.) probe placed on the skin of the quiescent animal either resting in air (dogs) or in water (mink, sea lion and dolphin). The probes were calibrated against a digital thermometer in an insulated water bath before and after the experimental period. The digital thermometer had been calibrated against a National Bureau of Standards mercury thermometer and agreed to within $0.1\,°C$ of the probe values over the experimental range of temperatures. Dolphins and sea lions were trained to remain in a stationary floating position next to a deck while the skin temperatures were taken on submerged sites. All measurements were made at acclimation ambient temperatures for the animals.

Resting metabolic rate

With the exception of the dolphins and sea lions, values for resting metabolic rate were obtained from the literature. Criteria for data selection were based on conformation to physiological conditions outlined by Kleiber (1975) for interspecific comparisons of metabolic rate. Specifically, the animals were mature, post-absorptive and quiescent and were measured in a metabolically neutral environmental temperature.

Metabolic measurements on California sea lions ($n = 3$) and bottlenose dolphins ($n = 2$) were carried out as described by Williams *et al.* (1991) and (1992), respectively. The dolphins were housed in salt water pens exposed to ambient ocean temperatures in San Diego (mean $T_{water} = 15.4\,°C$) during the experimental period. The sea lions were housed in salt water pools at the

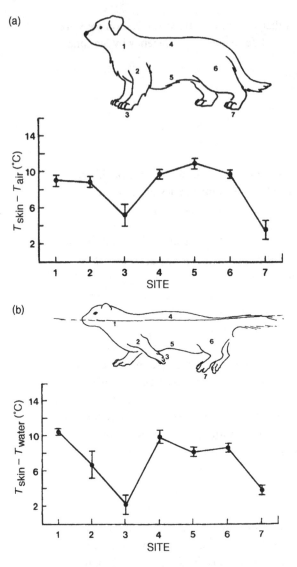

Figure 2.2. Regional skin temperatures for (a) dogs resting in air and for (b) mink, (c) sea lions and (d) dolphins resting in water. All values are presented as the differential between skin temperature and ambient air or water temperatures. Symbols and vertical lines are mean ±1 SE. Numbers on the abscissa correspond to the anatomical area measured as shown in the illustration. See the text for ambient temperatures and *n* for each species.

Scripps Institution of Oceanography (La Jolla, CA). The pools were maintained at local ocean temperatures (mean $T_{water} = 18.0\,°C$). Briefly, the two adult bottlenose dolphins were trained to rest in a water-filled metabolic box. Water temperature in the box was controlled by a salt water heat exchanger and ranged from 3.6 to 17.3 °C. Resting metabolic rates of the sea lions were determined while the animals floated beneath a metabolic hood in a water channel. Water temperature in the channel was varied from 5.0 to 20.0 °C.

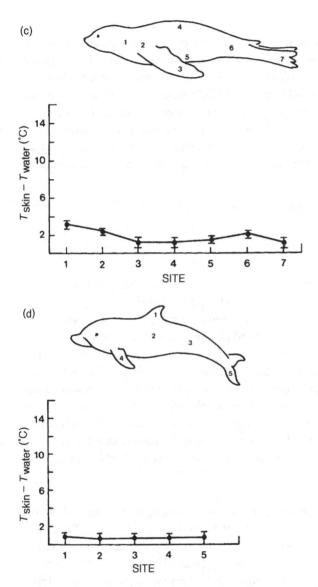

Oxygen consumption ($\dot{V}O_2$) for both species was measured using open flow respirometry systems (Williams, 1987) and calibrated using the nitrogen dilution method described by Fedak *et al.* (1981). Maintenance costs were calculated as the lowest resting metabolic rates maintained for at least 10 min at thermally neutral water temperatures. These costs represent the energy expended for endothermy and basal functions in the alert, resting animal at thermal neutrality.

Cost of transport

The total cost of transport (COT) for terrestrial, semi-aquatic and marine mammals was obtained from the literature where $COT = \dot{V}O_2/locomotor$ speed. Allometric relationships relating COT to body mass were from Taylor *et al.* (1982) for runners, Williams (1989) for semi-aquatic birds and mammals and Williams (1996) for marine mammals. Locomotor costs (LC) were calculated from the difference between COT and maintenance costs. LC is equivalent to the net cost of transport presented by Schmidt-Nielsen (1972) and represents the additional energy expended by animals for running or swimming.

Results

Regional heterothermy

Skin temperature patterns of the mammals varied with degree of aquatic specialization (Fig. 2.2). Regional skin temperatures for mink resting in water were nearly identical in pattern to temperatures measured for dogs standing in air. Like the mink, the dogs showed plantar skin temperatures 3–4 °C above ambient temperature while the more central sites averaged 10 °C above ambient temperature. Conversely, there was little evidence of regional heterothermy in the marine mammals. Skin temperatures of California sea lions and bottlenose dolphins were nearly uniform across peripheral and central sites at acclimation water temperatures. Rather than maintaining skin temperatures near core temperature, these aquatic mammals showed skin temperatures within 1–3 °C of T_{water} for all body sites measured (Fig. 2.2(c) and (d)).

Resting metabolic rate

The metabolic rates of many aquatic mammals resting in water are higher than predicted from allometric regressions for terrestrial mammals (Table 2.1). Mean oxygen consumption of the sea lions and dolphins resting in water was 6.33 (± 1.05) mlO_2 kg^{-1} min^{-1} at $T_{water} = 20.0\,°C$ and 6.99 (± 0.15) mlO_2 kg^{-1} min^{-1} at $T_{water} = 15.6\,°C$, respectively. These values were 1.5 to 2.4 times the predicted values from Kleiber (1975). The comparatively high metabolic rate of these marine mammals was similar to results reported for semi-aquatic mammals. Except for the beaver measured during the summer (MacArthur & Dyck, 1990), metabolic rates for six species of semi-aquatic mammal were 1.7 – 4.5 times predicted levels (Table 2.1). In general, the greatest discrepancies between measured and predicted values were reported for semi-aquatic mammals examined at water temperatures representing winter conditions. Among

Table 2.1. *The ratio between measured[†] and predicted[‡] metabolic rates for semi-aquatic and marine mammals resting in water*

Species	Body mass (kg)	Measured metabolic rate / Predicted metabolic rate	Reference
Semi-aquatic			
Muskrat (*Ondatra zibethicus*)	0.87	1.9 (summer)	Fish (1979)
Mink (*Mustela vison*)	0.97	1.8	Williams (1986)
Water rat (*Hydromys chryogaster*)	1.0	2.3 (summer) 4.2 (winter)	Dawson & Fanning (1981)
River otter (*Lutra lutra*)	8.5	1.7 (summer) 4.5 (winter)	Kruuk *et al.* (1994)
Beaver (*Castor canadensis*)	6.4–12.2	1.2 (summer) 1.8 (winter)	MacArthur & Dyck (1990)
Sea otter (*Enhydra lutris*)	20.0	2.4	Williams (1989)
Marine			
California sea lion (*Zalophus californianus*)	67.3	2.1–3.7	Liao (1990); and present study
Harp seal (*Phoca groenlandica*)	141.0	1.0	Gallivan & Ronald (1979)
Bottlenose dolphin (*Tursiops truncatus*)	145.0	1.9	Present study

[†] Measured values are from the cited references.
[‡] Predicted values are from Kleiber (1975) and Schmidt-Nielsen (1984) and are based on body mass.

marine mammals only the harp seal showed the predicted metabolic rate for a resting mammal (Gallivan & Ronald, 1979).

Cost of transport
The total cost of transport (COT) for swimming in semi-aquatic mammals was four to five times higher than costs determined for obligate terrestrial and marine mammals (Fig. 2.3). Like marine mammals, semi-aquatic mammals showed high maintenance costs in comparison to terrestrial mammals. As a result, a large fraction of total transport costs consisted of maintenance costs in both groups of swimming mammals. Maintenance costs accounted for less

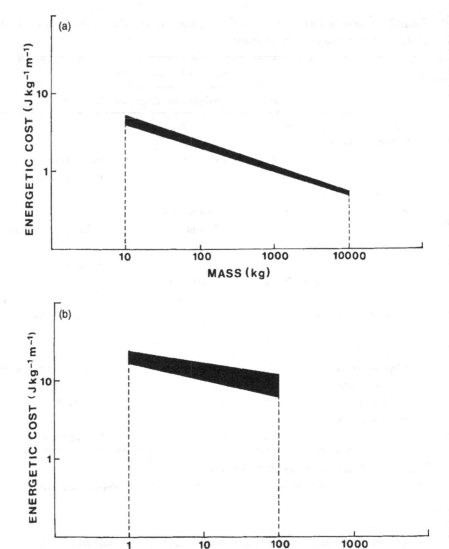

Figure 2.3. Comparison of energetic costs for terrestrial (a), semi-aquatic (b) and marine (c) mammals. Total transport costs, delineated by the top of each box, are from allometric regressions by Taylor *et al.* (1982) for runners, Williams (1989) for semi-aquatic birds and mammals and Williams (1996) for marine mammals. Total costs are subdivided into locomotor costs (open area) and maintenance costs (shaded areas). Maintenance costs were determined from resting metabolic rates (see the text). Note the elevation in both locomotor and maintenance costs for semi-aquatic mammals in comparison to terrestrial mammals.

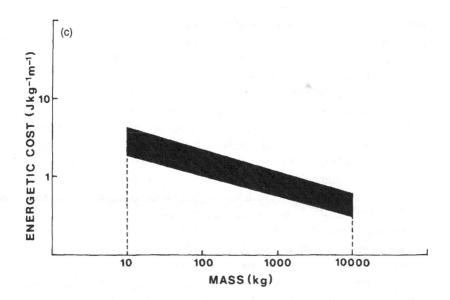

than 22% of COT in similarly sized terrestrial mammals. These maintenance costs, in addition to comparatively high locomotor costs, contributed to the elevated total cost of transport observed in semi-aquatic mammals (Fig. 2.3).

Discussion

Energetic costs of semi-aquatic swimming

The lifestyle of semi-aquatic mammals necessitates physiological mechanisms that can respond to the diverse physical demands of the terrestrial and aquatic environments. To account for this, the thermoregulatory and locomotor responses of semi-aquatic mammals share characteristics found in both obligate terrestrial and marine mammals (Figs. 2.2–2.3). One consequence of maintaining dual roles for physiological mechanisms, however, appears to be elevated energetic costs when in water.

Increased heat production is a well-documented thermoregulatory response for many aquatic mammals resting in water (Irving, 1973; Whittow, 1987; Estes, 1989). A survey of semi-aquatic and marine mammals revealed a general trend towards elevated resting metabolic rates regardless of level of aquatic specialization (Table 2.1). With few exceptions, both semi-aquatic and marine mammals demonstrated metabolic rates that were 1.7–4.5 times the predicted level for terrestrial mammals (Kleiber, 1975). Although basal metabolic rates for submerged marine mammals may approach the levels predicted for terrestrial mammals (Lavigne *et al.*, 1986; Innes & Lavigne, 1991; Hurley, 1996), it appears that many aquatic mammals show elevated metabolic rates when breathing at the water's surface. As a result, maintenance costs form a larger proportion of the total energetic cost of transport for semi-aquatic and marine mammals than they do for terrestrial mammals (Fig. 2.3).

In addition to high maintenance costs, semi-aquatic mammals demonstrate high locomotor costs (Fig. 2.3). This is due largely to surface swimming postures. One of the most important factors for reducing the cost of aquatic locomotion is position of the swimmer relative to the water surface. Swimming position has a profound effect on body drag, and hence energetic cost. Theoretically, if a mammal swims on or near the water surface body drag will be four to five times greater than if it is able to swim three body diameters below the surface (Hertel, 1966; Blake, 1983). Comparative measurements of total drag on surfaced and submerged sea otters (Williams, 1989), humans and harbour seals (Williams & Kooyman, 1985) confirm the importance of remaining submerged for minimizing body drag. Consequently, mammals specialized for aquatic activity often display adaptations for prolonging the period of submergence during swimming.

Many semi-aquatic mammals, like their terrestrial counterparts, swim primarily on the water surface. This results in elevated body drag and energetic costs to the swimmer. For example, surface swimming mink (Williams, 1983), humans (Holmer, 1972) and sea otters (Williams, 1989) incur transport costs that are four to five times the predicted values for marine mammals (Fig. 2.3(b)). This elevation in cost is nearly identical to the theoretical four- to five-fold increase in drag associated with a surface swimming position. Propulsive efficiency and streamlining undoubtedly contribute to the differences in transport costs between semi-aquatic and marine mammals (Fish, 1993). However, the ability to swim submerged appears to be a key adaptation for reducing energetic costs during aquatic activity in mammals.

Regional heterothermy can act as an energy-conserving mechanism for semi-aquatic mammals by allowing the skin temperature of thinly furred body regions to approach ambient temperatures. This in turn reduces the thermal gradient for heat loss (Irving & Krog, 1955). The present study shows that the use of regional heterothermy depends on the degree of aquatic specialization in mammals. At acclimation temperatures heterothermy was pronounced in terrestrial and semi-aquatic mammals, but absent in marine mammals (Fig. 2.2). Differences in insulation coincident with aquatic specialization most probably contributed to these patterns.

The thick blubber layer typical of marine mammals would make locomotion and heat transfer difficult for semi-aquatic mammals, especially during periods of high activity on land. To avoid this, semi-aquatic mammals balance thermal demands by relying on fur insulation coupled to thermal windows in sparsely furred areas. As found for river otters (Kruuk, 1995), muskrats (Fish, 1979), beavers (MacArthur & Dyck, 1990), Australian water rats (Fanning & Dawson, 1980) and mink (Williams, 1986), the resulting regional hetero-thermy enables the animal to balance the rate of heat loss against a labile core body temperature so as to prolong the duration of aquatic activity.

In general, the results from this study demonstrate that aquatic activity is energetically expensive for semi-aquatic mammals when compared to mammals specialized for either terrestrial or marine living. Flexible physiological responses enable semi-aquatic mammals to respond to the thermoregulatory and locomotor demands of terrestrial and aquatic environments. The energetic price for this versatility includes elevated maintenance and locomotor costs leading to a cost of transport that is the highest measured for any vertebrate group.

ACKNOWLEDGEMENTS

This work was supported by a UCSC Core Research grant and an Office of Naval Research grant R&T 4101-282. All experimental procedures were evaluated and approved according to animal welfare regulations specified under NIH guidelines. I thank the graduate students of UCSC and J. Estes for insightful discussions and comments on the manuscript. I also thank the many people and their animal charges who assisted on this project including D. Goley (Lucy), M. Zavanelli (Frodo), D. Costa (Sam) and D. Cepeda (Gunner). W. Hurley, J. Hurley and the Diving Physiology Group at Long Marine Laboratory provided invaluable patience and assistance with the sea lions and dolphins. Finally, the staff of the Dolphin Experience provided a unique research opportunity for taking temperature measurements on diving dolphins.

References

Blake, R. W. (1983). Energetics of leaping in dolphins and other aquatic animals. *J. mar. biol. Ass. U. K.* **63**: 61–70.

Dawson, T. J. & Fanning, F. D. (1981). Thermal and energetic problems of semi-aquatic mammals: a study of the Australian water rat, including comparisons with the platypus. *Physiol. Zool.* **54**: 285–296.

Dejours, P. (1987). Water and air physical characteristics and their physiological consequences. In *Comparative physiology: life in water and on land*: 3–11. (Eds Dejours, P., Bolis,

L., Taylor, C. R. & Weibel, E. R.). Springer Verlag, New York.

Estes, J. A. (1989). Adaptations for aquatic living by carnivores. In *Carnivore behaviour, ecology and evolution*: 242–282. (Ed. Gittleman, J. L.) . Chapman & Hall, London and Cornell University Press, Ithaca, NY.

Fanning, F. D. & Dawson, T. J. (1980). Body temperature variability in the Australian water rat, *Hydromys chrysogaster*, in air and water. *Aust. J. Zool.* **28**: 229–238.

Fedak, M. A., Rome, L. & Seeherman, H. J. (1981). One-step-N_2 dilution technique for calibrating open-circuit O_2 measuring systems. *J. appl. Physiol.* **51**: 772–776.

Fish, F. E. (1979). Thermoregulation in the muskrat (*Ondatra zibethicus*): the use of regional heterothermy. *Comp. Biochem. Physiol. (A).* **64**: 391–397.

Fish, F. E. (1993). Influence of hydrodynamic design and propulsive mode on mammalian swimming energetics. *Aust. J. Zool.* **42**: 79–101.

Gallivan, G. J. & Ronald, K. (1979). Temperature regulation in freely diving harp seals (*Phoca groenlandica*). *Can. J. Zool.* **57**: 2256–2263.

Hertel, H. (1966). *Structure, form, movement.* Reinhold Publishing Corporation, New York.

Holmer, I. (1972). Oxygen uptake during swimming in man. *J. appl. Physiol.* **33**: 502–509.

Hurley, J. A. (1996). *Metabolic Rate and Heart Rate during Trained Dives in Adult California Sea Lions.* PhD thesis: University of California, Santa Cruz.

Innes, S. & Lavigne, D. M. (1991). Do cetaceans really have elevated metabolic rates? *Physiol. Zool.* **64**: 1130–1134.

Irving, L. (1973). Aquatic mammals. In *Comparative physiology of thermoregulation. 3. Special aspects of thermoregulation*: 47–96. (Ed. Whittow, G. C.) . Academic Press, London and New York.

Irving, L. & Krog, J. (1955). Temperature of the skin in the Arctic as a regulator of heat. *J. appl. Physiol.* **7**: 355–364.

Kenyon, K. W. (1969). *The sea otter in the eastern Pacific Ocean.* Government Printing Office, Washington, DC. (*N. Am. Fauna* No. 69: 1–352).

Kleiber, M. (1975). *The fire of life: an introduction to animal energetics.* (2nd edn). R. E. Krieger Publ. Co., Huntington, NY.

Kruuk, H. (1995). *Wild otters: predation and populations.* Oxford University Press, Oxford.

Kruuk, H., Balharry, E. & Taylor, P. T. (1994). Oxygen consumption of the Eurasian otter *Lutra lutra* in relation to water temperature. *Physiol. Zool.* **67**: 1174–1185.

Lavigne, D. M., Innes, S., Worthy, G. A. J., Kovacs, K. M., Schmitz, O. J. & Hickie, J. P. (1986). Metabolic rates of seals and whales. *Can. J. Zool.* **64**: 279–284.

Liao, J. A. (1990). *An Investigation of the Effects of Water Temperature on the Metabolic Rate of the California Sea Lion.* MSc thesis: University of California, Santa Cruz.

MacArthur, R. A. & Dyck, A. P. (1990). Aquatic thermoregulation of captive and free-ranging beavers (*Castor canadensis*). *Can. J. Zool.* **68**: 2409–2416.

Pabst, D. A., Rommel, S. A., McLellan, W. A., Williams, T. M. & Rowles, T. K. (1995). Thermoregulation of the intra-abdominal testes of the bottlenose dolphin (*Tursiops truncatus*) during exercise. *J. exp. Biol.* **198**: 221–226.

Scholander, P. F. & Schevill, W. E. (1955). Counter-current vascular exchange in the fins of whales. *J. appl. Physiol.* **8**: 279–282.

Schmidt-Nielsen, K. (1972). Locomotion: energy cost of swimming, flying, and running. *Science* **177**: 222–228.

Schmidt-Nielsen, K. (1984). *Scaling: why is animal size so important?* Cambridge University Press, Cambridge.

Taylor, C. R., Heglund, N. C. & Maloiy, G. M. O. (1982). Energetics and mechanics of terrestrial locomotion. I. Metabolic energy consumption as a function of speed and body size in birds and mammals. *J. exp. Biol.* **97**: 1–21.

Webb, P. W. (1975). Hydrodynamics and energetics of fish propulsion. *Bull. Fish. Res. Bd Can.* **190**: 1–158.

Whittow, G. C. (1987). Thermoregulatory adaptations in marine mammals: interacting effects of exercise and body mass. A review. *Mar. Mammal Sci.* **3**: 220–241.

Williams, T. M. (1983). Locomotion in the North American mink, a semi-aquatic mammal. I. Swimming energetics and body drag. *J. exp. Biol.* **103**: 155–168.

Williams, T. M. (1986). Thermoregulation of the North American mink during rest and activity in the aquatic environment. *Physiol. Zool.* **59**: 293–305.

Williams, T. M. (1987). Approaches for the study of exercise physiology and hydrodynamics in marine mammals. In *Approaches to marine mammal energetics*: 127–145. (Eds Huntley, A. C., Costa, D. P., Worthy, G. A. J. & Castellini, M. A.) . Society for Marine Mammalogy, Lawrence, KS. (*Spec. Publ. Soc. mar. Mammal.* No. 1.)

Williams, T. M. (1989). Swimming by sea otters: adaptations for low energetic cost locomotion. *J. comp. Physiol.* (*A*) **164**: 815–824.

Williams, T. M. (1996). Physiological specialization dictates cost efficient locomotion in mammals. *Physiologist* **39**: A-27 (Abstract).

Williams, T. M. & Kooyman, G. L. (1985). Swimming performance and hydrodynamic characteristics of harbor seals, *Phoca vitulina. Physiol. Zool.* **58**: 576–589.

Williams, T. M., Kooyman, G. L. & Croll, D. A. (1991). The effect of submergence on heart rate and oxygen consumption of swimming seals and sea lions. *J. comp. Physiol.* (*B*) **160**: 637–644.

Williams, T. M., Haun, J. E., Friedl, W. A., Hall, R. A. & Bivens, L. W. (1992). Assessing the thermal limits of bottlenose dolphins: a cooperative study by trainers, scientists, and animals. *IMATA Soundings* **3**: 16–17.

3

Diving capacity and foraging behaviour of the water shrew (*Neomys fodiens*)

P. Vogel, C. Bodmer, M. Spreng and J. Aeschimann

Introduction

With regard to semi-aquatic mammals, Schröpfer & Stubbe (1992) distinguished three riparian guilds: the herbivores with the water vole and the beaver; the megacarnivores with the mink and the otter; and the macrocarnivores with water shrews and desmans. Among water shrews, the evolution of aquatic foraging behaviour occurred several times: *Nectogale* and *Chimarrogale* in Asia, several species of the genus *Sorex* in America, and *Neomys* in Eurasia (Churchfield, 1990). The fairly common European water shrew *N. fodiens* is the best known. However, the reports on the degree of adaptation to the water habitat are conflicting. Therefore some important findings from the literature are reviewed in this introduction, whereas new data are presented in the following sections.

The swimming locomotion of water shrews was analysed by Ruthardt & Schröpfer (1985) and Köhler (1991), and the related morphological adaptation were reviewed by Hutterer (1985) and Churchfield (this volume pp. 49–51). They obviously present a compromise between the requirements for activity on land and in the water.

Thermoregulation is a major problem for semi-aquatic mammals, because heat conductance in water is 25-fold greater than in air (Calder, 1969). According to this author, the body temperature of immersed American *Sorex palustris* dropped by a rate of 2.8 °C per min. However, this may be an experimental artefact, because *Neomys fodiens* can maintain its body temperature at 37 °C during an immersion of 6 min (Vogel, 1990). If heat loss is increased in water, then heat production should also increase to compensate in a well-adapted mammal. Sparti (1989, 1990, 1992) studied the basal metabolic rate, the maximum thermogenic capacity and the minimal heat conduction in air in 13 shrew species. Compared to two other syntopic soricine shrews (*Sorex minutus* and *S. araneus*, Table 3.1), the basal metabolic rate of the water shrew *Neomys fodiens* is not as high and the maximum metabolic rate is not increased, but the minimal thermal conductance in air is clearly lower, revealing

Table 3.1. *Comparison of some parameters (size, metabolic rate, hairs) of two terrestrial shrews (Sorex minutus and S. coronatus) and the water shrew (Neomys fodiens)*

Species	Mass (g)	HBL (mm)	BMR (ml O_2 g^{-1} h^{-1})	BMR % expect.	ADMR (J g^{-1} d^{-1})	ADMR % expect.
Sorex minutus	3.6±0.8	53.2±5.2	8.6±0.5	339%	7227	106%
S. coronatus	9.0±1.0	73.0±3.4	5.7±0.3	289%	4412	102%
Neomys fodiens	13.5±2.4	80.9±3.7	2.9±0.2	163%	3272	92%

Species	Hair density (hairs/4 mm^2)	Hair length (mm)	$\dot{V}O_2$ max. (ml O_2 g^{-1} h^{-1})	$\dot{V}O_2$ max. % expect.	Conductance (ml O_2 g^{-1} h^{-1} $°C^{-1}$)	Conductance % expect.
Sorex minutus	364.0	3.3	28.7±1.4	147%	0.52±0.05	99%
S. coronatus	†442.7	†4.2	20.3±1.2	137%	0.31±0.02	101%
Neomys fodiens	467.4	6.5	17.5±0.7	137%	0.23±0.01	84%

† Only data from the sibling species S. araneus were available.

Mass and head-body length (HBL) from Niethammer & Krapp (1990) and P. Vogel (unpublished results), BMR (basal metabolic rate), $\dot{V}O_2$ max. (maximal metabolic rate) and conductance from Sparti (1989), ADMR (average daily metabolic rate) from Hanski (1984) and Genoud (1985), hair density and hair length from Ivanter (1994). The deviation from the size-related expected value was calculated from the following sources: BMR of mammals from Kleiber (1961), ADMR of Insectivora from Grodzinski & Wunder (1975), $\dot{V}O_2$ max. from Lechner (1978) and conductance from Herreid & Kessel (1967).

Table 3.2. *The percentage frequency of aquatic prey items in the diet of* Neomys fodiens

Author	Aquatic prey	Habitat	Sample size
Niethammer (1977)	91%	Brook	13 stomachs
Niethammer (1978)	89%	Brook	45 stomachs
Illing *et al.* (1981)	94%	Brook	2 stomachs
Churchfield (1984)	50%	Watercress bed	160 faecal pellets
Kuvikova (1985)	53%	Brook	100 digestive tracts
DuPasquier & Cantoni (1992)	92%	Brook	107 faecal pellets

Taxonomic groups that were not clearly attributable to the terrestrial or aquatic habitat were omitted.

better insulation by the fur. This may be due to increased hair length and hair density (Table 3.1) as assessed by Ivanter (1994).

Even in very cold water, heat conductance in *Neomys fodiens* is clearly lower than heat production. This is explained by the fact that the fur of a healthy water shrew remains completely water-repellent, even after half an hour of aquatic activity (Vogel, 1990). This peculiarity, not found in other aquatic mammals, requires an explanation. The special grooved structure of the terminal segment of the curly overhair may play a role (Appelt, 1973; Vogel & Koepchen, 1978; Hutterer & Hürter, 1981; Ducommun *et al.*, 1994), but Köhler (1991) cited the tribo-electrical charge of the fur as a crucial element in keeping the fur dry.

If morphology and physiology give some indications of the degree of specialization, the most important evidence of the adaptation to a riparian habitat comes from field work (Churchfield, 1985). Diet analyses (Table 3.2) suggest that the water shrew catches 50 – 95% of its food in the water. The foraging activity, revealed by radioactive marking, occurs mainly along the watercourse (Lardet, 1987, 1988; Cantoni, 1990, 1993). This is explained by the concentration of food resources in the water habitat (DuPasquier & Cantoni, 1992).

Tracking of radioactively marked shrews does not allow quantification of the ratio between aquatic and terrestrial foraging. According to visual observations in nature, reported by Illing *et al.* (1981), 95% of the time spent in activity occurred on land, though stomach analyses showed that 95% of the prey was aquatic. On the other hand, Schloeth (1980) observed sensational sequences of diving activity, e.g. 96 dives during an activity bout of 50 min in icy water at an ambient temperature of −15 °C, and on another occasion a probable diving depth of 2 m. It would therefore be interesting to know more about the diving capacity of the water shrew. We investigated this problem in captivity in order to answer three questions, namely (1) how deep will a water shrew dive, (2) how often may a shrew dive per day in order to equilibrate its energy budget and (3) which parameters guide a water shrew in its patch choice?

With regard to this last question, shrews seem to be a good experimental model. The capacity of shrews to discriminate between food size has been shown by Barnard & Brown (1981) in tests with the terrestrial common shrew (*Sorex araneus*). The discrimination of patch richness was investigated by Arditi *et al.* (1983) using the terrestrial greater white-toothed shrew (*Crocidura russula*) and by Hanski (1989) with the common shrew. Finally, the ability of the water shrew to learn the position of prey underwater has been shown recently by Köhler (1993). According to optimal foraging theory (Charnov, 1976; Krebs & McCleery, 1984), it can be expected that the water shrew will develop a foraging strategy that takes into account prey profitability. We therefore conducted four foraging experiments in order to test the choices made by water shrews faced with food patches characterized by different prey size, prey density and prey accessibility.

Material and methods

Captivity conditions
The water shrews were captured using Longworth live-traps in the vicinity of Lausanne, Switzerland. They were kept either in wooden cages (50 × 50 × 30 cm) or in PVC containers (100 × 50 × 45 cm) in an outdoor animal room. In indoor conditions, the pelt in the shrews loses its water-repellent characteristics within a few days (Vogel, 1990). Seven centimetres of forest humus and a thick cover of moss served as substrate. The basal diet per day consisted of our standard food for shrews of 7.5 g minced horse meat mixed with some carrots, complemented with the polyvitamin product 'Fleischfresser-Zusatz' (Nafag, Gossau, CH) and about 10 mealworms. The cage contained a small water dish or was connected to a diving aquarium in which some *Gammarus* sp. were introduced daily.

Diving depth
In order to determine maximal diving depth, the cage was set on the first floor of a building, connected to the upper part of a perspex tube (length, 5 m; diameter 7 cm) fixed on the façade (Fig. 3.1) and filled with tap water by a

Figure 3.1. Diving pipe of 5 m length, connected to the case on the first floor. a, cage; b, cylinder for regulation of diving depth; c, tap for water change.

hose. The closed lower end contained two taps, a big one to clean the system every second day and a small one to inject air to furnish oxygen for the *Gammarus* released as prey in the system. The diving level, measured vertically, was regulated by a suspended cylinder closed by a disc on which the *Gammarus* concentrated. The slope of the tube was about 60°. In *Neomys*, the preferred diving angle is not known, but it is certainly not vertical. In natural conditions the water shrew generally starts on the river bank and follows the natural, highly variable slope to the bottom.

The access to the diving tube was opened every evening for about 1 h. The attraction of *Gammarus* stimulated, without delay, an activity period for the shrews. Starting with a depth of 20 cm, the level of the cylinder was lowered every day by 10–30 cm. The achieved diving depth was determined by direct observation or by video recording.

Monitoring the number of dives per day
In this experiment, the wooden cage was connected to an aquarium of $50 \times 30 \times 30$ cm (45 l) by a special perspex tunnel (Fig. 3.2). This tunnel was set under the water level of the aquarium and gave the shrew access to the water body after passing through the infrared beam of an electric photo cell reflected

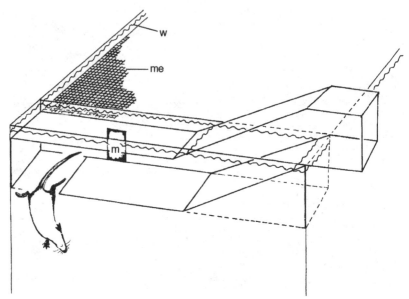

Figure 3.2. Connecting tunnel between the cage and the aquarium. For every diving event, the shrew passes in front of the immersed mirror, m, of the photo cell. The wire mesh, me, prevents the shrew from breathing on the water surface, w, and forces it to swim back through the tunnel.

by a mirror. A wire mesh placed 2 cm under the water surface obliged the shrew to pass through the immersed connecting tunnel to breathe and eat. Entry and exit, separated by a few seconds, were monitored using an event recorder and an appropriate chart speed in order to recognize every diving event by a double signal. A perforated suspended brick gave a certain infra-structure that was accessible for both prey and predator. Prey items (live *Gammarus*) were offered *ad libitum* (> 1000/day).

In order to determine the exact number of *Gammarus* caught in 24 h, one trial using one of the experimental animals was performed. At the onset, 800 *Gammarus* were offered, and in the second half of the day a further 200 *Gammarus* were added twice. For an estimation of the energy gain, a calorimetric equivalent of the *Gammarus* was determined in a bomb calorimeter for this sample. Three replicates of 50 *Gammarus* had a mean wet weight of 1225 mg, and a dry weight of 304 mg with an energetic value of 17.693 kJ/g dry mass, or 107.7 J/*Gammarus*. This value is close to the mean reported by Cummins & Wuycheck (1971).

Foraging strategy

In order to offer choice between two different food patches, the shrew's cage was connected to two aquaria, similar to the one described above, and the number of dives was monitored on both sides by photo cells. For experiments I and II, two aquaria of 45 l (depth 30 cm) were used, and in experiments III and IV, one of these was replaced by an aquarium of 75 l with the same ground surface area (50 × 30 cm), but with a depth of 50 cm. All choice experiments were conducted during April and May 1995 in outdoor conditions with fluctuating ambient temperature. The two water tanks did not show any significant difference in water temperature.

Selection of prey size

The *Gammarus* were caught under stones in Lake Geneva and set on a sieve with a mesh width of 2 mm. *Gammarus* that passed through the mesh were considered to be small prey, *Gammarus* that were retained were considered to be large prey.

Experimental design

EXPERIMENT I The foraging water shrew could choose between an aquarium with 200 large and an aquarium with 200 small *Gammarus*. The test should show whether, at equal prey density, *Neomys* prefer larger, energetically richer prey or smaller prey for which handling time may be shorter.

EXPERIMENT II The foraging water shrew could choose between an aquarium with 150 large *Gammarus* and an aquarium with 350 small *Gammarus*. Both quantities represented the same food volume, hence the same energy available. However, the encounter frequency was more than doubled in the aquarium with the smaller prey.

EXPERIMENT III The foraging water shrew could choose between a shallow (30 cm) and a deep (50 cm) aquarium. At the outset, both patches had a volume of 19 ml *Gammarus* of non-selected prey size. This volume was equivalent to about 400 *Gammarus*. As most of the *Gammarus* were concentrated on the bottom, which was of equal area in both aquaria, the prey density and the offered energy were identical. The patch difference consisted of the diving depth, hence the foraging effort.

EXPERIMENT IV The foraging shrew had to choose between a shallow (30 cm) and a deep (50 cm) aquarium. In the shallow patch a volume of 9.5 ml *Gammarus* was introduced, in the deep patch a volume of 28.5 ml *Gammarus*. The question was whether the increased diving depth was compensated for by a threefold higher encounter rate.

All experiments relating to foraging strategy were conducted with seven or eight shrews with a mean weight of 16.9 g (range 15.3–21.5 g). Every shrew was used for one run in all four experiments. Every run lasted 14 to 24 h. During the experiment, the stock of prey was not renewed and the shrew had access in the terrestrial part of the cage to minced meat.

Data analyses

In order to compare the number of dives per day, the results are expressed as the mean and the standard deviation. The number of animals is expressed as N, the total number of data is expressed as n. In the foraging strategy experiments, the number of dives per patch was summed after 7 and 14 h and at the end of the experiment. The hypothesis H_0 is that the water shrew makes no discrimination between the two patches. According to optimal foraging theory, significant differences should reveal higher profitability of the preferred patch. The number of dives in each patch was compared using a non-parametric Wilcoxon rank test. The limit of significance admitted was $p = 0.05$.

Results

Diving depth

The maximal diving depth of six shrews is shown in Table 3.3. The shrews with wet fur (hair tips wet and stuck together) reached a mean depth of 53 cm, the shrews with dry fur a mean depth of 213 cm. The deepest dive achieved was 260 cm. In this case, the shrew hesitated during the hour of direct observation. According to the video tape, it reached the bottom after midnight and then stopped diving. Obviously, the big diving effort was no longer rewarded by the gain. Dive time (DT), recorded for 25 dives of between 135 and 200 cm depth, gave the following regression: $DT = 1.92 + 0.075 \times depth$ (DT in s, depth in cm, $r = 0.635$, $p = 0.001$). Descent, search time and ascent were monitored during 10 diving events for one shrew to a depth of 150 cm. The descent lasted 5.5 (± 1.5) s, the catching of a prey at the bottom 1.3 (± 0.5) s, the return to the surface 4.2 (± 0.4) s. The mean total duration is thus 12 (± 1.5) s.

Number of dives per day

For these observations three water shrews with good fur quality (fur remaining dry) were used over three periods of 24 h. The mean number of dives per day was 545.6 (± 161.0) ($N = 3$, $n = 9$) with a minimum of 373 and a maximum of 890. This highest score was produced by a female on the day of parturition of a litter of six young.

Table 3.3. *Maximal diving depth achieved by water shrews in the diving pipe*

Shrew	Depth (cm)	State of fur
1	40	Wet
2	50	Wet
3	70	Wet
4	160	Dry
5	220	Dry
6	260	Dry

From the experiments of foraging strategy, lasting from 14 to 24 h, it is possible to extrapolate the number of dives to a period of 1 day. By this estimation, the theoretical daily mean number of dives is 524.9 ($\pm$145.9) ($N = 7$, $n = 28$).

Foraging strategy

The data are presented in Table 3.4. In the first experiment, with the choice being between the same number of large and small *Gammarus*, the preference for the larger prey was significant after 14 h. In the second experiment, with the choice being between the same volume of prey on both sides, but 150 larger versus 350 small prey items, the preference for the higher prey density was significant at the end of the experiment.

It is interesting to compare the mean diving success in the first two experiments, during which the number of prey items was counted at the beginning and at the end of each session (Table 3.5). The data show that the mean success ranged from 0.4 ($\pm$0.3) to 0.8 ($\pm$0.4) with individual scores from 0.1 to 1.4 *Gammarus* per dive. It is significantly higher on the side with small *Gammarus*, even if the number of dives in experiment I is higher on the side with larger prey.

In the third experiment, with the same volume and same number of prey items, but different depth of the aquaria, the shrews showed significant preference, after 7 h, for the shallow aquarium. In the fourth experiment, with three times more prey on the deeper side, the shrews generally started by diving more often on the shallower side. However, all but two of the seven shrews ended up diving more often on the deeper side, although the difference was not statistically significant.

Table 3.4. *Number of dives/shrew after 7 and 14 h and at the end of experiments I to IV in relation to large or small prey and deep or normal aquaria.*

	Experiment I						Experiment II					
	7 h		14 h		End		7 h		14 h		End	
Shrew	Large	Small	Large	Small	Large	Small	Large	Small	Large	Small	Large	Small
1	27	39	102	*113*	171	*191*	36	32	90	*120*	184	*308*
2	14	13	*173*	122	263	206	13	97	99	*390*	307	*602*
3	79	16	*181*	110	213	130	12	18	81	*184*	235	*298*
4	144	96	*207*	134	246	155	105	88	142	*148*	172	*207*
5	171	22	*252*	123	294	184	49	197	80	*269*	155	*317*
6	70	38	*231*	95	367	160	38	27	83	*87*	117	*132*
7	74	70	*214*	142	256	201	43	16	*186*	130	321	259
8							38	23	111	*132*	111	*132*
P	0.06		0.03		0.03		0.89		0.07		0.05	
	N S		S		S		N S		N S		S	

	Experiment III						Experiment IV					
	7 h		14 h		End		7 h		14 h		End	
Shrew	Deep	Norm.	Deep	Norm.	Deep	Norm.	Deep	Norm.	Deep	Norm.	Deep	Norm.
1	51	184	186	*309*	320	*371*	107	78	*240*	106	*330*	147
2	13	18	53	*161*	96	*294*	31	46	86	*198*	127	*215*
3	29	94	97	*227*	114	*300*	27	142	46	*172*	143	*255*
4	66	158	118	*193*	201	*261*	*91*	50	*229*	93	*317*	122
5	25	102	49	*203*	147	*269*	45	109	*126*	124	*230*	131
6	86	91	*181*	167	*284*	270	*175*	132	*373*	171	*483*	246
7	51	62	100	*216*	197	*284*	*111*	71	*305*	191	*479*	219
P	0.02		0.03		0.03		1.00		0.18		0.09	
	S		S		S		N S		N S		N S	

The highest score for each animal under each condition is given in italics. A significant difference in the number of visits (Wilcoxon test) is indicated by S.

Table 3.5. *Mean catching success (prey per dive) at the end of experiment I and II*

	Experiment I		Experiment II	
Prey no.	200	200	350	150
Prey size	Small	Large	Small	Large
Shrew				
1	0.34	0.11	0.36	0.11
2	0.73	0.60	0.45	0.41
3	0.88	0.83	0.48	0.15
4	1.18	0.69	1.35	0.67
5	0.71	0.56	1.00	0.83
6	0.84	0.43	1.47	0.66
7	0.66	0.65	0.73	0.24
8			1.44	0.18
Mean	0.76	0.55	0.83	0.44
S.D.	0.25	0.23	0.45	0.28
P	0.018		0.012	
	S		S	

Prey size and initial number of prey are indicated at the head of the columns. The differences are significant (S) in both experiments (Wilcoxon test).

Discussion

Diving depth

According to the literature, the reported diving depths of the water shrew in captivity are not very great (20–30 cm), but were probably, in most cases, limited by the technical possibilities. Ruthardt & Schröpfer (1985) offered a container with running water and depth increasing from 7 to 25 cm. Köhler (1991) worked with a tank of 28 cm depth. Field observations are therefore more relevant. Illing *et al.* (1981) observed diving behaviour in a creek 40 cm deep. Sanden-Guja (1957) observed, in a year of high population density, shrews diving to the bottom of a river 60–80 cm deep. This observation was erroneously transformed into 8 m in the *Handbuch der Säugetiere Europas* (Spitzenberger, 1990). Schloeth (1980), who observed numerous diving shrews in a mountain river, estimated a maximum diving depth of 2 m. Our experimental investigation shows that the capacity can go beyond this reported maximum. We believe that the record of 2.6 m is not the physiological limit of diving capacity, but it would be difficult to find an attractive stimulus to motivate the shrew for a better performance. The weak maximum of three of the shrews is explained by their wet fur. These experiments were carried out at a time when the optimum maintenance conditions were not known (Vogel, 1990).

Dive times in *Neomys fodiens* depend strongly on experimental or natural conditions. In the usual laboratory conditions, the duration is generally from 3 to 6 s (Churchfield, 1985; Ruthardt & Schröpfer, 1985) with individual maxima of 8.3–15.6 s (Köhler, 1991). In nature, the dive times are generally from 3–10 s (water depth: 30–50 cm) with a maximum of 24 s in water of depth 2 m (Schloeth, 1980). The results from the dive pipe show a linear correlation between dive time and depth, well-known from other semi-aquatic mammals such as the otter (Nolet *et al.*, 1993), mink (Davies, 1988) and the platypus (Kruuk, 1993). However, the dive time will be modulated by catching success. With high prey density in the dive pipe, search time was about 1 s, which is short compared to 11 s for descent and ascent at a depth of 1.5 m, showing the importance of underwater effort to reach the food patch.

Number of dives per day

It is obvious that the number of dives per day will be dependent on the diving conditions and on the feeding conditions. The water tank of 45 l did not present a limiting condition for a shrew's physiological capacity. The fact that the complement of laboratory food (minced meat) presented on land re-mained untouched clearly shows the high preference for the aquatic prey, *Gammarus*.

As a consequence, the mean of 546 dives per day with *Gammarus* offered *ad libitum* in the water may closely reflect the effort required to satisfy the daily budget of the shrew in conditions of captivity. After 24 h with a score of 582 dives, 834 *Gammarus* were missing, which means a catching success of 1.4 *Gammarus* per dive.

According to Hanski (1984) the daily energy need of a water shrew weighing 17.8 g is 58 248 J. With an assimilation efficiency of 70%, our shrew should eat 772 *Gammarus*. Hence the energy budget of our shrew is positive, without any supplementary terrestrial food. The budget is also positive when the mean number of dives (546) is considered, if the diving success was identical on the other days, resulting in a total of 780 *Gammarus* eaten per day.

According to Lardet (1987), the daily energy requirement in the field is 3 753 J/g. A shrew weighing 18 g with an assimilation efficiency of 70% (Hanski, 1984) and a diving success of 70% (Schloeth, 1980) should dive 1266 times for food items of energetic value equivalent to one *Gammarus* in order to satisfy 100% of the budget using aquatic prey. If diving frequency is limited to 600 dives per day, then either half of the food should be foraged for on land, or the available prey should be of larger size (e.g. Trichoptera, Plecoptera).

Foraging strategy

Optimization of foraging strategy in diving mammals was shown for the mink (Dunstone & O'Connor, 1979*a,b*) and the otter (Nolet *et al.*, 1993). Our foraging experiments with the water shrew were not conducted to test optimal foraging theory, since some important parameters e.g. energetic equivalent of the prey, search and handling time, energetics of diving activity, were not recorded. However, the results illustrate some of the problems a small mammal faces when foraging in water.

The first experiment, with the same number of large as of small prey, shows clearly that the water shrew discriminates between the two. But the discrimination is statistically significant only after 14 h. This long period may be explained by two facts. At the onset of the experience, both patches are saturated. Moreover, the size difference was compensated for by catching the smaller prey at a higher rate (0.76 *Gammarus* per dive), compared to the rate for the larger prey (0.55 *Gammarus* per dive). As the prey density decreases, the searching effort increases, and the patch selection becomes significant.

The second experiment, with a much higher density in the patch of small prey, increases the frequency of encounter, and the mean success rate went up from 0.76 to 0.83 catches per dive. The probability of double catches is greater, since three of the seven shrews have a mean success rate greater than one prey per dive. This may energetically compensate for the smaller prey size. In good conditions, the water shrew is able to catch several prey in one dive.

In the last two experiments, the searching effort was raised in one patch due to an increase of 60% in the diving depth. In this situation, with approximately identical prey density on the bottom, the shrews foraged in the shallower patch (experiment III). This is consistent with the observation of Lardet (1987), who showed that in the natural environment the water shrew generally prefers to forage in water not exceeding 30 cm in depth. The important buoyancy due to air trapped in the fur (Köhler, 1991) may be responsible for a probably high energy demand during the diving activity. This increased diving effort should be compensated for in experiment IV, since prey density was three times higher in the deeper patch. Five of the seven shrews did in fact forage more on the richer side after 14 h, but the difference was not statistically significant even at the end of the test.

During the experiments, the shrews always had access to terrestrial laboratory food. This factor may result in a slower speed of decision, but does not fundamentally change the behaviour. The observation that minced meat is only consumed when prey density and catching success of *Gammarus* become low is a very important accessory result. Meat has a higher energy content than *Gammarus* and a higher digestibility (Hanski, 1984), thus should be much more attractive from an energetic point of view. This result shows that the taste of a potential prey, probably selected over the long history of evolution and of habitat adaptation, may occasionally have a more important influence on food preference than energetic content. This is in agreement with the observation of Churchfield (1994) who also stated that prey choice is influenced by palatability. It is interesting to compare this situation with data from Genoud & Vogel (1981). Experimental availability of minced meat in the field during the winter period induced in the terrestrial *Crocidura russula* practically exclusive feeding on the artificial source with a dramatic reduction in foraging activity on natural food.

Terrestrial shrews have been successfully used as a model for optimal foraging theory (Barnard & Brown, 1981, 1982, 1985*a,b*; Arditi *et al.*, 1983; Barnard *et al.*, 1983; Hanski, 1989). From our results we can conclude that the European water shrew would be an excellent model for experimental optimal foraging theory studies of riparian foraging strategy. High energetic demand, willingness to enter the water and small body dimension all predestine this shrew for future investigations on foraging in terrestrial versus aquatic habitats.

ACKNOWLEDGEMENTS
We are grateful to M. Besson, N. Di Marco, C. Koenig, P. Moratal and O. Schneider for their technical assistance. Useful suggestions during the experiments were given by H. Brünner, Ph. Christe, C. Cretegny and A. Oppliger. We are also indebted to S. Churchfield and L. Rychlik for helpful comments on the manuscript.

References

Appelt, H. (1973). Fellstrukuruntersuchungen an Wasserspitzmäusen. *Abh. Ber. naturk. Mus. "Mauritianum"* **8**: 81–87.

Arditi, R., Genoud, M. & Küffer, P. (1983). Méthode d'étude de la stratégie de recherche de la nourriture chez les musaraignes. *Bull. Soc. vaud. Sci. nat.* **76**: 283–294.

Barnard, C. J. & Brown, C. A. J. (1981). Prey size selection and competition in the common shrew (*Sorex araneus* L.). *Behav. Ecol. Sociobiol.* **8**: 239–243.

Barnard, C. J. & Brown, C. A. J. (1982). The effect of prior residence, competitive ability and food availability on the outcome of interactions between shrews (*Sorex araneus* L.). *Behav. Ecol. Sociobiol.* **10**: 307–312.

Barnard, C. J. & Brown, C. A. J. (1985a). Risk-sensitive foraging in common shrews (*Sorex araneus* L.) . *Behav. Ecol. Sociobiol.* **16**: 161–164.

Barnard, C. J. & Brown, C. A. J. (1985b). Competition affects risk-sensitivity in foraging shrews. *Behav. Ecol. Sociobiol.* **16**: 379–382.

Barnard, C. J., Brown, C. A. J. & Gray-Wallis, J. (1983). Time and energy budgets and competition in the common shrew (*Sorex Araneus* L.) . *Behav. Ecol. Sociobiol.* **13**: 13–18.

Calder, W. A. (1969). Temperature relations and underwater endurance of the smallest homeothermic diver, the water shrew. *Comp. Biochem. Physiol.* **30**: 1075–1082.

Cantoni, D. (1990). *Etude en Milieux Naturel de l'Organisation Sociale de Trois Espèces de Musaraignes,* Crocidura russula, Sorex coronatus *et* Neomys fodiens (*Mammalia, Insectivora, Soricidae*). PhD thesis: University of Lausanne, Switzerland.

Cantoni, D. (1993). Social and spatial organization of free-ranging shrews, *Sorex coronatus* and *Neomys fodiens* (Insectivora, Mammalia). *Anim. Behav.* **45**: 975–995.

Charnov, E. L. (1976). Optimal foraging: attack strategy of a mantid. *Am. Nat.* **110**: 141–151.

Churchfield, S. (1984). Dietary separation in three species of shrews inhabiting water-cress beds. *J. Zool., Lond.* **204**: 211–228.

Churchfield, S. (1985). The feeding ecology of the European water shrew. *Mammal Rev.* **15**: 13–21.

Churchfield, S. (1990). *The natural history of shrews.* Christopher Helm/A & C Black, London.

Churchfield, S. (1994). Foraging strategies of shrews, and the evidence from field studies. In *Advances in the biology of shrews*: 77–87 (Eds Merritt, J. F., Kirkland, G. L. & Rose, R. K.) Carnegie Museum of Natural History, New York. (*Carnegie Mus. nat. Hist. Spec. Publ.* No. 18).

Cummins, K. W. & Wuycheck, J. C. (1971). Calorific equivalents for investigations in ecological energetics. *Mitt. int. Verein. theor. angew. Limnol.* **18**: 1–158.

Davies, S. W. (1988). *An Investigation of the Effects of Various Environmental Parameters on the Underwater Foraging Behaviour of the American Mink,* Mustela vison *Schreber.* PhD thesis: University of Durham.

Dunstone, N. & O'Connor, R. J. (1979a). Optimal foraging in an amphibious mammal. I. The aqualung effect. *Anim. Behav.* **27**: 1182–1194.

Dunstone, N. & O'Connor, R. J. (1979b). Optimal foraging in an amphibious mammal. II. A study using principal component analysis. *Anim. Behav.* **27**: 1195–1201.

Ducommun, M.-A., Jeanmaire-Besancon, F. & Vogel, P. (1994). Shield morphology of curly overhair in 22 genera of Soricidae (Insectivora, Mammalia). *Revue suisse Zool.* **101**: 623–643.

DuPasquier, A. & Cantoni, D. (1992). Shifts in benthic macroinvertebrate community and food habits of the water shrew, *Neomys fodiens* (Soricidae, Insectivora). *Acta oecol.* **13**: 81–99.

Genoud, M. (1985). Ecological energetics of two European shrews: *Crocidura russula* and *Sorex coronatus* (Soricidae: Mammalia). *J. Zool., Lond. (A)* **207**: 63–85.

Genoud, M. & Vogel, P. (1981). The activity of *Crocidura russula* (Insectivora, Soricidae) in the field and in captivity. *Z. Säugetierk.* **46**: 222–232.

Grodzinski, W. & Wunder, B. A. (1975). Ecological energetics of small mammals. In *Small mammals: their productivity and population dynamics*: 173–204. (Eds Golley, F. B., Petrusewicz, K. & Ryskowski, L.) . Cambridge University Press, Cambridge. (*Int. biol. Progm.* No. 5).

Hanski, I. (1984). Food consumption, assimilation and metabolic rate in six species of shrew (*Sorex* and *Neomys*). *Annls. zool. fenn.* **21**: 157–165.

Hanski, I. (1989). Habitat selection in a patchy environment: individual differences in common shrews. *Anim. Behav.* **38**: 414–422.

Herreid, C. F. & Kessel, B. (1967). Thermal conductance in birds and mammals. *Comp. Biochem. Physiol.* **21**: 405–414.

Hutterer, R. (1985). Anatomical adaptations of shrews. *Mammal Rev.* **15**: 43–55.

Hutterer, R. & Hürter, T. (1981). Adaptative Haarstrukturen bei

Wasserspitzmäusen (Insectivora, Soricinae). *Z. Säugetierk.* **46**: 1–11.

Illing, K., Illing, R. & Kraft, R. (1981). Freilandbeobachtungen zur Lebensweise und zum Revierverhalten der europäischen Wasserspitzmaus, *Neomys fodiens* (Pennant, 1771). *Zool. Beitr.* **27**: 109–122.

Ivanter, E. V. (1994). The structure and adaptive peculiarities of the pelage in soricine shrews. In *Advances in the biology of shrews*: 441–454. (Eds Merritt, J. F., Kirkland, G. L. & Rose, R. K.). Carnegie Museum of Natural History, New York. (*Carnegie Mus. nat. Hist. Spec. Publ.* No. 18).

Kleiber, M. (1961). *The fire of life: an introduction to animal energetics.* John Wiley, New York and London.

Köhler, D. (1991). Notes on the diving behaviour of the water shrew. *Neomys fodiens* (Mammalia, Soricidae). *Zool. Anz.* **227**: 218–228.

Köhler, D. (1993). Zum Erlernen der Lage aquatischer Futterquellen durch *Neomys fodiens* (Mammalia, Soricidae). *Zool. Anz.* **231**: 73–81.

Krebs, J. R. & McCleery, R. H. (1984). Optimization in behavioural ecology. In *Behavioural ecology: an evolutionary approach*: 91–121. (2nd rev. edn). (Eds Krebs, J. R. & Davies, N. B.) Blackwell, Oxford.

Kruuk, H. (1993). The diving behaviour of the platypus. (*Ornithorhynchus anatinus*) in waters with different trophic status. *J. appl. Ecol.* **30**: 592–598.

Kuvikova, A. (1985). Zur Nahrung der Wasserspitzmaus, *Neomys fodiens* (Pennant, 1771) in der Slowakei. *Biológia, Bratisl.* **40**: 563–572.

Lardet, J. P. (1987). *Contribution à l'Étude de Quelques Aspects de la Stratégie Énergétique de la Musaraigne Aquatique, Neomys fodiens (Mammifères, Insectivores).* PhD thesis: University of Lausanne, Switzerland.

Lardet, J. -P. (1988). Spatial behaviour and activity patterns of the water shrew *Neomys fodiens* in the field. *Acta theriol.* **33**: 293–303.

Lechner, A. J. (1978). The scaling of maximal oxygen consumption and pulmonary dimension in small mammals. *Respir. Physiol.* **34**: 29–44.

Niethammer, J. (1977). Ein syntopes Vorkommen der Wasserspitzmäuse *Neomys fodiens* und *Neomys anomalus. Z. Säugetierk.* **42**: 1–6.

Niethammer, J. (1978). Weitere Beobachtungen über syntope Wasserspitzmäuse der Arten *Neomys fodiens* und *N. anomalus. Z. Säugetierk.* **43**: 313–321.

Niethammer, J. & Krapp, F. (Eds) (1990). *Handbuch der Säugetiere Europas.* 3(1) *Insektenfresser–Insectivora.* Aula-Verlag, Wiesbaden.

Nolet, B. A., Wansink, D. E. H. & Kruuk, H. (1993). Diving of otters (*Lutra lutra*) in a marine habitat: use of depths by a single-prey loader. *J. Anim. Ecol.* **62**: 22–32.

Ruthardt, M. & Schröpfer, R. (1985). Zum Verhalten der Wasserspitzmaus *Neomys fodiens* (Pennant, 1771) unter Wasser. *Z. angew. Zool.* **72**: 49–57.

Sanden-Guja, von W. (1957). Wasserspitzmausjahre. *Beitr. Naturk. Niedersachs.* **10**: 73–75.

Schloeth, R. (1980). Freilandbeobachtungen an der Wasserspitzmaus, *Neomys fodiens* (Pennant, 1771) im Schweizerischen Nationalpark. *Revue suisse Zool.* **87**: 937–939.

Schröpfer, R. & Stubbe, M. (1992). The diversity of European semiaquatic mammals within the continuum of running water systems–an introduction to the symposium. In *Semiaquatische Säugetiere*: 9–14. (Eds Schröpfer, R., Stubbe, M. & Heidecke, D.). *Martin-Luther-Univ. Halle-Wittenberg, Wiss. Beitr.*

Sparti, A. (1989). *Etude Comparative de la Thermorégulation et du Budget de l'eau chez les Soricinés et les Crociduriés (Insectivora, Mammalia).* PhD thesis: University of Lausanne, Switzerland.

Sparti, A. (1990). Comparative temperature regulation of African and European shrews. *Comp. Biochem. Physiol. (A)* **97**: 391–397.

Sparti, A. (1992). Thermogenic capacity of shrews (Mammalia, Soricidae) and its relationship with basal rate of metabolism. *Physiol. Zool.* **65**: 77–96.

Spitzenberger, F. (1990). *Neomys fodiens* (Pennant, 1771)–Wasserspitzmaus. In *Handbuch der*

Säugetiere Europas. 3(1) In *Insektenfresser–Insectivora*: 335–374. (Eds Niethammer, J. & Krapp, F.). Aula-Verlag, Wiesbaden.

Vogel, P. (1990). Body temperature and fur quality in swimming water-shrews, *Neomys fodiens* (Mammalia, Insectivora). *Z. Säugetierk.* 55: 73–80.

Vogel, P. & Koepchen, B. (1978). Besondere Haarstrukturen der Soricidae (Mammalia, Insectivora) und ihre taxonomische Deutung. *Zoomorphologie* 89: 47–56.

4

Habitat use by water shrews, the smallest of amphibious mammals

S. Churchfield

Introduction

Water shrews are the smallest of amphibious mammals. They are insectivores belonging to the family Soricidae and, although the aquatic species are larger than most terrestrial species, they have a body length of only 70–130 mm and a body mass of 8–56 g. Thirteen species of water shrews are currently recognized (Wilson & Reeder, 1993), belonging to four genera (Table 4.1), and they have a wide geographical distribution through the Nearctic, Palaearctic and parts of south-east Asia where they are associated with freshwater streams and marshlands. Shrews have excited considerable interest because of their large energy demands, high levels of activity and voracious appetites. Water shrews, with their habit of diving in cold waters for aquatic prey, and the consequent energy costs of this, provide an additional dimension to the study of the physiology and ecology of very small mammals. However, compared with their terrestrial counterparts, remarkably little is known about water shrews, and many species have rarely been sighted or captured. This review investigates the geographical and habitat occurrence of water shrews and the use they make of the aquatic mode of life.

Anatomical adaptations to a semi-aquatic mode of life

Convergent evolution has occurred several times in the Soricidae, and adaptations for a semi-aquatic mode of life have developed in four different genera (*Sorex, Neomys, Chimarrogale* and *Nectogale*) and in two different continents. It is noteworthy that the genus *Crocidura* (to which some 125 of the 233 shrew species are ascribed) possesses no aquatic examples. Water shrews show a graded series of progressively greater adaptation to a semi-aquatic existence, from *Sorex* and *Neomys* to *Chimarrogale* and culminating in *Nectogale*, the web-footed water shrew. Their anatomical adaptations have been described in detail by Hutterer (1985), but are briefly summarized below.

Table 4.1. *Species of water shrew* (Insectivora: Soricidae), *with body dimensions where available*

	Head and body Length (mm)	Tail length (mm)	Body mass (g)	Source
American water shrews				
Sorex alaskanus	—	—	—	
S. bendirii	75–86	62–81	7.5–18.0	van Zyll de Jong (1983)
S. palustris	69–74	61–89	8.5–17.9	van Zyll de Jong (1983)
Eurasian water shrews				
Neomys				
anomalus	67–87	40–52	8.0–17.0	Pucek (1981)
N. fodiens	70–96	51–72	11.4–17.5	Pucek (1981)
N. schelkovnikovi	87–102	60–75	—	Gureev (1971)
Oriental water shrews				
Chimarrogale				
hantu	90–117	85–98	—	Hoffmann (1987)
C. himalayica	91–110	75–90	—	Hoffmann (1987)
C. phaeura	91–110	81–95	—	Hoffmann (1987)
C. platycephala	100–130	94–108	24–56	Koyasu (1993)
C. styani	96–108	61–85	—	Hoffmann (1987)
C. sumatrana	—	—	—	
Asian web-footed water shrew				
Nectogale elegans	90–128	89–110	c. 45[†]	Gureev (1971) [†]Hutterer (1992)

While water shrews greatly resemble their terrestrial counterparts in general appearance, several features set them apart. The pelage of water shrews, besides being darker than terrestrial species (black on the dorsal surface) is also denser and the awn hairs possess an **H**-shaped profile with lateral grooves filled with numerous ridges that increase in numbers and complexity from *Sorex* to *Neomys, Chimarrogale* and *Nectogale* (Hutterer & Hürter, 1981; Hutterer, 1985; Ivanter, 1994). This probably serves to retain air and provide better insulation and buoyancy. The terminal parts of the awn hairs are also flattened, which may aid water repellence. All water shrews possess a fringe of stiff hairs on the lateral edges of the hind feet, and most species possess it on the fore feet too. In *Sorex palustris, Neomys* spp. and *Chimarrogale* spp. these stiff hairs are also found on the lateral edges of each toe. These hairs increase the surface area of the foot when swimming and so aid propulsion. Only *Nectogale* possesses webbed feet: both fore and hind feet are webbed to the base of the terminal phalanges. *Nectogale* also possesses disc-like pads on the base of each foot, thought to aid adhesion on wet rocks, and small scales graduating into transverse scutes on the dorsal surface (Hutterer, 1985). A keel of stiff hairs is present on the ventral surface of the tail of all species (best-developed in *Neomys*), which may help to prevent rolling while swimming. *Nectogale* possesses two ventral, two lateral and one dorsal keel of stiff hairs on the tail. The lengths of the tibia and the digits of the hind limbs of water shrews are slightly greater than in terrestrial shrews (Hutterer, 1985).

The ears of water shrews are smaller and situated more posteriorly than in terrestrial species (Hutterer, 1985). In *Chimarrogale* and *Nectogale* the ears are reduced and possess a valvular antitragus that seals the opening when underwater. The external ears of *Nectogale* are reduced to just a slender opening with no pinna. The eyes of water shrews show no adaptation except in *Nectogale* where they are covered by skin with the apparent consequential loss of visual sense (Hutterer, 1985). The olfactory system is also rather reduced compared with terrestrial shrews, and there are modifications of the rhinarium for aquatic life (Hutterer, 1980, 1985). In *Neomys*, the lateral lobes of the rhinarium are very large and probably close the nostrils while underwater. In *Chimarrogale* and *Nectogale*, the nostrils are placed behind the rhinarium shield (as in water tenrecs), which may prevent water entering.

There is some enlargement in aquatic shrews (with an increase in brain size) of the trigeminal system, which innervates the vibrissae of the muzzle and may replace the olfactory system in underwater foraging (Hutterer, 1985). *Neomys* alone possesses narcotizing saliva, which probably helps to subdue large prey items, a most unusual feature amongst mammals (Pucek, 1959, 1969).

Geographical distribution and habitat occurrence

Although several individual species have a very wide geographical distribution, water shrews are generally more restricted globally than their terrestrial counterparts. For example, although there are many genera and species of shrews in Africa, none is aquatic (Gureev, 1971; Corbet & Hill, 1991; Wilson & Reeder, 1993). In continental Africa their ecological role is taken by tenrecs, namely species of *Potamogale* and *Micropotamogale*, and in Madagascar by the water tenrec (*Limnogale*). The paucity of aquatic insectivores in Africa, and their very limited distribution, may reflect the limited availability of suitable habitats and their geographical isolation. Although terrestrial shrew species are numerous and widespread in central and southern Asia, the seven species of water shrew are more contained and localized (Corbet & Hill, 1991; Wilson & Reeder, 1993).

The three aquatic members of the genus *Sorex* are restricted to North America, and are the only water shrews to be found on that continent (Whitaker, 1980; Hall, 1981). Most widespread of these is *S. palustris*, whose range extends from south-east Alaska southwards to New Mexico and east to Nova Scotia. It mostly inhabits boreal areas and montane habitats below the tree line, but it also occurs in more open, marshy, lowland habitats (Buckner & Ray, 1968; Clark, 1973; Benenski & Stinson, 1987). Typically it occurs beside small, swift-flowing streams bordered by grasses, sedges and willows, however sometimes it inhabits ephemeral creek beds, but rarely far from open water (Conaway, 1952; Banfield, 1974). Clark (1973) never found it further than 9 m from water and, sampling a range of xeric, mesic and hydric habitats, the present author encountered it only on stream banks.

In contrast to the wide distribution of *S. palustris*, *S. bendirii* inhabits only a narrow coastal area from south-west British Columbia southwards to north-west California (Pattie, 1973; Whitaker, 1980; Hall, 1981). It, too, is found along stream sides and in marshy areas, but it may extend up to 1 km from open water in moist forests. *S. alaskanus* is known only from a few specimens collected from the Glacier Bay area of south-east Alaska, and is of doubtful status (Hall, 1981; Corbet & Hill, 1991; Wilson & Reeder, 1993).

The Eurasian water shrews of the genus *Neomys* have a wide distribution in the Palaearctic (Gureev, 1971; Corbet, 1978; Hoffmann, 1987; Wilson & Reeder, 1993). Best known (and best studied) of these is *Neomys fodiens*, which occurs through most of Europe (including the British Isles) and eastwards through western and central Siberia to the River Yenesei and Lake Baikal, southwards to northern Asia Minor, the montane forests of the Tien Shan and north-west Mongolia. It is thought to have had a continuous distribution to the Pacific in the past, giving it the largest range of any shrew species, but today there is a gap of some 2000 km before it recurs in the mixed forests of the Pacific coast region of Russia, in north-east China and north Korea (where it has a disjunct distribution), and the islands of Sakhalin and Hokkaido. It is absent from the taiga zone of eastern Siberia. It extends well beyond the Arctic Circle in Europe but not so far north in Asia where it is confined to forest and does not inhabit tundra.

It is typically a forest species, inhabiting temperate deciduous forest and coniferous taiga where it is usually found associated with streams, rivers, marsh and bog habitats (Aulak, 1970; Stein, 1975; Pucek, 1981; Schröpfer, 1983; Kuvikova, 1985; Lardet, 1988a). It makes use of seasonally flooded habitats, such as alder forests in river valleys, but also extends up to the alpine zone in mountains. It does occur in more open habitats, including hedgerows, old fields, scrub, drainage ditches through meadows, fish ponds and watercress beds (Buchalczyk & Pucek, 1963; Wolk, 1976; Kraft & Pleyer, 1978; Churchfield, 1984a). Occasionally it is found several kilometres from open water.

Neomys anomalus is sympatric with *N. fodiens* over much of its range, but has a much more limited distribution. It mostly occurs in the temperate forests of Europe, from Portugal to Poland, Ukraine and southern Russia as far east as the River Don (Aulak, 1970; Gureev, 1971; Corbet, 1978; Pucek, 1981; Niethammer & Krapp, 1990). It lives in much the same habitats as *N. fodiens* but shows a greater affinity for mountainous areas, such as the Pyrenees, Alps and West Carpathians, where it reaches an altitude of 2000 m (Pucek, 1981). Although it is closely associated with water (Dehnel, 1950; Pucek, 1981), it is less well-adapted for an aquatic existence (the fringes of hairs on the feet and the keel on the tail are less prominent than in *N. fodiens*), and it rarely, if ever, dives underwater for food (Rychlik, 1995).

The distribution of *N. schelkovnikovi* is limited to the Caucasus and possibly adjacent Turkey and Iran (Gureev, 1971; Nowak & Paradiso, 1983; Corbet & Hill, 1991; Wilson & Reeder, 1993). Its ecology is poorly known.

The web-footed *Nectogale elegans* is the largest and most specialized of all the water shrews, with a body mass reaching approximately 45 g (Hutterer, 1992). Its distribution is centred on the Himalayan region, including Nepal, Tibet, Sikkim and Bhutan, but extending into west and central China and northern Burma (Nowak & Paradiso, 1983; Hoffmann, 1987; Corbet & Hill, 1991; Wilson & Reeder, 1993). It has been found only beside fast-flowing streams in mountain forests on southern and eastern slopes at altitudes of 900–4600 m (Pruitt, 1957; Hutterer, 1992). The few observations made of this species suggest that it rarely leaves its aquatic habitat.

Species of *Chimarrogale* are distributed in different parts of the Oriental region, and are the only water shrews whose distribution encompasses tropical areas. Like *Nectogale*, they seem to be associated with mountainous areas and they are mostly found by small forest streams (Harrison, 1958; Jones & Mumford, 1971; Abe, 1982; Nowak & Paradiso, 1983; Arai et al., 1985; Hoffmann, 1987; Wilson & Reeder, 1993). *Chimarrogale hantu, C. phaeura* and *C. sumatrana* occur along streams in tropical forests of the Malayan peninsula, Borneo and Sumatra, respectively. *Chimarrogale himalayica* is found along alpine streams in Kashmir, through much of south-east Asia and south and central China, and Taiwan. Over part of its range it is sympatric and syntopic with *Nectogale* in montane, riparian habitats. *Chimarrogale styani* occurs only in the highlands of western-central China and, sympatrically with *C.*

himalayica, in northern Burma. *Chimarrogale platycephala* is restricted to the islands of Japan, notably Honshu, Kyushu and Shioku.

The Oriental water shrews, in particular, are found at high altitudes. Both *C. himalayica* and *C. styani* have been found at 3060 m in China (Hoffmann, 1987). In Burma, where they are sympatric, some altitudinal segregation of these two species may occur, since *C. himalayica* has been found at 918–1928 m and *C. styani* at 1591–3152 m (Hoffmann, 1987).

Accounts of the habitats where water shrews have been located frequently describe small, swiftly-flowing streams or rivers and occasionally provide details of stream widths and bank cover (e.g. Conaway, 1952; Arai *et al.*, 1985). Unfortunately, water depth and substrate type are seldom mentioned, yet these may be important factors in habitat selection by water shrews because of their limited diving ability. Although captive *N. fodiens* can dive up to 260 cm in still water (see Chapter 3), wild shrews utilize stream depths of only 20–80 cm and they seem to prefer to forage in water not exceeding 30–40 cm in depth (Illing *et al.*, 1981; Weissenberger *et al.*, 1983; Churchfield, 1984*a*; Lardet, 1988*a*). Stony substrates are often implied, if not mentioned.

Use of the aquatic foraging mode

Variation amongst water shrew species
Water shrews vary considerably in their use of aquatic foraging. *Nectogale elegans*, the web-footed water shrew, is the most specialized and best-adapted of the water shrews fore aquatic foraging and is reported to swim and dive very well, rarely leaving its aquatic habitat (Pruitt, 1957; Nowak & Paradiso, 1983; Hutterer, 1992). It is thought to feed exclusively on aquatic prey, including small fish and invertebrates, but there have been very few observations of the foraging behaviour of this species, either in captivity or in the wild. *Chimarrogale* spp. are also reported to swim well underwater and are occasionally caught in fish traps (Harrison, 1958; Jones & Mumford, 1971; Novak & Paradiso, 1983). They have a diet of aquatic insect larvae, crustaceans and fish but, again, observations of these species are few (Abe, 1982).

Sorex palustris makes major use of aquatic foraging, although not exclusively. Only 49% of stomachs ($n = 87$) from Montana examined by Conaway (1952) contained aquatic organisms, mostly Plecoptera and Ephemeroptera nymphs, and Trichoptera larvae, totalling about 27% of dietary occurrences. They have been observed to catch and eat small fish, fish eggs and larval salamanders (Nussbaum & Maser, 1969; Buckner, 1970; Banfield, 1974). Major

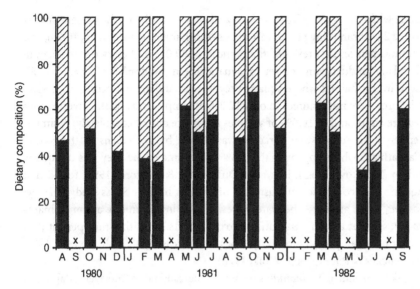

Figure 4.1. The monthly proportions of aquatic (black bars) and terrestrial (hatched bars) prey in the diet of *N. fodiens* in southern England. A cross denotes a month not sampled. (After Churchfield, 1984a).

terrestrial items were insects, their larvae, gastropods and earthworms (Conaway, 1952; Whitaker & Schmeltz, 1973; Whitaker & French, 1984). However, analysis of 15 stomachs and of faecal middens from shrews inhabiting stream banks in California during August by the present author suggested that aquatic prey may assume greater importance: a mean of 77.2% of dietary occurrences (74% by volume) were aquatic in origin, mostly Plecoptera nymphs and Trichoptera larvae.

Sorex bendirii, which is slightly less well-adapted to swimming than *S. palustris* (the fringes of stiff hairs on the feet and the keel of the tail being relatively poorly developed: Pattie, 1973; Hall, 1981), seems to make much less use of aquatic foraging. Only 29.8% ($n = 24$) of dietary occurrences were aquatic (mostly Ephemeroptera nymphs) in forested habitats studied by Whitaker & Maser (1976) but, compared with *S. palustris*, this species is more often found in marshy areas than at stream-sides and so may be more terrestrial in habits. The bulk of its diet comprised terrestrial gastropods, insect larvae and earthworms (Whitaker & Maser, 1976).

Neomys fodiens, although very similar morphologically to *S. palustris*, generally takes about equal proportions of aquatic and terrestrial prey in habitats where both are available (Fig. 4.1). Although it can survive on terrestrial prey alone when living away from streams and ponds, studies by the present author over a 2 year sampling period in a stream-side habitat (Churchfield, 1984*a*, and unpublished results) revealed that a mean of 49.0% of dietary occurrences in faecal and midden samples were of aquatic origin (range 33.3–67.2%, *n* = 166). Even in this habitat where aquatic prey were abundant and accessible, terrestrial prey were not ignored. In only 12.7% of samples did aquatic invertebrates feature alone, and in 10.8% of samples they were absent completely. A similar tendency for underwater foraging was shown by *N. fodiens* in the Slovak Carpathians where approximately 45% by volume of the prey was of aquatic origin (Kuvikova, 1985). However, DuPasquier & Cantoni (1992) found that aquatic prey comprised at least 80% of the diet in their Swiss study. *Neomys fodiens*, alone amongst the water shrews (and almost unique amongst mammals), secretes a mildly narcotizing toxin into its saliva that may permit it to feed on larger prey, commonly frogs and occasionally fish (Pucek, 1959; Buchalczyk & Pucek, 1963; Wolk, 1976).

Like *S. bendirii*, *N. anomalus* is morphologically rather less well-adapted to aquatic life than its congener, in this case *N. fodiens*. But both species have similar feeding habits and *N. anomalus* feeds on a range of aquatic invertebrates (Niethammer, 1977, 1978). However, it dives much less readily than *N. fodiens* and forages mostly in shallow water (< 5 cm deep), in muddy substrates beside running or stagnant water, and along stream banks (Rychlik, 1995). Differences in the foraging strategies of the two species in sympatry help to reduce competition (Rychlik, 1995).

With the possible exception of *Nectogale* and *Chimarrogale*, all water shrews can subsist entirely on terrestrial prey both in the wild, when they are sometimes found far from water (Corbet & Harris, 1991; personal observations), and in captivity (personal observations). But their reduced olfactory system (Hutterer, 1985) and agility compared with terrestrial species may affect their success at terrestrial foraging.

Seasonal use of aquatic foraging

The occurrence of water shrews at high latitude and altitude and their strong association with cold mountain streams appear to be at odds with the high energetic demands of shrews generally, particularly their unfavourable surface-area-to-volume ratios and potential for high heat loss. Their problems would seem to be exacerbated by exposure to cold water and frequent wettings. On the basis of the prey requirements of captive *N. fodiens* of 2.2 g dry weight of food per day (Churchfield, 1984*a*) and the biomass of aquatic prey (Churchfield, 1984*a*, and unpublished results), a mixed diet of aquatic invertebrates taken in proportion to their abundance could necessitate up to 2017 dives per day or 84 per hour. This approximates to the diving frequency of wild *N. fodiens* observed by Schloeth (1980) over a limited period of some 50 min. Moreover, many dives are unsuccessful, resulting in no prey capture or in collection of inappropriate items such as twigs and small stones mistaken for prey (Wolk, 1976; Schloeth, 1980; personal observation), increasing the number of submergences required to meet daily food requirements. The number of dives could be reduced by selectivity for the largest and most profitable aquatic invertebrates available, namely crustaceans and Trichoptera larvae, which indeed were the commonest dietary items (Churchfield, 1984a), and by supplementing the diet with terrestrial prey. With prey selection and terrestrial prey comprising 50% of the diet, the number of daily dives could be reduced to approximately 200. Another solution would be to restrict underwater foraging to the warmest seasons when exposure to cold can be minimized and rapid drying of the pelage can be achieved. Are there seasonal differences in the use of the aquatic feeding mode?

In the absence of detailed information on the seasonal habits of *Nectogale* and *Chimarrogale*, we may assume that these shrews forage underwater throughout the year since they have never been found in any habitat except stream-sides. *Sorex palustris*, too, is highly associated with this habitat type, and has been observed to swim underwater beneath the ice in winter (van Zyll de Jong, 1983). It has enzymes that function well at low temperatures and is apparently capable of reducing its metabolic demands, which allows it to dive year-round in cold mountain streams (Boernke, 1977), a feature probably common to all water shrews.

Neomys fodiens forages in water all year round. In a study of *N. fodiens* in southern England, Churchfield (1984a, and unpublished results) found that, although the proportion of aquatic prey taken varied from month to month (see Fig. 4.1), there was no significant difference between summer and winter foraging (aquatic dietary occurrences: April–October = 51.0%; November–March = 46.2%). Even in the northern and continental parts of its range, such as Poland, this species employs aquatic foraging in winter (Wolk, 1976; Schloeth, 1980; DuPasquier & Cantoni, 1992).

It is clear, then, that cold, winter conditions do not curb the aquatic foraging habit. It is notable that many of these water shrews, even those from lower latitudes (e.g. *Chimarrogale*), inhabit cold, fast-flowing streams, often at high altitude. The coefficient of heat transfer of captive *S. palustris* in water averaged 4.6 times that in air, resulting in a cooling rate in water of 10–12 °C (equivalent to typical stream temperature in summer) of 1.43 °C per 30 s of submergence (Calder, 1969). Lardet (1988b) recorded lower rates of temperature decrease in *N. fodiens*, but this depended upon water temperature, being

0.5 °C per min at 19 °C and 0.8 °C per min at 5 °C. This suggests that aquatic foraging in summer, let alone in winter, incurs significant energy costs. Lardet (1988*b*) proposed that the increased cost of foraging in cold water may explain observations of the disappearance of water shrews from normal stream-side home ranges in winter as recorded by Dehnel (1950), Price (1953), Weissenberger *et al.* (1983) and Lardet & Vogel (1985). However, as we have seen, field data demonstrate aquatic foraging by both species in winter.

In fact, the differential between air and water temperatures in winter is often small, implying that foraging in water incurs no greater heat loss than would foraging on land on a cold day. But a crucial factor in energy-saving is the condition of the pelage, particularly its tendency to get waterlogged, since this will lead to rapid heat loss. The pelage of water shrews is better adapted to retain air and repel water than that of their terrestrial counterparts (Hutterer & Hürter, 1981; Ivanter, 1994). Air trapped in the fur is a very effective insulator: heat loss of *S. palustris* with a well-developed air layer was reduced by 50% compared with fur from which the air had been expelled (Calder, 1969). The fur of captive shrews is often not in good condition and is prone to wetting, which may exacerbate heat loss (Vogel, 1990: personal observation). Vogel (1990) found that shrews with wet fur suffered a 1.1 °C decrease in body temperature per min during swimming, but shrews whose pelage was in good condition, with a well-developed air layer, were able to keep dry during swimming and maintain their normal body temperature of about 37 °C in a variety of test situations.

Another factor affecting heat loss is the period of submergence. Captive *S. palustris* are capable of surviving forced submergences of 31–48 s (Calder, 1969), which could result in a significant reduction in body temperature, particularly if the pelage has a poorly-developed air layer. However, periods of submergence by wild and captive water shrews able to dive freely are frequent, but of much shorter duration, which helps to minimize waterlogging. Observations of wild water shrews revealed that dives mostly lasted 3–10 s, once up to 24 s, in water depths generally less than 30 cm, occasionally up to 200 cm (Schloeth, 1980), and 10–20 s as recorded by Lardet (1988*a*). Studies of captive *N. fodiens* by Köhler (1991) in large aquaria with 12 cm depth of water showed that maximum diving duration per individual was 8–16 s, and that even trained shrews did not remain submerged for more than 16 s. In aquaria with 75 cm depth of water where *N. fodiens* were permitted to dive freely for prey amongst gravel at the base, dives lasted only 4 s (Churchfield, 1985). So, provided the coat is in good condition and is groomed quickly to remove surplus water, aquatic foraging should incur no greater thermoregulatory cost than foraging on land. There is no information on the diving ability of *Chimarrogale* or *Nectogale*, which are rather better adapted for swimming.

Advantages of aquatic foraging: food availability

Aquatic invertebrates provide an abundant and dependable food supply that is not utilized by other shrews, and only to a limited extent by other vertebrates, with insectivorous fish and birds being the major potential competitors. Moreover, the activity of many invertebrate and vertebrate predators of aquatic invertebrates is highly seasonal, being lowest in winter. Quantitative samples of macro-invertebrates inhabiting the stony substratum of streams supplying water-cress beds in southern England (for details, see Churchfield, 1984a) showed that the total number of potential prey in five samples taken at approximately monthly intervals ranged from 2750 to 16 275/m^2 (mean 7060 ($\pm$ 1053)), with biomasses of 3.0–16.1 g dry weight/m^2. Not all these macro-invertebrates were found in the diets of shrews but even known prey taxa, revealed by faecal analysis, numbered 828–5957/m^2 (mean 3357($\pm$ 552)). Invertebrates were abundant throughout the year (Fig. 4.2), total numbers in summer and winter months showing no significant difference. However, the abundance of known prey was significantly greater in winter (November – March) than in summer (April–October), with means of 4427/m^2 and 2466/m^2, respectively ($t = 4.34$, $p < 0.02$). Numbers of insect nymphs and larvae, particularly, decline in summer as they emerge as adults. Similarly, the abundance of the major prey items (small freshwater crustaceans such as *Gammarus* and *Asellus*) was greater in winter (mean $= 3949$/m^2) than in summer (mean $= 1964$/m^2; $t = 3.34$, $p < 0.01$). Their occurrence in summer diets also declined, but not significantly. There was no correlation between the monthly abundance of aquatic prey (all invertebrates, known prey, or crustaceans) and their occurrence in the diets.

The abundance of terrestrial invertebrates at this site was not measured but data from a grassland/scrub area in central England using a similar methodology are available for comparison (for details, see Churchfield, 1982). Figure 4.2 shows the relative abundance of known prey items in the two study areas over an equivalent period of months (although not in the same years). Though detailed comparison on this basis may be spurious, the data do show some interesting trends. Biomass estimates of terrestrial and aquatic invertebrates prey in the two habitats were similar (2.5–16.5 g dry weight/m^2 of terrestrial prey), the major contributor to terrestrial biomass being earthworms. But the abundance (and hence encounter rate) of aquatic prey was, in most months,

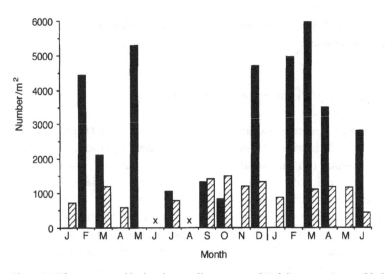

Figure 4.2. The mean monthly abundances of known prey of *N. fodiens*: aquatic prey (black bars) in 1980–81 and terrestrial prey (hatched bars) in 1976–77. A cross denotes a month not sampled. See the text for details.

much greater than that of terrestrial prey (mean monthly number of aquatic prey = 3358/m^2, terrestrial prey = 1043/m^2; $t = 14.12$, $p < 0.001$).

While it could be argued that the aquatic habitat exemplified here is atypical in providing particularly good conditions for water shrews, it is worth noting that most species of water shrew are associated with similar habitats, namely cold, fast-flowing streams and rivers with stony substrata that typically support large populations of aquatic invertebrates. DuPasquier & Cantoni (1992) found that aquatic prey were plentiful in all seasons in a stony-bottomed Swiss river inhabited by *N. fodiens*. Quantitative sampling on one occasion in each of the four seasons produced similar orders of abundance and biomass to those reported here.

The abundance and availability of freshwater invertebrates raises the question of why water shrews are not exclusively aquatic foragers in suitable habitats, which can only be speculated upon. Water shrews may need a more varied diet than can be provided by aquatic prey alone, and it is notable that even so-called specialist feeders like the earthworm-eating *Sorex roboratus* augment their diet with a range of other invertebrates (Churchfield & Sheftel, 1994). Water shrews move from one aquatic feeding ground to another by land (see below), during which they encounter terrestrial invertebrates upon which they opportunistically prey. The energy costs of swimming compared with running on land for shrews are not known, but may be relevant to their foraging strategy.

Home ranges and habitat use

With an abundance of aquatic prey available to water shrews, what implications does this have for the size and use of their home ranges? There have been relatively few studies of home range use by water shrews simply because they have proved elusive and difficult to catch and track. Our knowledge of the ecology of *Chimarrogale* spp. and *Nectogale* is practically non-existent because these species have rarely been caught. Even live-trapping studies of *S. palustris* are few and limited in scope: mark–recapture of just two individuals produced a crude estimate of home range area of 0.2–0.3 ha (Buckner & Ray, 1968). Most of our information about the ecology of water shrews, including home ranges, comes from studies on *N. fodiens*.

Studies by Lardet (1988a) provide an interesting insight into the behaviour of this species. He tracked *N. fodiens* marked with radioactive ear tags as they traversed their stream-side habitat in Switzerland. Home ranges were small, with a mean of 207 m^2 in April–September, and only 106 m^2 in October–December. This is consistent with the greater abundance of aquatic prey in winter compared with summer found by the present author (see above). Similar home-range sizes have been recorded in equivalent habitats by other workers: 60–80 m^2 (Illing *et al.*, 1981), 118–276 m^2 (van Bemmel & Voesenek, 1984). However, Cantoni (1993) found daily home ranges in a canal-side habitat were more extensive (260–509 m^2), and largest in winter.

Home-range sizes of aquatic and terrestrial shrews are not easily compared because of the different dimensions involved, but those of water shrews are generally significantly smaller than those of terrestrial species. Home ranges of *S. araneus*, for example, are 500–2800 m^2 (Churchfield, 1990). In her canal-bank study, Cantoni (1993) recorded very small home ranges for *S. coronatus*, which were comparable with syntopic *N. fodiens* for most of the year (except in winter when they were even smaller) but also found that they had a higher total daily activity than the water shrews. Water shrews spent more time resting. This suggests that aquatic prey do indeed provide a more abundant, predictable and accessible food source than terrestrial prey.

Home-range studies by Illing *et al.* (1981) revealed the extent to which aquatic and terrestrial habitats were used. Of the total range size occupied in a stream-side habitat (60–80 m^2), the terrestrial area occupied 22–30 m^2. Individuals had range lengths of 20–24 m, which overlapped with neighbours only

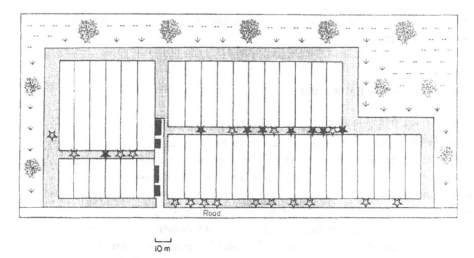

Figure 4.3. The study area of water-cress beds showing captures of seven marked individuals of *N. fodiens* (black stars) and 12 *S. araneus* (open stars) in May 1981. The bar represents 10 m.

at the periphery. Ninety-five percent of above-ground activity was spent along the stream bank searching for terrestrial invertebrates; only 5% was spent in swimming and diving. Shrews preyed upon both terrestrial and aquatic invertebrates but Illing *et al.* (1981) surmised from diet analysis that aquatic foraging was most successful.

Lardet (1988*a*) confirmed that most of the shrews' time was spent confined to the stream and its banks, although occasional forays were made into nearby woodland. Different foraging sites along the stream were used, the shrews passing from one to the other *via* rodent burrows within the banks. Periods of inactivity were spent mostly in one underground nest, although up to four secondary nests were also used.

Water shrews exhibit preferences not only for certain habitats, but also for particular parts of the same habitat, and they can be very site-specific. Studies by Churchfield (1984*b*, and unpublished results) of *N. fodiens* inhabiting an area of water-cress beds revealed a preference for inhabiting linear, grassy banks in the centre of the study area that were relatively isolated and had ready access to the water on both sides, compared with the grass banks at the periphery which provided access to the water on one side and to hedgerows, scrub and marsh on the other side (Fig. 4.3). Over 2 years, a mean of 57.2% of water shrews were found on the central banks, although these sites only occupied 27% of the total bank area. Often, several shrews lived in close proximity on the central banks, while peripheral banks had few or no shrews

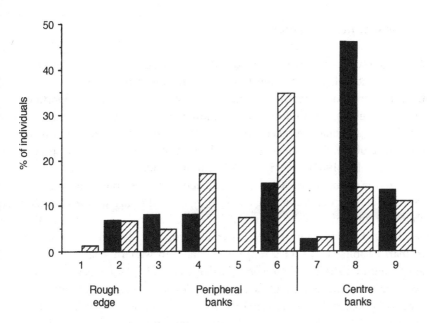

Figure 4.4. Site occurrence of *N. fodiens* (black bars) and *S. araneus* plus *S. minutus* (hatched bars) at different bank sites in the study area of water-cress beds (see the text for details).

(Fig. 4.3). The main central bank (110 m long and 3 m wide) was inhabited by a mean of 2.3 water shrews but up to 6 individuals were trapped there at any one time, more than occurred on any other banks, either separately or collectively. Range overlap there was greater in summer (when shrews were breeding), with a mean of 4.0 individuals overlapping in their home ranges, than in winter (mean 2.3 individuals overlapping).

Coexisting in the same study area were large numbers of *S. araneus* and some *S. minutus*. Although all three species had partially overlapping home ranges on many banks, the central banks so favoured by *N. fodiens* were inhabited by a mean of only 28.1% of *S. araneus* and *S. minutus* per sampling session, these terrestrial species occurring predominantly on the peripheral banks (Fig. 4.4). A mean of 1.3 *S. araneus*/*S. minutus* inhabited the main central bank (110 m long) compared with 4.1 individuals on the most-preferred peripheral bank (160 m long) where a mean of only 0.7 *N. fodiens* existed. Overlap (Percentage Similarity: Southwood, 1978) in site use between *N. fodiens* and the terrestrial species was 62.3%. This suggests that these species were tending to avoid each other, or that the different sites predominantly occupied by each species favoured their respective modes of life and access to prey.

Abundance of water shrews

Although they are widely distributed and can be locally abundant in selected habitats, water shrews have much lower population densities than most of their terrestrial counterparts, and generally constitute only a small proportion of the population in multi-species communities of shrews. In a range of Russian taiga habitats sampled over a 13 year period, *N. fodiens* constituted only 0.5% of the shrew catch and was the second rarest of the nine species present (Sheftel, 1989). Even in its peak year of abundance it constituted only 0.9%. In various biotopes in Bialowieza, Poland, *N. fodiens* constituted 6.5% of the catch of five species, and *N. anomalus* only 1.6%, although they were locally abundant and numerically dominant in hydric habitats with > 30% of the catch (Aulak, 1970). Similarly, *N. fodiens* constituted 8.5% of the shrew catch in marsh habitat (3.7% in a range of habitats) in western France and was greatly outnumbered by *S. coronatus* (Yalden *et al.*, 1973). In a 3 year study of a Swiss canal-side habitat, monthly captures of *S. coronatus* again outnumbered those of *N. fodiens* (Cantoni, 1993). Stein (1975) recorded proportions of 6% *N. fodiens* and 94% *S. araneus* in hydric habitats over an 11 year period and found that the proportions of trapped animals were similar to those found in barn owl pellets, concluding that *N. fodiens* is not an abundant species. In the favoured habitat of water-cress beds in England (Churchfield, 1984*b*), *N. fodiens* constituted a much higher proportion of the shrew catch over a 2 year period (31%), but was always outnumbered by *S. araneus* (52%) (Fig. 4.5).

   Sorex palustris has similarly low levels of abundance. In an extensive survey of 1000 sites in Manitoba where *S. palustris* is common, it constituted only 13.6% of the total catch of six species in hydric habitats (swamp, bog and fen), and 1.4% in mesic habitats (forest, shrub and grassland) (Wrigley *et al.*, 1979). It was most abundant in marsh and fen but, even here, it constituted only 15.9% of shrew captures and ranked third in abundance after two common terrestrial species. Trapping beside, and in close proximity to, streams running through forest habitats in New Brunswick, Whitaker & French (1984) found this water shrew to be the rarest of the six shrew species present, constituting approximately 1.6% of the catch.

   Captures of *Chimarrogale* and *Nectogale* have been very few, suggesting that population density and abundance of these water shrews is also low, and that they are relatively rare species. For example, during an extensive survey of small mammals in various habitats in Nepal, Abe (1982) collected only two specimens of *Chimarrogale* (0.77% of total shrew catch).

   Such low abundance relative to many terrestrial shrew species may simply be evidence of inefficiency in capturing water shrews or, bearing in mind their

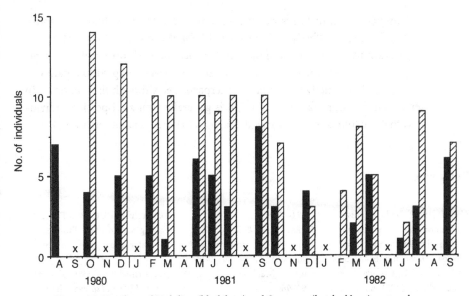

Figure 4.5. Numbers of *N. fodiens* (black bars) and *S. araneus* (hatched bars) captured per trapping period (after Churchfield, 1984*b*).

site specificity, failure to locate their favoured habitats or sites for sampling. The first reason seems unlikely since the use of live-traps, snap-back traps and pitfalls have given similar results. Owl pellet analysis also suggests that these shrews are generally not abundant. It seems that, with few exceptions, water shrews are outnumbered by common terrestrial species, even where their favoured sites have been located for study.

With the advantages of aquatic foraging techniques, the abundance of aquatic prey, the need for very small home ranges, and the advantage of relatively large body size providing superiority in interference competition with smaller shrews, why it is that they are not more abundant? The reasons can only be speculated upon.

Given that water shrews are generally more habitat-specific than other shrews, the availability of suitable habitats may be limited (today, if not palaeo-historically). Their favoured habitats cover smaller areas than larger-scale habitats such as forests and grasslands, and are more isolated. Although stream systems offer dispersal corridors over large distances, water shrews are highly selective in their habitat use and, as we have seen, show considerable site specificity. This may be related to their limited diving ability and the availability of sites with suitable water depths to permit foraging amongst the substratum. So, their populations may be more isolated and restricted than their less specialized terrestrial counterparts.

Competition with terrestrial species, coupled with an inability to survive well and compete effectively in terrestrial habitats, may also be a factor, and this in turn may limit their population densities and their dispersal. Anatomical features (see above) suggest that they are not as well-adapted to conditions on land as their strictly terrestrial counterparts are. Although they do occur in terrestrial situations far from water, it is generally young, dispersing individuals that are found there (Corbet & Harris, 1991; personal observations).

References

Abe, H. (1982). Ecological distribution and faunal structure of small mammals in central Nepal. *Mammalia* **46**: 477–503.

Arai, S., Mori, T., Yoshida, H. & Shiraishi, S. (1985). A note on the Japanese water shrew, *Chimarrogale himalayica platycephala*, from Kyushu, *J. mammal. Soc. Japan* **10**: 193–203.

Aulak, W. (1970). Small mammal communities of the Bialowieza National Park. *Acta theriol.* **15**: 465–515.

Banfield, A. F. W. (1974). *The mammals of Canada*. University of Toronto Press, Toronto.

Benenski, J. T. & Stinson, D. W. (1987). *Sorex palustris. Mammal. Sp.*, No. 296: 1–6.

Boernke, W. E. (1977). A comparison of arginase maximum velocities from several poikilotherms and homeotherms. *Comp. Biochem. Physiol.* **56**: 113–116.

Buchalczyk, T. & Pucek, Z. (1963). Food storage of the European water-shrew, *Neomys fodiens* (Pennant, 1771). *Acta theriol.* **7**: 376–379.

Buckner, C. H. (1970). Direct observation of shrew predation on insects and fish. *Blue-Jay* **28**: 171–172.

Buckner, C. H. & Ray, D. G. H. (1968). Notes on the water shrew in bog habitats of south-eastern Manitoba. *Blue-Jay* **26**: 95–96.

Calder, W. A. (1969). Temperature relations and underwater endurance of the smallest homeothermic diver, the water shrew. *Comp. Biochem. Physiol.* **30**: 1075–1082.

Cantoni, D. (1993). Social and spatial organization of free-ranging shrews, *Sorex coronatus* and *Neomys fodiens* (Insectivora, Mammalia). *Anim. Behav.* **45**: 975–995.

Churchfield, S. (1982). Food availability and the diet of the common shrew. *Sorex araneus,* in Britain. *J. Anim. Ecol.* **51**: 15–28.

Churchfield, S. (1984a). Dietary separation in three species of shrew inhabiting water-cress beds. *J. Zool., Lond.* **204**: 211–228.

Churchfield, S. (1984b). An investigation of the population ecology of syntopic shrews inhabiting water-cress beds. *J. Zool., Lond.* **204**: 229–240.

Churchfield, S. (1985). The feeding ecology of the European water shrew. *Mammal Rev.* **15**: 13–21.

Churchfield, S. (1990). *The natural history of shrews*. Christopher Helm/A. & C. Black, London.

Churchfield, S. & Sheftel, B. I. (1994). Food niche overlap and ecological separation in a multi-species community of shrews in the Siberian taiga. *J. Zool., Lond.* **234**: 105–124.

Clark, T. W. (1973). Distribution and reproduction of shrews in Grand Teton National Park, Wyoming. *NW. Sci.* **47**: 128–131.

Conaway, C. H. (1952). Life history of the water shrew (*Sorex palustris navigator*). *Am. Midl. Nat.* **48**: 219–248.

Corbet, G. B. (1978). The mammals of the Palaearctic region: a taxonomic review. *Publs Br. Mus. (Nat. Hist.)*, No. 788: 1–314.

Corbet, G. B. & Harris, S. (Eds) (1991). *The handbook of British Mammals*. (3rd edn). Blackwell Scientific, Oxford.

Corbet, G. B. & Hill, J. E. (1991). *A world list of mammalian species*. (3rd edn). Natural History

Museum Publications, London and Oxford University Press, Oxford.

Dehnel, A. (1950). Studies on the genus *Neomys* Kaup. *Annls Univ. Mariae Curie-Sklodowska (C)* **5**: 1–63.

DuPasquier, A. & Cantoni, D. (1992). Shifts in benthic macroinvertebrate community and food habits of the water shrew. *Neomys fodiens* (Soricidae, Insectivora). *Acta oecol.* **13**: 81–99.

Gureev, A. A. (1971). [*Shrews (Soricidae) of the world.*] Nauka Press, Leningrad. [In Russian].

Hall, E. R. (1981). *The mammals of North America* 1. John Wiley & Sons, New York.

Harrison, J. L. (1958). *Chimarrogale hantu* a new water shrew from the Malay Peninsula, with a note on the genera *Chimarrogale* and *Crossogale* (Insectivora, Soricidae). *Ann. Mag. nat. Hist.* (13) **1**: 282–290.

Hoffmann, R. S. (1987). A review of the systematics and distribution of Chinese red-toothed shrews (Mammalia: Soricinae). *Acta theriol. sin.* **7**: 100–139.

Hutterer, R. (1980). Das Rhinarium von *Nectogale elegans* und anderen Wasserspitzmäusen (Mammalia, Insectivora). *Z. Säugetierk.* **45**: 126–127.

Hutterer, R. (1985). Anatomical adaptations of shrews. *Mammal Rev.* **15**: 43–55.

Hutterer, R. (1992). Ein Lebensbild der Tibetanischen Wasserspitzmaus (*Nectogale elegans*). In *Semiaquatische Säugetiere*: 39–51. (Eds Schröpfer, R., Stubbe, M. & Heidecke,

D.). *Martin-Luther-Univ. Halle-Wittenberg*, Wiss. Beitr.

Hutterer, R. & Hürter, T. (1981). Adaptive Haarstrukturen bei Wasserspitzmäusen (Insectivora, Soricinae). *Z. Saugetierk.* **46**: 1–11.

Illing, K., Illing, R. & Kraft, R. (1981). Freilandbeobachtungen zur Lebensweise und zum Revierverhalten der europäischen Wasserspitzmaus, *Neomys fodiens* (Pennant, 1771). *Zool. Beitr.* **27**: 109–122.

Ivanter, E. V. (1994). The structure and adaptive peculiarities of the pelage in soricine shrews. In *Advances in the biology of shrews. Carnegie Mus. nat. Hist. Spec. Publ.* No. 18: 441–454.

Jones, G. S. & Mumford, R. E. (1971). *Chimarrogale* from Taiwan. *J. Mammal.* **52**: 228–231.

Köhler, D. (1991). Notes on the diving behaviour of the water shrew. *Neomys fodiens* (Mammalia, Soricidae). *Zool. Anz.* **227**: 218–228.

Koyasu, K. (1993). *Tracks, trails and signs of the Japanese land mammals.* Nikkei Science Inc., Japan.

Kraft, R. & Pleyer, G. (1978). Zur Ernährungsbiologie der Europäischen Wasserspitzmaus, *Neomys fodiens* (Pennant, 1771), an Fischteichen. *Z. Säugetierk.* **43**: 321–330.

Kuvikova, A. (1985). Zur Nahrung der Wasserspitzmaus, *Neomys fodiens* (Pennant, 1771) in der Slowakei. *Biológia, Bratisl.* **40**: 563–572.

Lardet, J. -P. (1988a). Spatial behaviour and activity patterns of the water shrew *Neomys fodiens*

in the field. *Acta theriol.* **33**: 293–303.

Lardet, J. -P. (1988b). Evolution de la température corporelle de la musaraigne aquatique (*Neomys fodiens*) dans l'eau. *Revue suisse Zool.* **95**: 129–135.

Lardet, J. -P. & Vogel, P. (1985). Evolution démographique d'une population de musaraignes aquatiques (*Neomys fodiens*) en Suisse Romande. *Bull. Soc. vaud. Sci. nat.* **368**: 353–360.

Niethammer, J. (1977). Ein syntopes Vorkommen der Wasserspitzmäuse *Neomys fodiens* und *Neomys anomalus. Z. Säugetierk.* **42**: 1–6.

Niethammer, J. (1978). Weitere Beobachtungen über syntope Wasserspitzmäuse der Arten *Neomys fodiens* und *N. anomalus. Z. Säugetierk.* **43**: 313–321.

Niethammer, J. & Krapp, F. (Eds) (1990). *Handbuch der Säugetiere Europas.* 3 (1) Insektenfresser–Insectivora Aula-Verlag, Wiesbaden.

Nowak, R. M. & Paradiso, J. L. (Eds) (1983). *Walker's mammals of the world,* 1 (4th edn). Johns Hopkins University Press, Baltimore & London.

Nussbaum, R. A. & Maser, C. (1969). Observations of *Sorex palustris* preying on *Dicamptodon ensatus. Murrelet* **50**: 23–24.

Pattie, D. (1973). *Sorex bendirii. Mammal. Sp.* No. 27: 1–2.

Price, M. (1953). The reproductive cycle of the water shrew, *Neomys fodiens bicolor. Proc. zool. Soc. Lond.* **123**: 599–621.

Pruitt, W. (1957). A survey of the mammalian family Soricidae (shrews). *Säugetierk. Mitt.* 5: 18–27.

Pucek, M. (1959). The effect of the venom of the European water shrew (*Neomys fodiens fodiens* Pennant) on certain experimental animals. *Acta theriol.* 3: 93–104.

Pucek, M. (1969). *Neomys anomalus* Cabrera, 1907 – a venomous mammal. *Bull. Acad. Pol. Sci.* (*Sér. Sci. biol.*) 17: 569–573.

Pucek, Z. (Ed.) (1981). *Keys to vertebrates of Poland. Mammals.* Polish Scientific Publishers, Warsaw.

Rychlik, L. (1995). *Differences in Foraging Behaviour between Neomys anomalus and N. fodiens.* PhD thesis: Mammal Research Institute, Polish Academy of Sciences, Bialowieza.

Schloeth, R. (1980). Freilandbeobachtungen an der Wasserspitzmaus, *Neomys fodiens* (Pennant, 1771) im Schweizerischen Nationalpark. *Revue suisse Zool.* 87: 937–939.

Schröpfer, R. (1983). Die Wasserspitzmaus (*Neomys fodiens* Pennant 1771) a/s Biotopgutanzeiger fur Uferhabitate an Fliessgewassern. *Verh. Dt. zool. Ges.* 76: 137–141.

Sheftel, B. I. (1989). Long-term and seasonal dynamics of shrews in Central Siberia. *Annls. zool. Fenn.* 26: 357–369.

Stein, G. H. W. (1975). Uber die Bestandsdichte und ihre Zusammenhänge bei der Wasserspitzmaus, *Neomys fodiens* (Pennant, 1771). *Mitt. zool. Mus. Berl.* 51: 187–198.

Southwood, T. R. E. (1978). *Ecological methods: with particular reference to the study of insect populations.* (2nd rev. edn) Chapman & Hall, London.

Van Bemmel, A. C. & Voesenek, L. A. C. J. (1984). The home range of *Neomys fodiens* (Pennant, 1771) in the Netherlands. *Lutra* 27: 148–153.

Van Zyll de Jong, C. G. (1983). *Handbook of Canadian mammals. 1. Marsupials and insectivores.* National Museums of Canada, Ottawa.

Vogel, P. (1990). Body temperature and fur quality in swimming water-shrews, *Neomys fodiens* (Mammalia, Insectivora). *Z. Säugetierk.* 55: 73–80.

Weissenberger, T., Righetti, J. -F. & Vogel, P. (1983). Observations de populations marquées de la musaraigne aquatique *Neomys fodiens* (Insectivora, Mammalia). *Bull. Soc. vaud. Sci. Nat.* 76: 381–390.

Whitaker, J. O. Jr. (1980). *The Audubon Society field guide to North American mammals.* Alfred A. Knopf, New York.

Whitaker, J. O. Jr. & French, T. W. (1984). Foods of six species of sympatric shrews from New Brunswick. *Can. J. Zool.* 62: 622–626.

Whitaker, J. O. Jr. & Maser, C. (1976). Food habits of five western Oregon shrews. *NW. Sci.* 50: 102–107.

Whitaker, J. O. Jr. & Schmeltz, L. L. (1973). Food and external parasites of *Sorex palustris* and food of *Sorex cinereus* from St. Louis County Minnesota *J. Mammal.* 54: 283–285.

Wilson, D. E. & Reeder, D. M. (Eds) (1993). *Mammal species of the world. A taxonomic and geographic reference.* (2nd edn). Smithsonian Institution Press, Washington & London.

Wolk, K. (1976). The winter food of the European water-shrew. *Acta theriol.* 21: 117–129.

Wrigley, R. E., Dubois, J. E. & Copland, H. W. R. (1979). Habitat, abundance, and distribution of six species of shrews in Manitoba. *J. Mammal.* 60: 505–520.

Yalden, D. W., Morris, P. A. & Harper, J. (1973). Studies on the comparative ecology of some French small mammals. *Mammalia* 37: 257–276.

5

The importance of the riparian environment as a habitat for British bats

P. A. Racey

Introduction

This review has three main aims: to examine the distribution of bat roosts in relation to rivers and other water bodies, to assess the evidence that some bat species forage preferentially along rivers, and to consider what resources are provided by the rivers for the bats. Recent studies on the effects of eutrophication on bats will also be described. The review will be based principally on information obtained in Britain and will include the published and unpublished results obtained by myself and my colleagues on five rivers in Scotland: the Dee, Don, Ythan, Spey and Tay. The conclusions of these studies together with those selected from mainland Europe and North America are thought to apply equally to rivers in England and Wales or point to the need for similar research in these countries.

Fourteen species of vespertilionid bats breed in the UK and nine are referred to in this review (Table 5.1).

Are bat roosts concentrated in river valleys?

Roosts in buildings

Over the last 20 years, data have been collected of the locations of bat roosts in north-east Scotland (ca. 57 °N), where four bat species commonly occur (Speakman *et al.*, 1991). The pipistrelle is the commonest species and accounted for 147 of 184 bat roosts known in 1989, located close to the Rivers Dee and Don from source to sea (Fig. 5.1). Thirty-four of the roosts were occupied by brown long-eared bats, most of which were concentrated along a well-wooded stretch of the River Dee (Entwistle, 1994). Seven roosts of Daubenton's bats were located, mostly adjacent to rivers. In north-east Scotland only three roosts of Natterer's bat have been found, also close to rivers, although in Tayside, Swift (1997) found 11 roosts of this species either close to rivers or in river

Table 5.1.*British bat species referred to in the text, their distribution and status*

	Distribution			
Species	England	Wales	Scotland	Status
1. Daubenton's bat (*Myotis daubentoni*)	+	+	+	Common
2. Brandt's bat (*Myotis brandti*)	+	+		Scarce
3. Whiskered bat (*Myotis mystacinus*)	+	+		Scarce
4. Natterer's bat (*Myotis nattereri*)	+	+	+	Frequent
5. Pipistrelle bat (*Pipistrellus pipistrellus*)	+	+	+	Common
6. Noctule bat (*Nyctalus noctula*)	+	+	+	Frequent
7. Leisler's bat (*Nyctalus leisleri*)	+			Rare
8. Brown long-eared bat (*Plecotus auritus*)	+	+	+	Common
9. Serotine (*Eptesicus serotinus*)	+	+		Frequent

Data given in the table have been modified from Hutson (1993*a*), distribution is based on information in Arnold (1993).

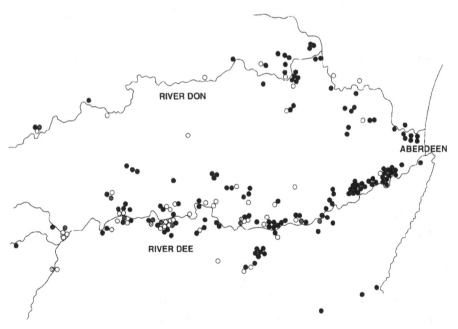

Figure 5.1. Bat roosts in north-east Scotland in relation to two of the major rivers. ●, pipistrelle; ○, brown long-eared; ⊛, Daubenton's; ◉ Natterer's.

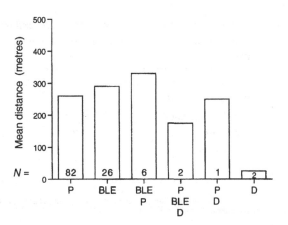

Figure 5.2. The mean distance from bat roosts in houses in north-east Scotland to the nearest water body shown on Ordnance Survey maps (from Park, 1988). Nine houses contained roosts of more than one bat species. N, number of houses containing bat roosts. P, pipistrelle; BLE, brown long-eared; D, Daubenton's.

valleys. A similar survey of roost distribution was conducted by the Durham Bat Group, and revealed an association between bat roosts and the rivers Derwent, Wear and Tees (Sargent, 1991).

The mean distance between the roosts in north-east Scotland and water (rivers and lochs) was about 300 m (Park, 1988) (Fig. 5.2). In Northumberland at ca. 54 °N, the majority of roosts of six bat species were located within 100 m of freshwater (Table 5.2; Fig. 5.3: Northumberland Bat Group, 1985). Although such roost distributions suggest an association between bats and rivers, some caution is required in the interpretation of these results. Bats are highly synanthropic and, during summer, most temperate zone species roost in buildings, behind soffits or weather boarding, in cavity walls and roof voids (Hutson, 1993b). Buildings are often concentrated in river valleys so that the observed distribution of roosts is hardly surprising. Furthermore, a study of roost site selection (Entwistle et al., 1997) and foraging behaviour (Entwistle et al., 1996) of brown long-eared bats has revealed their preference for large houses more than 100 years old with complex roof voids containing several compartments and with immediately adjacent deciduous woodlands in which the bats forage mainly by gleaning prey from foliage (Anderson & Racey, 1991). The abundance of such houses in river valleys explains the observed pattern of roost distribution in the valley of the River Dee and there is no evidence that the distribution of brown long-eared bats is related to the proximity of rivers themselves. The association between roosts of Natterer's bats and river valleys may also be explained by the presence of woodland in these valleys where this species also gleans at least part of its prey from foliage (Arlettaz, 1996; Swift, 1997).

Stronger inferences may be made about the association between roosts of other bat species and rivers by comparing the incidence of such roosts in

Table 5.2. *Distance (m) between bat roosts and fresh water in Northumberland*

Bat species	Distance (m)	n
Pipistrelle	141±116	18
Brown long-eared	178±187	10
Whiskered/Brandt's	105± 53	4
Natterer's	140±162	3
Daubenton's	50	1
Noctule	25	1

Data collected by the Northumberland Bat Group (1985).

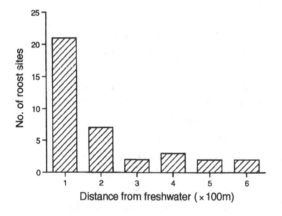

Figure 5.3. The distance between bat roosts and freshwater in Northumberland (Northumberland Bat Group, 1985).

buildings situated in river valleys and those elsewhere. A survey conducted by Pritchard & Murphy (1987) of the buildings in Glen Lyon and Glen Rannoch, in Perthshire, Scotland (ca. 56°N), is unique in this respect. The area is lightly wooded, the majority of the buildings are of stone, with slate roofs over rough pine underboarding, and modern housing development is minimal. Of the 450 buildings present, 360 were inspected and bats had roosted in 82, 61 of which were occupied in the year of the survey. Of the dwelling houses occupied by humans, 26% were also used by bats. The incidence of bat roosts (one in six buildings) is far higher than in areas of countryside without river valleys (Fig. 5.1) and points to an association between the roosts of some bat species and rivers.

Roosts in bridges

In recent years, considerable attention has been paid to bats roosting in bridges, out of concern that repointing work was entombing bats that had

Table 5.3. *Bat species roosting in North Yorkshire bridges*

Bat species	Summer roosts	Winter roosts
Daubenton's	16	3
Natterer's	5	2
Noctule	1	1
Brown long-eared	—	1
Unidentified	6	—

Data from Roberts (1989).

Table 5.4. *Bat species roosting in bridges in Cork and Waterford*

Bat species	Bridges n	Bats/roost mean	Bats/bridge mean
Daubenton's	38	1.39	1.76
Brown long-eared	8	2.76	3.0
Natterer's	4	1	1
Whiskered	3	1	1
Pipistrelle	3	1–2	1–2

Data from Smiddy (1991).

taken up residence (Mitchell-Jones, 1989). Roberts (1989) surveyed 306 bridges in North Yorkshire and considered that 78, almost all of which were built of stone, contained potential roost sites. Roosting was confirmed by the discovery of bats, or their faeces, in a third of these. Many of these bridges had been built in stages, and had been widened, sometimes twice, to meet increased traffic demand, and the commonest roosting site identified by Roberts (1989) was in the gaps between the stages on the underside of the bridges. The commonest bat species roosting in these bridges was Daubenton's (Table 5.3), and Roberts (1989) concluded that bridges were important roost sites for this species in North Yorkshire.

A similar survey was carried out by Smiddy (1991) in the west of Ireland, where 14% of 366 bridges contained roosting bats, 11% showed evidence of roosting, 26% had suitable roosting sites but no evidence of occupancy and the remainder were considered unsuitable for bats. Daubenton's bat was again the commonest species encountered (Table 5.4) although the numbers of individuals in each bridge was low. These are likely to have been males or non-reproductive females, since reproducing females form large nursery colonies during pregnancy and lactation (Speakman, 1991).

In an extensive survey of canal bridges and tunnels in England and Wales, Richards (1992) reported a total of 45 Daubenton's bats in eight bridges and five tunnels, the largest colony consisting of 19 bats in a bridge. In contrast, a survey of 90 bridges in Fife (Fife Bat Group, 1988) revealed no bats and only two individual Daubenton's bats were found in a survey of 72 bridges in north-east Scotland (Speakman *et al.*, 1991).

Roosts in trees

Bats are thought to have evolved in association with trees and several European species such as noctules (Racey, 1991), Daubenton's bats (Childs & Aldhous, 1995) and pipistrelles (F. Mathieson, personal communication) still roost in holes in trees in the UK, as well as in buildings. In Switzerland, 50 roosts of Daubenton's bats were located in 1.5 km^2 of woodland, mainly in beech and oak trees (Rieger, 1996) and a radiotracking study of this species foraging over the River Rhine located 67 roosts, 60 of which were in trees (Rieger & Alder, 1995). A study currently in progress in the valleys of the Rivers Wye, Ithon and Usk in Wales has revealed that Daubenton's bat breeds in bat boxes installed close to the river (J. Messenger, personal communication). Roosts have also been found in tree holes but not in buildings.

Do bats forage preferentially in river valleys?

Autecological studies

In a study of Daubenton's bat in the valley of the River Spey (250 m above sea level (a.s.l.)), in the Highlands of Scotland, bats foraged almost entirely over or adjacent to the river, over riparian vegetation (Swift & Racey, 1983). They also foraged over pools and drainage ditches that connected pools and river.

In a similar study of the foraging behaviour of pipistrelles in the lowland agricultural valley of the River Don (100 m a.s.l.), with a range of land uses, marked individuals followed the river, feeding over water and within 100 m of the river, in areas where there were trees and thick undergrowth (Racey & Swift, 1985). The only bats seen further from the river were over two large ponds, about 600 m from the river. No bats were sighted where the river passed through open fields with no riparian trees nor were they seen in or near coniferous plantations on the valley sides. In an upland area in Perthshire, Scotland (350–750 m a.s.l.) containing a stretch of the River Blackwater and its catchment, pipistrelles foraged over all areas containing deciduous trees and water, including the river itself and a small loch, and were not observed in open fields or moorland (Racey & Swift, 1985).

Similar results were found by Sargent (1991) in County Durham. Daubenton's bats fed almost exclusively over rivers and showed a significant preference for wide rivers with calm water and adjacent deciduous woodland. Pipistrelles also favoured wide rivers but showed no preference for the presence of calm or turbulent water or for a particular level of bankside tree cover. Pipistrelle activity was also recorded around deciduous trees and hedges away from the river. Whiskered bats selected narrower rivers with abundant bankside vegetation up to 5 m high.

From these initial studies, it was not clear whether pipistrelles feeding in riparian situations were associated with the vegetation, mainly riparian trees, or with the water itself or with a combination of the two. This question was addressed by Rydell et al. (1994) who studied 84 sites in the valleys of the Rivers Dee and Ythan on 19 nights between June and August, representing the principal foraging habitats available in the two valleys – woodland, open farmland, ponds and rivers. The Dee is the larger of the two rivers with a catchment area of 2100 km². It is fed mainly by mountain streams and woodland occurs along 40% of its banks. In contrast, the Ythan is a much smaller river, with a catchment area of 690 jm² consisting mainly of farmland. It is entirely treeless for much of its length, with woodland occurring along only 15% of its banks. The Dee is oligotrophic (Jenkins, 1985) whereas the Ythan is eutrophic (Macdonald et al., 1995). Each of the woodland and farmland sites was 100–1000 m from the river and each was 0.5 ha in size, and 100 m of river or pond bank were surveyed. Each site was visited once and monitored by using a bat detector for 15 min, sometime between 1 h after sunset and midnight, avoiding the dusk period, since the time at which bats start to feed varies with species. The occurrence of pipistrelles and Daubenton's bats differed significantly between woodlands, open farmlands, ponds and rivers (Fig. 5.4). Pipistrelles were observed in every river and pond site, but also occurred frequently in woodlands away from water, and occasionally in open farmland. Daubenton's bats were observed in most river and pond sites, but were observed only once in woodland and never in open farmland. Brown long-eared bats were observed only in woodland, and drinking from a pond. The number of bat species detected in each site differed significantly between the sampled habitats and the modal numbers of species was 2 over water, 1 in woodland, and none in farmland (Rydell et al., 1994).

If the distribution of bats along the rivers was affected by the presence of riparian woodland, more bats would be expected over the river Dee where the

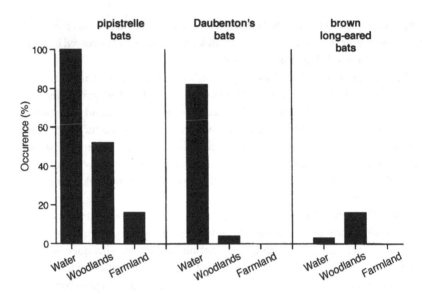

Figure 5.4. Frequency of occurrence (%) of common pipistrelles, Daubenton's and brown long-eared bats in 0.5 ha sites located over water ($n = 40$), in woodlands ($n = 25$) and in open farmland ($n = 19$). Each site was monitored for 15 min using a bat detector (Rydell *et al.*, 1994). Reproduced by courtesy of the publishers of *Folia Zoologica*.

riparian woodlands are more extensive than over the river Ythan, which is bordered mainly by treeless farmland. However, the abundance of foraging pipistrelles and Daubenton's bats did not differ significantly between the two (Fig. 5.5), and Rydell *et al.* (1994) concluded that the river itself is more attractive to foraging bats than riparian vegetation.

The habitat preferences of European bats may vary with latitude. In a comparative study of Daubenton's and whiskered bats in Finland (ca. 60 °N), Nyholm (1965) showed that both species foraged in woodland during the first half of the summer, where they were less conspicuous to avian predators, before moving to open hunting areas which, for both species, included ponds, rivers and lakes as well as meadows. More recently, Taake (1984) reported the preference of whiskered bats for habitats containing flowing water and riparian vegetation. Sixty-nine percent of sightings were within 200 m of ditches, brooks and small rivers with flowing water. Similar observations were made by P.W. Richardson (personal communication) in Northamptonshire where whiskered bats forage over dense willow and alder scrub on the edges of all types of water body.

In Central Scotland, 17% of sightings of foraging Natterer's bats were over sheltered areas of water surrounded by vegetation (Swift, 1997). In the

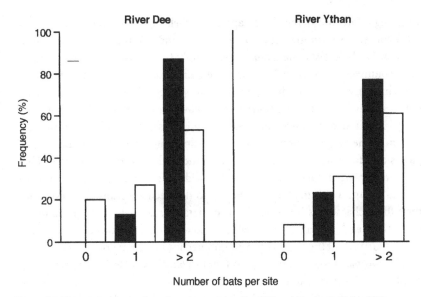

Figure 5.5. The minimum number of common pipistrelles (■) and Daubenton's bat (□) encountered in each 0.5 ha site over the River Dee ($n = 15$) and the River Ythan ($n = 13$). Each site was monitored for 15 min using a bat detector (Rydell *et al.*, 1994). Reproduced by courtesy of the publishers of *Folia Zoologica*.

Bialowieza Primeval Forest in eastern Poland (52 °N), Rachwald (1992) recorded the highest activity of the noctule bat over a river, and Kronwitter (1988) reported that the most frequently used foraging habitat of noctules radiotracked in Germany was over a lake. Similarly in Northamptonshire, P.W. Richardson (personal communication) observed noctules feeding over all water bodies, often following the line of rivers and canals and turning at features such as bridges and telegraph poles. This raises the question of the extent to which bats use linear landscape features such as rivers for navigation as well as sources of food (Limpens *et al.*, 1989; Limpens & Kapteyn, 1991).

The UK National Bat Habitat Survey

The scale of the studies reviewed in the previous section was greatly expanded during the National Bat Habitat Survey (Walsh *et al.*, 1995; Walsh & Harris, 1996). This adopted a random stratified sampling system based on the Institute of Terrestrial Ecology's land classification scheme, which assigns every 1 km square in Britain to one of 32 land classes (Bunce *et al.*, 1981*a,b*, 1983). Squares in each land class have a similar climate, physiogeography and pattern of land use, and within each land class a sample of 1 km squares was selected at random, to avoid observer bias in the selection of sites and ensure a standard sampling effort in different landscape types. Field work was carried out over three consecutive summers from 1990 to 1992, and involved professional and amateur bat biologists, the latter drawn mainly from Britain's 90 bat groups. Each volunteer was allocated one or more 1 km squares and walked a transect in each square four times during predetermined periods in summer, avoiding nights when weather conditions were unfavourable to bats. They carried tuneable bat detectors (mainly QMC Mini 2) set at 45 khz in order to maximize the range of species encountered, and noted the total number of bat passes and feeding buzzes in each square and within each habitat type. The more experienced surveyors were able to identify some bats to species or species groups by tuning the bat detector to their echolocation calls and from their flight characteristics. Analysis of the data revealed relationships between bat activity and habitat variables within and across the 32 land classes, which for the purposes of the analysis were combined into seven major groups – three arable, two pastoral, marginal upland and upland. Avoidance or selection of habitat types was examined by constructing Bonferroni confidence intervals around the observed use of each habitat type (Neu *et al.*, 1974), and regression analyses were employed to evaluate habitats of critical importance in determining high bat activity in each land-class group.

Of the 1030 1 km squares surveyed, 910 provided data suitable for analysis, and involved 2700 h of search effort with 30 000 bat passes recorded in the 9000 km walked. Only 24% of bat passes were identified to a particular species or species group, and 71.0% of these were *Pipistrellus pipistrellus*, 17.0% *Myotis* spp., 7.6% *Nyctalus noctula*, 2.7% *Plecotus* spp. and 1.7% *Eptesicus serotinus* (Serotine bat). Since a similar proportion of the unidentified bat passes were probably *P. pipistrellus* (the most abundant bat in the UK, Harris *et al.*, 1995), the habitat relationships described apply to *P. pipistrellus* in particular.

The incidence of bat passes was strongly related to land class, with the greatest bat activity occurring in pastoral land classes. Across all land-class groups, bats tended to forage selectively in edge and linear habitats and avoided more open and intensively managed habitat types. They showed a far stronger preference for all water bodies and woodland edge than for any other habitat type, emphasizing the importance of these habitats as key foraging sites. Linear vegetation corridors, particularly tree-lined hedgerows and covered ditches, were also selected by bats, emphasizing the importance of landscape connectivity.

Bats foraged preferentially in habitats that were comparatively rare within each land-class group. For example, the percentage availability in each land-class group of the preferred habitats of woodland edge, treelines, hedgerows and water bodies ranged from 14 to 31%, with a mean of 25%. In contrast, the availability of habitats that were consistently avoided (moorland, arable and most grassland categories) ranged from 40 to 54% with a mean of 47%. Optimum habitats tend to be at the perimeters of other habitat types or linear strips, and so in comparison with contiguous blocks of pasture or arable land, for instance, their area is proportionally smaller. Water bodies generally represent less than 1% of the available habitat, and broadleaved woodland edge ranges from 3 to 4%. Because the selection patterns were consistent between individual land classes, the results of Walsh & Harris's (1996a) analysis can be summarized by habitat type rather than by land class (Table 5.5) and reveal the importance of all water bodies for foraging bats.

Analysis of habitat factors affecting high bat activity resulted in equations with high predictive power and of particular value in forecasting the effects of changes in land use on bats. Although vespertilionids use a diversity of habitats, these regression models identify riparian and woodland habitats as particularly important.

The only woodland category that was avoided by bats was coniferous plantations in upland landscapes, which often occur in large blocks (Walsh & Harris, 1996a). However, in a study of coastal Scots pine (*Pinus sylvestris*) and lodgepole pine (*P. contorta*) plantations in north-east Scotland, Neville (1986) showed that bat foraging activity was concentrated over ponds and water-courses, even when insect biomass was low. In a similar study in mixed lowland woodland with a range of habitat types, greatest bat activity occurred over ponds and woodland rides and more species (pipistrelles, Daubenton's, noctules and Leisler's) were found over ponds than over other habitat types (Walsh & Mayle, 1991).

What resources are available to bats foraging over rivers and other water bodies?

Food

By analysing the faeces of Daubenton's and pipistrelle bats, and also sampling the insects available to the latter species using a suction trap, Swift & Racey (1983) and Swift et al. (1985) showed that in north-east Scotland both species fed mainly on Trichoptera and Nematocera throughout pregnancy and lactation (Fig. 5.6), and the composition of pipistrelle diet reflected the availability of these insects over a wide range of abundances. Although Ephemeroptera and Neuroptera accounted for a small proportion of the diet in both bat species, they were significantly over-represented in pipistrelle diet, indicating that

Table 5.5. *Summary of habitats significantly selected by bats in Britain, used in proportion to availability or avoided in 19 discrete land classes*

Selected in all land classes	Selected in some land classes, never avoided	Selected in some and avoided in other land classes	Avoided in some land classes, never selected	Avoided in all land classes
Treeline	Hedgerow	Stone wall	Improved grassland	Arable
Broadleaved woodland edge	Coniferous woodland edge	Coniferous woodland opening	Semi-improved grassland	Moorland
Lake and reservoir	Mixed woodland edge	Scrub	Lowland unimproved glassland	Upland unimproved grassland
	Broadleaved woodland opening	Parkland		
	Mixed woodland opening	Urban land		
	Felled woodland	**Open ditch**		
	River and canal	**Covered ditch**		
	Stream			
	Pond			

Data from Walsh & Harris (1996a). Habitat types that include water bodies are listed in bold.

(a)

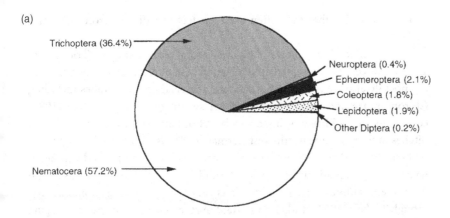

(b)

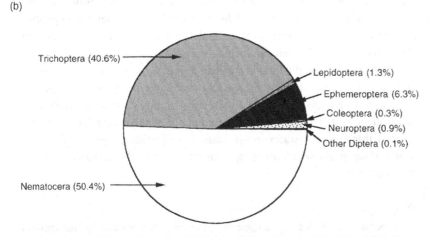

Figure 5.6. The diet of (a) Daubenton's and (b) pipistrelle bats (from Swift, 1981) established by faecal analysis. Daubenton's bat faeces were collected in roosts from May to August and the results are expressed as percentage frequency. Pipistrelle faeces were collected from pipistrelles mist-netted from July to mid-August and the results are derived from the actual numbers of insect fragments recovered.

when they occurred they were selected (Swift *et al.*, 1985). The close similarity between the diets of pipistrelles and Daubenton's bats was further emphasized by Sullivan *et al.* (1993) who analysed faeces collected from roosts in Ireland. Trichoptera, Ephemeroptera and several nematoceran families (Chironomidae and Culicidae) have aquatic larval stages and others (such as Ceratopogonidae and Psychodidae) are found in damp habitats. Although pipistrelles take flying prey (Kalko, 1995), Daubenton's bats frequently break the water surface and Jones & Rayner (1988) and Kalko & Schnitzler (1989), using photographic methods, revealed that they often use their feet to gaff prey from the water surface.

Sullivan *et al.* (1993) showed that Leisler's bat feeds mainly on Trichoptera and nematoceran and cyclorrhaphan Diptera; Gloor *et al.* (1995) revealed that the noctule also feeds on Trichoptera and several dipteran families including Chironomidae and Anisopodidae, which swarm over water. Taake (1992) compared the incidence of insects in faeces of bats hunting over ponds and ditches in forests in the north-west German lowlands with those in a light trap and concluded that the most important prey taxa for Brandt's and whiskered bats were non-aquatic insects such as tipulids. Similarly Natterer's bats fed on Brachycera, although Daubenton's bats fed mainly on Chironomidae. Although Taake's (1992) study is important in revealing that bat species foraging close to water do not necessarily feed on insects with aquatic larvae stages, Swift (1997) showed that faeces of Natterer's bats collected from one roost contained a high proportion of Trichoptera and from another a high proportion of Nematocera (Chironomidae and Ceratopogonidae). Natterer's bat clearly has a broad diet that can include aquatic insects.

In North America, the little brown bat, *Myotis lucifugus*, occupies a similar niche to Daubenton's bat and forages low over water on aerial prey, especially chironomids (Barclay, 1991). In one of the few studies where insect availability, determined by sticky traps, was compared in different habitats, Barclay (1991) showed that the over-water insect mass was significantly greater than that found along paths or in the neighbouring forest in the Kananaskis valley, Alberta, Canada.

Drinking water

Bats spend up to 20 h a day in their roosts during summer and, in hot weather, may become dehydrated. Webb (1992) showed that the maximum urine-concentrating ability of brown long-eared and Daubenton's bats was at the lower end of the range found in bats, and was lower than that predicted for a terrestrial mammal of the same body mass. Evaporation from resting bats was also higher than that predicted for terrestrial mammals, but not significantly different to that in other bat species of the same size. At a relative humidity of less than 20%, Webb *et al.* (1995) showed that resting brown long-eared and Daubenton's bats could lose over 30% of their body mass per day through evaporation. In free-flying captive colonies, Daubenton's bats both drank and fed sooner upon emerging from the roost than brown long-eared bats, and the total daily water intake of Daubenton's bats was greater than that of either brown long-eared bats or pipistrelles (Webb, 1992). Evaporative water loss will increase during flight and the need to drink on emergence from the roost may be one of the factors determining the proximity of roost sites to water. Access to water may be even more important during winter and Speakman & Racey (1989) have suggested that the primary function of the winter emergence flights recorded for many bat species (e.g. Avery, 1985) is to drink.

Mates

Since the density of bats foraging along rivers is comparatively high compared with other habitat types, mating opportunities may arise in such situations during autumn when females first come into oestrus. To investigate this possibility, the activity and behaviour of bats was monitored by Rydell et al. (1994) at 42 different sites on three rivers – the Dee, the Don and the Ythan – during nights in September and October coinciding with the mating season of pipistrelles (Racey, 1974). At each site, two observers, each with a bat detector, monitored bat activity under or immediately adjacent to a bridge, and simultaneously over open water 200–250 m away from the bridge. Bat passes and attempted prey captures (feeding buzzes) were counted for 10 min each, with the detector set at 50 kHz. The detector was then retuned to 20 kHz and mating calls were detected for a further 10 min. Although significantly more bat passes were detected near the bridges than over open water away from the bridges, there was no significant difference in the incidence of feeding buzzes at bridges compared with open water. These results indicate that the bats' feeding activities were not concentrated near bridges, but that some other activity was. Pipistrelle mating calls were significantly more frequent near the bridges than over open water away from the bridges, indicating that bridges may be used as mating stations by pipistrelles. In this way, males may be able to attract or intercept females foraging over rivers, and at the same time have access to food in the immediate vicinity of the territory. In a subsequent study in which calls were analysed sonagraphically, Russ (1995) recorded a high incidence of mating calls and territorial song flights in the vicinity of bridges.

Factors affecting the attractiveness of rivers to bats

Eutrophication

Eutrophication refers to nutrient enrichment and its effects on water bodies (Harper, 1992) and despite its widespread occurrence in the second half of this century there have been few studies of its consequences for foraging bats.

In a major study of habitat use in the province of Uppland, Sweden (59 °N), de Jong (1994) found that relatively open deciduous forests and adjacent shallow eutrophic lakes were the only habitats attracting large numbers and high diversity of bats during early summer, as a result of the high chironomid productivity (de Jong & Ahlén, 1991). Bats foraged in more diverse habitats later in summer, although lakes, wetland and deciduous forest remained important.

Vaughan *et al.* (1996) monitored bat activity and feeding rates upstream and downstream from 19 sewage outfalls in south-west England at the same time each night. The two recently described phonic types of the pipistrelle, in one of which most of the energy of the echolocation call is at 46 kHz and in the other 55 kHz (Jones & van Parijs, 1993), were recorded separately. Bat species of the genus *Myotis* are more difficult to separate by using bat detectors and were referred only to genus. Overall bat activity, as measured by passes, was reduced by 11% downstream of sewage outfalls (Fig. 5.7(a)) whereas feeding buzzes (the increase in pulse repetition rate during the terminal phase of an attack on an insect) were reduced by 28% (Fig. 5.7(b)). Both phonic types of pipistrelles were less active at downstream than at upstream sites, and there was clear evidence that the 46 kHz pipistrelle concentrated its feeding at upstream sites. Bats of the genus *Myotis*, on the other hand, foraged at higher rates downstream from sewage outfalls than upstream (Fig. 5.7(c)). Vaughan *et al.* (1996) concluded that for the conservation of pipistrelles, the maintenance of high standards of water quality may be important, but that Daubenton's bat may benefit from eutrophication.

This is of particular interest in view of Kokurewicz's (1995) hypothesis that increases in the numbers of chironomids resulting from eutrophication and canalization of waterways has resulted in the rise in numbers of hibernating Daubenton's bats in several countries in mainland Europe (Daan, 1980; Voûte *et al.*, 1980; Bartá *et al.*, 1981; Cerveny & Bürger, 1990; Weinrich & Oude Voshaar, 1992; Harrje, 1994). This runs counter to the general trend of declining bat numbers throughout Europe (Stebbings, 1988), which has been the subject of widespread concern and has led to the Agreement on the Conservation of Bats in Europe (Hutson, 1991).

An opportunity to test Kokurewicz's (1995) hypothesis arose in north-east Scotland where the Dee is oligotrophic (Jenkins, 1985), but the Ythan has experienced significant increases in nitrate levels over the last 30 years (North East River Purification Board, 1993; Macdonald *et al.*, 1995) and has been considered for designation as a nitrate-sensitive area under the EU Nitrate Directive. Racey *et al.* (1998) made pairwise comparisons of bats and insects at 10 sampling sites on the Dee and Ythan for three periods of 10 nights each during pregnancy, lactation and post-lactation of the bats. Bat detectors were tuned to 55 khz and 45 khz for 5 min each and counts were made of the

Figure 5.7. (a) Bat passes and (b) feeding buzzes of all bats (mainly pipistrelles and *Myotis* spp.) and (c) feeding buzzes of *Myotis* spp., upstream and downstream of sewage treatment works (Vaughan *et al.*, 1996). Reproduced by courtesy of the publishers of *Biological Conservation*.

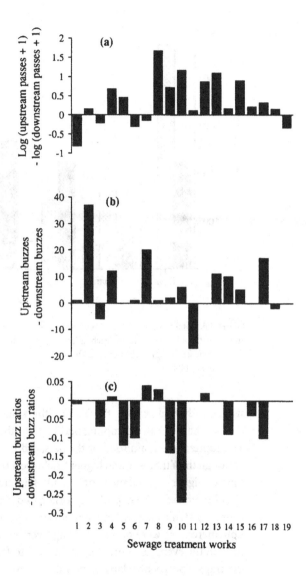

number of passes of the 55 khz and 45 khz pipistrelle phonic types, respectively. The detectors were then tuned to 35 khz for 5 min to count passes of Daubenton's bats. This sequence was then repeated. The less enriched of the two rivers, the Dee, supported higher numbers of bats than the Ythan in June, but this may have resulted from between-river differences in weather conditions (Fig. 5.8). In July and August/September, the abundances of the bat species did not differ significantly between the two rivers, but the density (i.e. abundance corrected for difference in river size) of pipistrelle bats was

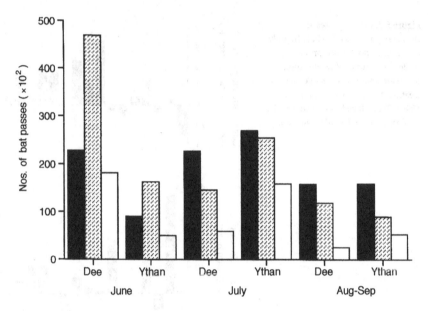

Figure 5.8. Total number of bat passes detected over each of the two rivers (the oligotrophic Dee and the eutrophic Ythan) during each sampling period (June, July and August/September). ■, *M. daubentonii;* ▨, *P. pipistrellus* of the 55 khz phonic type; □, *P. pipistrellus* of the 45 khz phonic type (Racey *et al.*, 1998).

significantly higher over the Ythan (the smaller of the two rivers) in July. During the latter two sampling periods, the abundance of Chironomidae and Trichoptera, the main food of the bats, did not differ between the two rivers, although the Ythan showed higher total insect biomass than the Dee, and also a much higher abundance of small Diptera, mainly Psychodidae and Cecidomyiidae, which probably originated from ditches draining the surrounding fertilized grassland and not from the main river (Fig. 5.9). Although Ephemeroptera were observed flying over both rivers, they were most active early in the evening before the bats started to feed, and were seldom caught in the traps. Species of relatively pollution-sensitive Trichoptera were caught in abundance at both rivers. Thus the current level of eutrophication of the Ythan appears to have a positive effect on insect biomass, and bat activity, and may be related to increased production of Diptera in the fertilized river valley. This supports Kokurewicz's (1995) hypothesis that eutrophication may have beneficial effects on Daubenton's bat. However, Racey *et al.* (1998) have shown that pipistrelles respond similarly to increased nutrient enrichment because they use the same feeding sites and eat the same insect taxa as Daubenton's bats.

Eutrophication can result in the production of algal blooms (Harper, 1992)

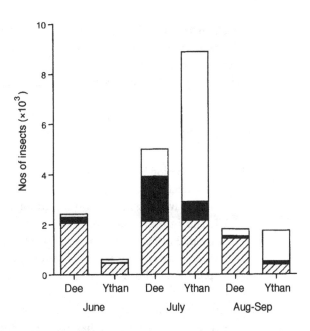

Figure 5.9. Total number of insects suction-trapped over each of the two rivers (the oligotrophic Dee and the eutrophic Ythan) during each sampling period (June, July and August/September). ▨, chironomids (Diptera: Chironomidae); ■, caddis flies (Trichoptera); □, all other insects. (From Racey *et al.*, 1998).

and there is a single report of the effects of such blooms on bats (Pybus *et al.*, 1986). Five hundred dead bats were counted on a lake in Alberta during one August, and mortality was estimated at twice that number. A toxic alkaloid (*Anabaena* Very Fast Death Factor) produced by the blue-green alga *A. flos-aquae* was extracted from the green slime on the carcasses of six *Myotis* spp. and one hoary bat (*Lasiurus cinereus*).

Turbulent water

Von Frenckell & Barclay (1987) showed that the little brown bat, *Myotis lucifugus*, feeding low over water, concentrated its activity over calm areas (ponds and stream pools) and avoided areas of turbulence or riffles. Differences in insect abundance, as measured by sticky traps, was not responsible for this preference. It was hypothesized that high-frequency sounds produced by running water may mask the weaker echoes of echolocation calls and thus reduce prey-detection efficiency. Greater levels of surface roughness (or clutter) may also increase the difficulty of detecting or capturing insects near the water surface by acting as an obstacle to flight and by producing extraneous background echoes that must be discriminated from prey echoes. In subsequent experiments, Mackey & Barclay (1989), showed that both artificial clutter and playbacks of the sound of turbulent water reduced the activity of *M. lucifugus*, while the activity of the big brown bat, *Eptesicus fuscus*, which feeds higher over the water surface, was reduced by playbacks but not by artificial clutter. In the only British study of this subject, Sargent (1991) recorded significantly higher activity levels of Daubenton's bats over pools compared to riffles, indicating that surface prey capture from disturbed water is difficult. There was no such difference in the activity levels of pipistrelle bats.

Future work

(1) The association between bat roosts and river valleys has been documented in few catchments in the northern half of the UK. More information is probably available in the roost records of the 90 amateur bat groups in the UK and in other bat roost data bases and is in need of collation and analysis. Further work on the distribution of bat roosts is needed in relation to river systems in southern England.

(2) At present, no clear relationship has been established between the use bats make of rivers and water quality. In particular, the extent to which eutrophication affects bats through changes in the abundance and diversity of their insect prey has received little attention. Further work is required on eutrophic and polluted rivers in England and Wales to determine the threshold levels of nutrient enrichment and pollution that affect bats through their insect prey, and to establish the extent to which Daubenton's bat is an indicator species as far as water quality is concerned.

(3) Many of the data on bats foraging over rivers and water bodies relate to Daubenton's bats and pipistrelles, and further studies are required to establish the importance of rivers in England and Wales to less common species such as noctules, Leisler's, whiskered and Brandt's bats.

ACKNOWLEDGEMENTS
Most of the work undertaken in north-east Scotland by the author and his research associates was supported by The Natural Environment Research Council. I am grateful to Drs A. C. Entwistle, J. Rydell, J. R. Speakman, S. M. Swift and A. L. Walsh, and to Mr A. M. Hutson, Ms G. Sargent and Miss N. Vaughan for comments on earlier drafts of the review. Dr J. R. Speakman jointly supervised some of the work reported here.

References

Anderson, M. E. & Racey, P. A. (1991). Feeding behaviour of captive brown long-eared bats, *Plecotus auritus. Anim. Behav.* **42**: 489–493.

Arlettaz, R. (1996). Foraging behaviour of the gleaning bat *Myotis nattereri* (Chiroptera, Vespertilionidae) in the Swiss Alps. *Mammalia* **60**: 181–186.

Arnold, H. (1993). *Atlas of mammals in Britain.* Her Majesty's Stationery Office, London.

Avery, M. I. (1985). Winter activity of pipistrelle bats. *J. Anim. Ecol.* **54**: 721–738.

Barclay, R. M. R. (1991). Population structure of temperate zone insectivorous bats in relation to foraging behaviour and energy demand. *J. Anim. Ecol.* **60**: 165–178.

Bartá, Z., Cerveny, J., Gaisler, I., Harnák, P., Horácek, L., Hurka, L., Miles, P., Nevrilý, M., Rumler, Z., Sklenár, J. & Salman, J. (1981). Results of winter census of bats in Czeckoslovakia 1969–1979. *Sb. oresniho Muz. Moste* **3**: 71–116.

Bunce, R. G. H., Barr, C. J. & Whitaker, H. A. (1981*a*). An integrated system of land classification. In *the Institute of Terrestrial Ecology report of 1980*: 28–33. Institute of Terrestrial Ecology, Cambridge.

Bunce, R. G. H., Barr, C. J. & Whitaker, H. A. (1981*b*). *Land classes in Great Britain: preliminary descriptions for users of the Merlewood method of land classification.* Institute of Terrestrial Ecology, Grange-over-Sands, Cumbria.

Bunce, R. G. H., Barr, C. J. & Whitaker, H. A. (1983). A stratified system for ecological sampling. In *Ecological mapping from ground, air and space*: 39–46. (Ed. Fuller, R. M.). Institute of Terrestrial Ecology, Abbots Ripton.

Cervený, J. & Bürger, P. (1990). Changes in bat population sizes in the Sumava Mountains (south-west Bohemia). *Folia zool.* **39**: 213–226.

Childs, J. & Aldhous, A. (1995). Chemiluminescent marking of Daubenton's bats at Stockgrove Country Park, Bedfordshire. *Bat News* No. 36: 3–4.

Daan, S. (1980). Long term changes in bat populations in the Netherlands: a summary. *Lutra* **22**: 95–118.

de Jong, J. (1994). *Distribution Patterns and Habitat Use by Bats in Relation to Landscape Heterogeneity and Consequences for Conservation.* Dissertation: Swedish University of Agricultural Sciences, Uppsala.

de Jong, J. & Ahlén, I. (1991). Factors affecting the distribution pattern of bats in Uppland, central Sweden. *Holarct. Ecol.* **14**: 92–96.

Entwistle, A. (1994). *The Roost Ecology of the Brown Long-eared bat Plecotus auritus.* PhD thesis: University of Aberdeen.

Entwistle, A. C., Racey, P. A. & Speakman, J. R. (1996). Habitat exploitation by a gleaning bat, *Plecotus auritus. Phil. Trans. R. Soc. (B)* **351**: 921–931.

Entwistle, A. C., Racey, P. A. & Speakman, J. R. (1997). Roost selection by the brown long-eared bat (*Plecotus auritus*). *J. appl. Ecol.* **34**: 399–408.

Fife Bat Group (1988). *Report.* The Bat Conservation Trust, London.

Gloor, S., Stutz, H. -P. & Ziswiler, V. (1995). Nutritional habits of the noctule bat *Nyctalus noctula* (Schreber, 1994) in Switzerland. *Myotis* **32–33**: 231–242.

Harper, D. (1992). *Eutrophication of fresh waters.* Chapman & Hall, London.

Harris, S., Morris, P., Wray, S. & Yalden, D. (1995). *A review of British mammals: population estimates and conservation status of British mammals other than cetaceans.* JNCC, Peterborough.

Harrje, C. (1994). Etho-ökologische Untersuchungen der ganzjährigen Aktivität von Wasserfledermäusen (*Myotis daubentoni* Kuhl, 1819) am winter quartier. *Mitt. naturf. Ges. Schaffhausen* **39**: 15–52.

Hutson, A. M. (1991). European Bats Agreement approved. *Bat News* No. 23: 4–5.

Hutson, A. M. (1993*a*). *Action plan for the conservation of bats in the United Kingdom.* The Bat Conservation Trust, London.

Hutson, A. M. (1993*b*). *Bats in houses.* The Bat Conservation Trust, London.

Jenkins, D. (1985). *The biology and management of the River Dee.* Institute of Terrestrial Ecology, Huntingdon.

Jones, G. & Rayner, J. M. V. (1988). Flight performance, foraging tactics and echolocation in free-living Daubenton's bats. *Myotis daubentoni* (Chiroptera: Vespertilionidae). *J. Zool., Lond.* **215**: 113–132.

Jones, G. & van Parijs, S. M. (1993). Bimodal echolocation in pipistrelle bats: are cryptic species present? *Proc. R. Soc. (B)* **251**: 119–125.

Kalko, E. K. V. (1995). Insect pursuit, prey capture and echolocation in pipistrelle bats (Microchiroptera). *Anim. Behav.* **50**: 861–880.

Kalko, E. K. V. & Schnitzler, H. -U. (1989). The echolocation and hunting behaviour of Daubenton's bat, *Myotis daubentoni. Behav. Ecol. Sociobiol.* **24**: 225–238.

Kokurewicz, T. (1995). Increased population of Daubenton's bat (*Myotis daubentonii* (Kuhl 1819)) (Chiroptera: Vespertilionidae) in Poland. *Myotis* **32–33**: 155–161.

Kronwitter, F. (1988). Population structure, habitat use and activity patterns of the noctule bat, *Nyctalus noctula* Schreb., 1774 (Chiroptera: Vespertilionidae) revealed by radio-tracking. *Myotis* **26**: 23–85.

Limpens, H. J. G. A. & Kapteyn, K. (1991). Bats, their behaviour and linear landscape elements. *Myotis* **29**: 39–48.

Limpens, H. J. G. A., Helmer, W., Van Winden, A. & Mostert, K. (1989). Vleermuizen (Chiroptera) en lint vormige landschapselementen. *Lutra* **32**: 1–20.

Macdonald, A. M., Edwards, A. C., Pugh, K. B. & Balls, P. W. (1995). Soluble nitrogen and phosphorus in the river Ythan system, U. K. : annual and seasonal trends. *Wat. Res.* **29**: 837–846.

Mackey, R. L. & Barclay, R. M. R. (1989). The influence of physical clutter and noise on the activity of bats over water. *Can. J. Zool.* **67**: 1167–1170.

Mitchell-Jones, A. J. (1989). Bridge surveys. *Batchat* No. 12:5.

Neu, C. W., Byers, C. R. & Peek, J. M. (1974). A technique for analysis of utilization – availability data. *J. Wildl. Mgmt* **38**: 541–545.

Neville, P. A. (1986). Factors affecting the distribution and feeding ecology of bats in Grampian forests. *Forest Res.* **1986**: 72–73.

North East River Purification Board (1993). *Water quality review.* North East River Purification Board, Aberdeen.

Northumberland Bat Group (1985). *Report.* The Bat Conservation Trust, London.

Nyholm, E. E. (1965). Zur Ökologie von *Myotis mystacinus* (Leisl.) und *Myotis daubentoni* (Leisl.) (Chiroptera). *Annls. zool. Fenn.* **2**: 77–123.

Park, M. E. (1988). *Distribution of Bat Roosts along River Systems in North-east Scotland.* BSc thesis: University of Aberdeen.

Pritchard, J. S. & Murphy, F. J. (1987). *A Highland bat survey 1986. Fortingall and Glen Lyon.* Fife Bat Group. The Bat Conservation Trust, London.

Pybus, M. J., Hobson, D. P. & Onderka, D. K. (1986). Mass mortality of bats due to probable blue-green algal toxicity. *J. Wildl. Dis.* **22**: 449–450.

Racey, P. A. (1974). Ageing and assessment of reproductive status of pipistrelle bats, *Pipistrellus pipistrellus. J. Zool., Lond.* **173**: 264–271.

Racey, P. A. (1991). Noctule bat *Nyctalus noctula.* In *The handbook of British mammals:* 117–121. (Eds. Corbet, G. B. & Harris, S. H.) Blackwell Scientific, Oxford.

Racey, P. A. & Swift, S. M. (1985). Feeding ecology of *Pipistrellus pipistrellus* (Chiroptera: Vespertilionidae) during pregnancy and lactation. 1. Foraging behaviour. *J. Anim. Ecol.* **54**: 205–215.

Racey, P. A., Rydell, J., Swift, S. M. & Brodie, L. (1998). Bats and insects over two Scottish rivers with contrasting nitrate status. *Anim. Cons.* (in press).

Rachwald, A. (1992). Habitat preference and activity of the noctule bat *Nyctalus noctula* in the Bialowieza Primeval Forest. *Acta theriol.* **37**: 413–422.

Richards, B. (1992). *The Importance of Canal Bridges and Tunnels as Roost Sites for Bats and the Environmental Factors that may Influence Roost Site Selection.* BSc thesis: University of Aberystwyth.

Rieger, I. (1996). Tagesquartiere von Wasserfledermäusen, *Myotis daubentonii* (Kuhl 1819), in hohlen Bäumen. *Schweiz. Z. Forstwes.* **147**: 1–20.

Rieger, I. & Alder, H. (1994). *Wasserfledermäuse in der Region Rheinfall.* Fledermäus-Gruppe Rheinfall, Dachsen und Schaffhausen.

Roberts, D. (1989). Bats under bridges in North Yorkshire. *Bat News* No. 16: 6–7.

Russ, J. M. (1995). *Bats, Bridges and Acoustic Signalling.* BSc thesis: University of Aberdeen.

Rydell, J., Bushby, A., Cosgrove, C. C. & Racey, P. A. (1994). Habitat use by bats along rivers in NE Scotland. *Folia zool.* **43**: 417–424.

Sargent, G. (1991). *The Importance of Riverine Habitats to Bats in County Durham.* MSc thesis: University of Durham.

Smiddy, P. (1991). Bats and bridges. *Ir. Nat. J.* **23**: 425–426.

Speakman, J. R. (1991). Daubenton's bat *Myotis daubentonii.* In *The handbook of British mammals*: 108–111. (Eds Corbet, G. B. & Harris, S. H.). Blackwell Scientific, Oxford.

Speakman, J. R. & Racey, P. A. (1989). Hibernal ecology of the pipistrelle bat: energy expenditure, water requirements and mass loss, implications for survival and the function of winter emergence flights. *J. Anim. Ecol.* **58**: 797–813.

Speakman, J. R., Racey, P. A., Catto, C. M. C., Webb, P. I., Swift, S. M. & Burnett, A. M. (1991). Minimum summer populations and densities of bats in N. E. Scotland, near the northern borders of their distributions. *J. Zool., Lond.* **225**: 327–345.

Stebbings, R. E. (1988). *The conservation of European bats.* Christopher Helm, London.

Sullivan, C. M., Shiel, C. B., McAney, C. M. & Fairley, J. S. (1993). Analysis of the diets of Leisler's *Nyctalus leisleri*, Daubenton's *Myotis dauben-*

toni and pipistrelle *Pipistrellus pipistrellus* bats in Ireland. *J. Zool., Lond.* **231**; 656–663.

Swift, S. M. (1981). *Foraging, Colonial and Maternal Behaviour of Bats in North-east Scotland.* PhD thesis: University of Aberdeen.

Swift, S. M. (1997). Roosting and foraging behaviour of Natterer's bats (*Myotis nattereri*) close to the northern border of their distribution. *J. Zool., Lond.* **242**: 375–384.

Swift, S. M. & Racey, P. A. (1983). Resource partitioning in two species of vespertilionid bats (Chiroptera) occupying the same roost. *J. Zool., Lond.* **200**: 249–259.

Swift, S. M., Racey, P. A. & Avery, M. I. (1985). Feeding ecology of *Pipistrellus pipistrellus* (Chiroptera: Vespertilionidae) during pregnancy and lactation. II. Diet. *J. Anim. Ecol.* **54**: 217–225.

Taake, K. -H. (1984). Strukturelle Unterschiede zwischen den Sommerhabitaten von kleiner und Grosser Bartfledermaus (*Myotis mystacinus* und *M. brandti*) in Westfalen. *Nyctalus* **2**: 16–32.

Taake, K. -H. (1992). Strategien der Ressourcennutzung an Waldgewässern jagender Fledermäuse (Chiroptera; Vespertilionidae). *Myotis* **30**: 7–74.

Vaughan, N., Jones, G. & Harris, S. (1996). Effects of sewage effluent on the activity of bats (Chiroptera: Vespertilionidae) foraging along rivers. *Biol. Conserv.* **38**: 337–343.

von Frenckell, B. & Barclay, R. M. R. (1987). Bat activity over calm and turbulent water. *Can. J. Zool.* **65**: 219–222.

Voûte, A. M., Sluiter, J. W. & van Heerdt, P. F. (1980). De vleermuizenstand in einige zuidlimburgse groeven sedert 1942. *Lutra* **22**: 18–34.

Walsh, A. L. & Harris, S. (1996a). Foraging habitat preferences of vespertilionid bats in Britain. *J. appl. Ecol.* **33**: 508–518.

Walsh, A. L. & Harris, S. (1996b). Factors determining the abundance of vespertilionid bats in Britain: geographic, land class and local habitat relationships. *J. appl. Ecol.* **33**: 519–529.

Walsh, A. L. & Mayle, B. A. (1991). Bat activity in different habitats in a mixed lowland woodland. *Myotis* **29**: 97–104.

Walsh, A. L., Harris, S. & Hutson, A. M. (1995). Abundance and habitat selection of foraging vespertilionid bats in Britain: a landscape-scale approach. *Symp. zool. Soc. Lond.* No. 67: 325–344.

Webb, P. I. (1992). *Aspects of the Ecophysiology of some Vespertilionid Bats at the Northern Borders of their Distribution.* PhD thesis: University of Aberdeen.

Webb, P. I., Speakman, J. R. & Racey, P. A. (1995). Evaporative water loss in two sympatric species of vespertilionid bat, *Plecotus auritus* and *Myotis daubentoni*: relation to foraging mode and implications for roost site selection. *J. Zool., Lond.* **235**: 269–278.

Weinrich, J. A. & Oude Voshaar, J. H. (1992). Population trends of bats hibernating in marl caves in the Netherlands (1943–1987) *Myotis* **30**: 75–84.

6

A preliminary study of the behaviour of the European mink *Mustela lutreola* in Spain, by means of radiotracking

S. Palazón and J. Ruiz-Olmo

Introduction

The European mink, *Mustela lutreola*, is one of the least-known mustelids in Europe (Youngman, 1982; Camby, 1990; Saint-Girons 1991). The species is, moreover, declining in number and many populations have disappeared (Romanowski, 1990; Saint-Girons 1991; Rozhnov, 1993). Formerly, mink were widely distributed through Europe, but today are found only in some eastern European countries (Sidorovich, 1991; Maran, 1992; Tumanov, 1992), in a small area of southern France (Braun, 1990; Camby, 1990; C. Maizeret, unpublished results) and in northern Spain (Ruiz-Olmo & Palazón, 1991; Palazón & Ruiz-Olmo, 1992; Gosàlbez *et al.*, 1995). Thus, *Mustela lutreola* is one of the five most endangered small carnivore taxa in the Palaearctic (Schreiber *et al.*, 1989).

The European mink was recorded in Spain during the middle of the twentieth century (Rodriguez de Ondarra 1955). Subsequently, the species expanded to the south and east, following the main river basins, principally the Ebro river (Palazón & Ruiz-Olmo, 1995). Today Spain is the only area in the world where *Mustela lutreola* is not in decline. The arrival of mink in certain areas of Spain has presumably meant adaptation to a new environment: the Mediterranean ecosystem. Here, most of the rivers have no polecats, *Mustela putorius*, American mink, *Mustela vison*, or otters, *Lutra lutra* (Delibes, 1990; Ruiz-Olmo & Delibes, 1998).

Other than studies of distribution and status very little has been published on the European mink. There is a little information related to diet in publications of a more general nature (Camby, 1990) but only one extensive study exists (Sidorovich, 1992). Similarly, there are few studies of their spatial organization and use of space and time (Novikov, 1975; Danilov & Tumanov, 1976; Maran, 1989; Palazón & Ruiz-Olmo, 1993; Sidorovich *et al.*, 1995).

Here we present the first radiotracking study of the behaviour of wild European mink.

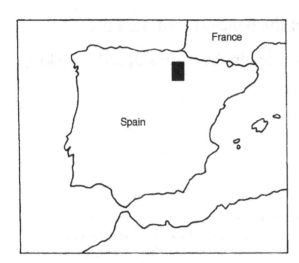

Figure 6.1. Location of study area in Spain.

Study area

The study area (Figs 6.1, 6.2 and 6.3) comprised about 20 km of the River Ega, south-west of Navarra, Spain (2°00' W, 42°55' N). The altitude ranges from 350 to 400 m. River breadth ranges from 10 to 25 m and depth exceeds 2 m.

The study area is in the Mediterranean biogeographic zone, with a dry ombroclimate. Mean annual precipitation is 400–600 mm. Generally, riparian forest is well developed but relatively sparse (between 2 and 10 m) and surrounded by extensive cultivations of cereals and asparagus.

Upstream of the study area, riparian vegetation is dominated by alders (*Alnus glutinosa*). Downstream, these give way to poplars (*Populus alba*), black poplars (*Populus nigra*), white willows (*Salix alba*), narrow leaf ash (*Fraxinus angustifolia*) and elm (*Ulmus minor*). Frequently associated with riparian forest were small, sparse woods of holm oak (*Quercus rotundifolia*), extending to the river in some places.

Fauna in the area is very rich and diverse. In the river, cyprinids (*Barbus graellsi, Chondrostoma toxostoma*), trout (*Salmo trutta*) and American crayfish (*Proambarus clarkii*) are very common. Other carnivore species include *Mustela nivalis, Martes foina, Genetta genetta, Meles meles* and *Vulpes vulpes*, while *Lutra lutra* and *Mustela putorius* have become extinct in the last 5–10 years.

The city of Estella and its industrial area are situated 1 km upstream from the study area. There is also a large paper factory in the Ega study stretch, as well as a poultry farm. These birds are unprotected and mink frequently enter in search of food (Fig. 6.3).

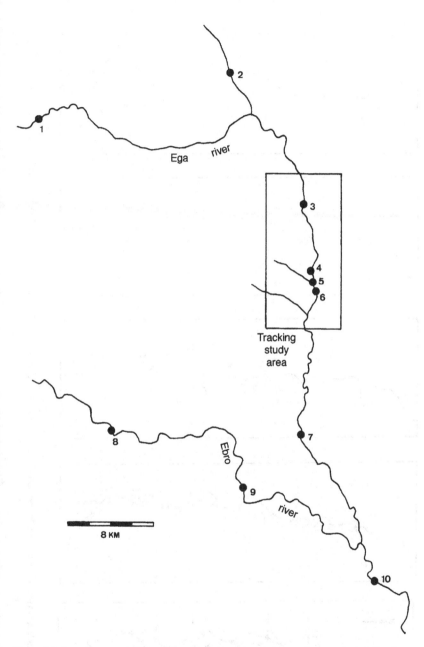

Figure 6.2. Area of capture of European mink and tracking study area. 1, one subadult female (ML05); 2, one adult and one subadult male (2A and 2B); 3, one adult female (3A); 4, one adult male (ML03); 5, three adult males (ML01, ML02 and 5A) and one adult female (ML04); 6, two adult males (ML06 and ML07) and one adult female (ML08); 7, one adult male (7A) and one of undetermined sex (7B); 8, two adult males (8A and 8B); 9, one adult male (9A); 10, three adult males (10A, 10B and 10C).

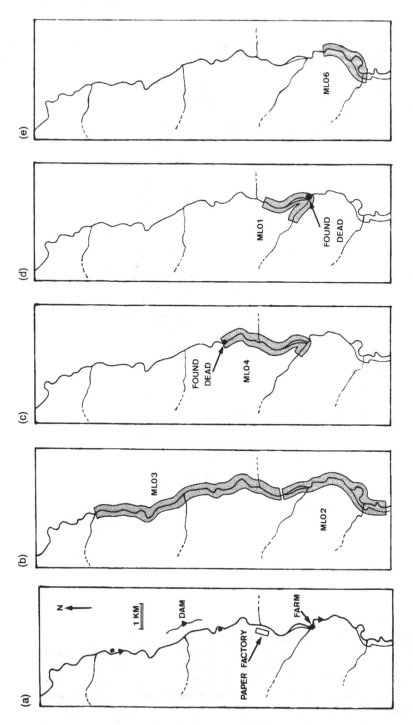

Figure 6.3. (a) Enlarged view of tracking study area shown boxed in Figure 6.2, and (b) to (e) home ranges of five European mink.

Fifteen kilometres to the west of the study area there is an American mink farm, equipped with good security measures to prevent escapes. No escaped American mink have been recorded to date.

Material and methods

European mink were captured in wooden-box live-traps ($15 \times 15 \times 60$ cm) baited with tinned sardines over a total of 615 trap-days from 1992 to 1995. Following capture they were anaesthetized with a combination of ketamine hydrochloride (0.1 ml/kg of body mass) and tiazine hydrochloride (0.03 ml/kg of body mass).

Each animal was measured, weighed and fitted with a radio-collar transmitter with an activity sensor attached to its neck (four radios from Urmeneta, Navarra, Spain, and four more from Biotrack, UK). Radiotracking was carried out by means of an antenna (AF Antronics Inc. model F151-3FB), and a receiver (Customs Electronics Inc. (Urbana, USA), model CE-12).

During the first study period (April–May, 1992) we radiotracked four mink (three males and one female). In the second, (December, 1993–January, 1994) four more animals (two males and two females) were tracked (Table 6.1).

Each day the position of all tracked animals was located. Each night a single animal was tracked. Triangulations were unnecessary because animals could be tracked closely enough. The activity of each tracked animal was noted every half hour.

Results

Trapping efficiency
We caught a total of 20 European mink (4.07% trap-days of effectiveness if we include five recaptures). During the first period we caught five different individuals in a 1 km stretch (Table 6.1(a)). During the second trapping period, the capture study area was larger and we caught 15 more individuals in a stretch of 11 km (Table 6.1(b)). Thus 1.36 European mink were captured per kilometre of river in the second trap period.

Characteristics of the captured animals and tracking
We radiotracked eight European minks (five males and three females). Three of them (ML05, ML07 and ML08) were lost during the first day due to transmitter failure or unknown causes. The remaining five (ML01, ML02, ML03, ML04 and ML06) (Table 6.1(a)) were tracked for 25, 58, 4, 13 and 5

Table 6.1. *Characteristics of European mink captured in the study area*

(a) European minks captured and tracked

Animal	Sex	Age	Weight (g)	Total length (mm)	Date tag fitted	Last tracking date	Number of recaptures	End anir
ML01	M	Adult	610	508	01-04-92	26-04-92	3	Dea
ML02	M	Adult	850	545	01-04-92	30-05-92	0	Los
ML03	M	Adult	720	540	13-04-92	17-04-92	0	Los
ML04	F	Adult	460	490	18-04-92	01-05-92	1	Dea
ML05	F	Adult	473	471	24-12-93	25-12-93	0	Los
ML06	M	Adult	825	550	31-12-93	05-01-94	1	Los
ML07	M	Adult	805	507	02-01-94	03-01-94	0	Los
ML08	F	Subadult	410	454	07-01-94	08-01-94	0	Los

(b) European minks captured but not tracked

Animal	Sex	Age	Weight (g)	Total length (mm)	Date capture	End animal
5A	M	Adult	795	551	30-03-92	Dead
7A	M	Adult	760	—	20-12-93	Free
7B	—	—	—	—	24-12-93	Escaped
3A	F	Adult	590	473	27-12-93	Dead
10A	M	Adult	750	530	03-01-94	Free
10B	M	Adult	950	566	06-01-94	Free
10C	M	Adult	720	550	07-01-94	Free
8A	M	Adult	920	550	14-01-94	Free
9A	M	Adult	880	560	14-01-94	Free
8B	M	Adult	830	535	22-01-94	Free
2A	M	Adult	718	536	23-03-94	Free
2B	M	Subadult	625	525	26-03-94	Dead

days, respectively. Female ML04 was found dead after this period, and male ML01 was found dead 51 days after the start of tracking. Both were killed deliberately by man. The female was pregnant with five embryos in the early stages of development. Tracking of the remaining individuals stopped when the transmitters failed.

Spatial use and movements

Adult males ML02 and ML03 had home ranges of 7.8 and 11.4 km, respectively, along the river. During the tracking period their home ranges did not overlap. Male ML06 had a home range of 2.9 km along the river during the 8 days of radiotracking. His range overlapped that of male ML02, tracking of the latter was, however, undertaken 2 years later (Fig. 6.3).

Male ML01 occupied a 2.94 km river section around the farm, between males ML02 and ML03 (which were heavier and had larger home ranges). On the night of 13 April, 1992, he was followed 0.7 km from the river (straight line distance) to a tributary (Fig. 6.3). Here he was located in the same den as female ML04 on three alternate days, where mating presumably occurred (the female was later captured with tooth wounds on the back of the neck). ML01 was found dead in front of the chicken farm 2 months later.

Female ML04's home range overlapped with that of male ML01, but was longer (5.06 km). The female spent most of her time around the farm but was found dead 4.9 km upstream (Fig. 6.3). A fisherman killed the female on the opening day of the fishing season.

All European mink were located in river and riparian habitats of the main river (River Ega) or in the secondary tributaries with plentiful helophytic vegetation. They were also found in the vegetation of some small parallel irrigation channels, 50 m from the river. We could verify that on one occasion ML01 left the river, crossing 700 m to other river (other specimens were found killed by road traffic near the study area but a considerable distance from the river).

During the study, we found a total of 16 resting sites (Table 6.2). Mink occupied an average of four resting sites (range 1–8).

Resting sites were located in helophytic vegetation, tree roots and blackberry thickets (Table 6.2); 81.2% of the dens were found in the bank of the main river and 18.7% in tributaries, within 200 m of the main river. Distances between the resting sites used by each animal (Table 6.3) were calculated in relation to the neighbouring den along the river.

No significant correlations were found between mink body mass and home range size ($r = 0.18$; $p = 0.771$), number of resting sites ($r = 0.89$; $p = 0.111$) and average distance travelled per active period ($r = 0.85$; $p = 0.34$), but a clear tendency was observed in the latter two cases.

Male ML02 made two long journeys, one of 4.8 km upstream (mean speed = 9.1 m/min) and a second of 1.92 km downstream (mean speed = 8.1 m/min). Also, male ML03 made one journey between two resting sites with a distance of 2.36 km upstream (mean speed = 21.1 m/min).

Table 6.2. *Location of the resting sites of four radiotracked European mink*

| | Individual | | | | | |
Type	ML01	ML02	ML03	ML04	Total	%
Roots	—	1	—	—	1	6.2
Blackberries	3	6	3	—	12	75.0
Helophytic veg.	1	1	—	1	3	18.7

Table 6.3. *Use of space by radiotracked European mink*

Animal	Sex	Age	Home range (km)	Number resting Sites	DNRS (km) Average	DNRS (km) SD	Distance per day (km)	SD distance per day (km)
ML01	M	Ad	2.94	4	0.39	0.196	0.85	0.48
ML02	M	Ad	7.8	8	1.06	0.561	2.12	1.24
ML03	M	Ad	11.4	3	5.70	0.334	0.74	0.94
ML04	F	Ad	5.14	1	0	—	—	—
ML06	M	Ad	2.9	—	—	—	—	—

DNRS, distance between neighbouring resting sites; SD, standard deviation.

During the period of radiotracking, the female ML04 made no long journeys, remaining between the farm and the den (150 m).

Activity
The average activity of the radiotracked animals measured by means of an activity sensor was 31.28% ($n = 227$ radio-locations). The greatest activity (42.15%) was recorded during the night (18:00–06:00 h). By day (06:00–18:00 h) activity was significantly lower (18.87%) ($\chi^2 = 35.16$; d.f. = 3; $p < 0.01$). Maximum activity was observed at the beginning and end of the night during the period of darkness (Fig. 6.4 and Table 6.4). Foraging occurred at night and at twilight and finished at a different resting site. Daytime activity was confined to short movements around the resting site or within it (Fig. 6.5 and Table 6.4).

Discussion

The data show that European mink were active throughout the 24 h period and coincide with observations carried out in captivity (Maran, 1989).

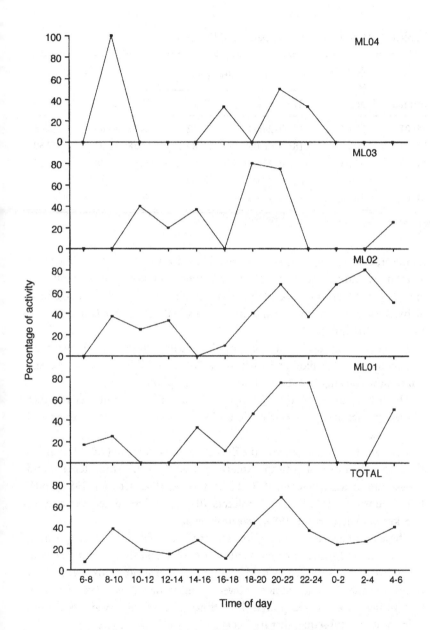

Figure 6.4. Activity of the European mink in relation to the time of day.

Table 6.4. *Main and secondary periods of European mink activity*

Animal	Average % activity	1st activity period		2nd activity period	
		Time of day	% act.	Time of day	% act.
ML01	27.46	18:00–24:00	53.12	04:00–10:00	23.08
ML02	40.12	18:00–06:00	57.89	08:00–12:00	27.59
ML03	33.33	18:00–22:00	44.00	10:00–16:00	40.00
		04:00–06:00	60.00		
ML04	12.50	16:00–18.00	16.70	20:00–24:00	27.27

act., activity.

However, clear crepuscular peaks were recorded with a lower level of activity maintained during the night. Daytime activity did not involve significant movements. American mink and polecats also present two activity maxima at night (the beginning and end) and little activity in the afternoon (Birks & Linn, 1982; Herrenschmidt, 1982).

The home range of the European mink in Spain may be much longer than that reported by Danilov & Tumanov (1976) in the Karelia region, Russia, using snow-tracking (2.4 km without sex specification).

The behaviour of *M. lutreola* appears to be different from that of other *Mustela* species of similar weight and ecology, such as the American mink and the polecat.

The size of the home range of the European mink in Spain (2.9–11.4 km) is also larger than that of both the American mink (1.8–2.2 km in females and 2.5–4.4 km in males: Gerell, 1970; Birks & Linn, 1982; Chanin, 1983; Lodé, 1991; Dunstone, 1993) and of polecats (0.65–1.65 km in females and 1–3.05 km in males: Weber, 1989; Brzezinski *et al.*, 1992).

These findings have certain implications for conservation and population estimation. The further mink move, the more vulnerable they become (two out of the five animals tracked were killed by man). European mink, particularly young and subadult individuals, showed a behaviour that was overly fearless. One young male was recaptured three times and was observed twice (once swimming and subsequently running at a distance from the river).

Density estimations using snow-tracking (1–3 days following snowfall), based on the concentration of tracks, where each 1–1.5 km of track is considered to belong to a different animal (Danilov & Tumanov, 1976; Sidorovich *et al.*, 1995) should be treated with caution. The mink studied here travelled up to 4.5 km in one day. Such density estimations might, therefore, be overestimat-

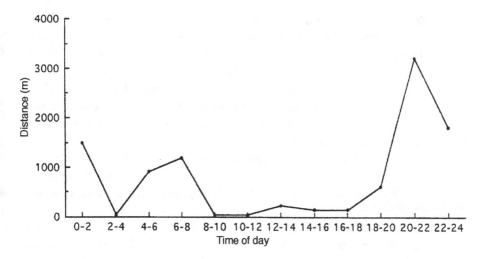

Figure 6.5. Distance travelled by European mink in relation to time of day.

ing mink numbers, with clear implications for conservation and the ecology of the species.

There is no overlap between the home ranges of resident males in European mink, American mink (Gerell, 1970) or polecat (Weber, 1989; Brzezinski *et al.*). In these three species, the female home range is smaller and overlaps with the male home range, while young males live at the extremities of adult male home ranges and travel across them during dispersal (Gerell, 1970; Brzezinski, *et al.*, 1992).

The type and number of resting sites used by European mink are similar to those used by American mink and polecat. American mink resting sites are highly varied (inside tree trunks, tree roots, heaps of stone, bridge bases, rabbit dens, walls, cattle barns, between reeds, scrub and blackberry bushes, wood poles, etc) (Gerell, 1970; Ternowskii, 1977; Birks & Linn, 1982; Dunstone & Birks, 1983) and two to five dens are used (Gerell, 1970). Polecats use mainly tree roots and rabbit dens with one male using up to eight dens (Brzezinski *et al.*, 1992). Gerell (1970) found a correlation between home range size and the number of dens in American mink. In polecats, the mean distance recorded in an activity period was 1.1 km during the day (Brzezinski *et al.*, 1992) and 0.6 km at night (Nilsson, 1978). In the European mink, maxima also seem to be higher (2.0–4.65 km).

Foraging speeds of the European mink are the same as those that Weber (1989) calculated for the polecat, 10 m/min inside the activity centre; speed over longer distances is 75 m/min. The speed of American mink is less than that of European mink inside the home range (activity centre) but greater when travelling from one activity centre to another (Gerell, 1969). Speeds during activity recorded on journeys between activity centres for the European mink were similar to those of American mink (8.1–21.1 m/min).

Because they are active in daylight, European mink can be observed and are highly vulnerable (as in the case of female ML04). As they enter chicken farms in search of food, they can be killed by dogs, shot or caught in traps, which greatly hinders conservation.

We can conclude that the European mink presents a behavioural pattern similar to that of the American mink and polecat, but with greater movement and greater spatial requirements. Such findings could have important ecological repercussions, as these animals could be successful colonists, but are at the same time highly vulnerable. More research is needed. In November, 1995 this study was renewed, tracking more animals and at different times of the year.

ACKNOWLEDGEMENTS

We are grateful to the Department of the Environment (Government of Navarra) for co-operation and permission for trapping.

We thank the many people who helped with field work, especially E. Castien, J. Ochoa, E. Camacho, O. Arribas, O. Bilbao, J. M. López-Martín, Y. Cortés, R. Baquero, E. Virgós, J. Casanova, M. A. Pasquina, S. Cahill, R. Hortelano, J. A. Lacunza, F. Ezcurra and E. Tomás.

Financial support was provided by ICONA (Ministry of Agriculture and Fish) and the Department of the Environment (Government of Navarra).

References

Birks, J. D. S. & Linn, I. J. (1982). Studies of the home range of the feral mink, *Mustela vison*. *Symp. zool. Soc. Lond.* No. 49: 231–257.

Braun, J. (1990). The European mink in France: past and present. *Mustelid Viverrid Conserv.* 3: 5–8.

Brzezinski, M., Jedrzejewski, W. & Jedrzejewska, B. (1992). Winter home ranges and movements of polecats *Mustela putorius* in Bialowieza Primeval

Forest, Poland. *Acta theriol.* 37: 181–191.

Camby, A. (1990). Le vison d'Europe (*Mustela lutreola* Linnaeus, 1761). In *Encyclopédie des carnivores de France* 13: 1–18. Société Française pour l'Etude et la Protection des Mammifères, Paris.

Chanin, P. (1983). Observations on two populations of feral mink in Devon, U. K. *Mammalia* 47: 463–476.

Danilov, P. I. & Tumanov, I. L. (1976). [*Mustelids of the northwestern USSR.*] Akademiya Nauk, Leningrad. [In Russian.]

Delibes, M. (1990). *La nutria* (Lutra lutra) *en España.* Instituto pera la Conservación de la Naturaleza (ICONA), Madrid.

Dunstone, M. (1993). *The mink.* T. & A. D. Poyser, London.

Dunstone, N. & Birks, J. D. S. (1983). Activity budget and habitat usage by coastal-living mink (*Mustela vison* Schreber). *Acta zool. Fenn.* **174**: 189–191.

Gerell, R. (1969). Activity patterns of the mink *Mustela vison* Schreber in southern Sweden. *Oikos* **20**; 451–460.

Gerell, R. (1970). Home ranges and movements of the mink *Mustela vison* Schreber in southern Sweden. *Oikos* **21**: 160–173.

Gosàlbez, J., Palazón, S. & Ruiz-Olmo, J. (1995). *Distribución del visón europeo* Mustela lutreola *en Navarra*. Gobierno de Navarra, Pamplona.

Herrenschmidt, V. (1982). Note sur les déplacement et le rythme d'activité d'un putois, *Mustela putorius* L., suivi par radiotracking. *Mammalia* **46**: 554–556.

Lodé, T. (1991). Les déplacements du vison américain *Mustela vison* Schreber suivi par radiotracking sur une rivière bretonne. *Mammalia* **55**: 643–646.

Maran, T. (1989). Einige Aspekte zum gegenseitigen verhalten des Europäischen *Mustela lutreola* und Amerikanischen Nerzes *Mustela vison* sowie zu ihrer Raum- und Zeitnutzung. In *Populationsökologie marderartiger Säugetiere:* 321–332. *Martin-Luther-Univ. Halle-Wittenberg wiss. Beitr.* **P39**.

Maran, T. (1992). The European mink *Mustela lutreola* in protected areas in the former Soviet Union. *Small Carniv. Conserv.* No. 7: 10–12.

Nilsson, T. (1978). Home range utilization and movements in polecat, *Mustela putorius* during autumn. *Congr. theriol. int., Brno, CSSR* **2**: 173. (Abstract).

Novikov, I. (1975). *Biologija Lesngh ptic i sverej.* Moscow.

Palazón, S. & Ruiz-Olmo, J. (1992). *Estatus actual del visón europeo* Mustela lutreola *y del visón americano* Mustela vison *en España.* Instituto para la Conservación de la Naturaleza (ICONA), Madrid.

Palazón, S. & Ruiz-Olmo, J. (1993). First data on the activity and use of space of the European mink (*Mustela lutreola*) revealed by radiotracking. *Small Carniv. Conserv.* No. 8: 6–8.

Palazón, S. & Ruiz-Olmo, J. (1995). *Estudi per determinar la presència del visó europeu a Catalunya.* Generalitat de Catalunya, Barcelona.

Rodriguez de Ondarra, P. (1955). Hallazgo, en Guipúzcoa, de un mamífero no citado en la "Fauna Ibérica" de Cabrera. *Munibe* **15**: 103–110.

Romanowsi, J. (1990). Minks in Poland. *Mustelid Viverrid Conserv.* **2**: 13.

Rozhnov, V. V. (1993). Extinction of the European mink: Ecological catastrophe or a natural process? *Lutreola* **1**: 10–16.

Ruiz-Olmo, J. & Delibes, M. (1998). *La nutria* (Lutra lutra) *en España.* Sondeo de 1994–96. SECEM-WWF-Adena, Madrid. (In press).

Ruiz-Olmo, J. & Palazón, S. (1991). New information on European and American minks in the Iberian Peninsula. *Mustelid Viverrid Conserv.* **5**: 13.

Saint-Girons, M. C. (1991). Wild mink (*Mustela lutreola*) in Europe. *Counc. Eur. Nat. Environ. Ser.* No. 54: 1–41.

Schreiber, A., Wirth, R., Rifel, M. & von Rompaey, H. (1989). *Weasels, civets, mongooses and their relations: an action plan for the conservation of mustelids and viverrids.* IUCN, Gland, Switzerland.

Sidorovich, V. E. (1991). Distribution and status of minks in Byelorussia. *Mustelid Viverrid Conserv.* **5**: 14.

Sidorovich, V. E. (1992). Comparative analysis of the diets of European mink (*Mustela lutreola*), American mink (*Mustela vison*) and polecat (*Mustela putorius*) in Byelorussia. *Small Carniv. Conserv.* **6**: 2–4.

Sidorovich, V. E., Savchenko, V. V. & Bundy, B. V. (1995). Some data about the European mink *Mustela lutreola* distribution in Lovat River Basin in Russia and Belarus: current status and retrospective analysis. *Small Carniv. Conserv.* No. 12: 14–18.

Ternovskii, D. V. (1977). *Biologija Kuniceobraznyh.* Nauka, Novosibirsk.

Tumanov, I. L. (1992). The numbers of the European mink (*Mustela lutreola*) in the east of area and its relation with American species. In *Semiaquatische Saügetiere:* 329–335. (Eds Schröpfer, R., Stubbe, M. & Heidecke, D.). *Martin-Luther-Univ. Halle-Wittenberg wiss. Beitr.*

Weber, D. (1989). *Beobachtungen zu Aktivität und Raumnutzung beim Iltis* (Mustela putorius L.). *Revue suisse Zool.* **96**: 841–862.

Youngman, P. M. (1982). Distribution and systematics of the European mink *Mustela lutreola* Linnaeus 1761. *Acta zool. Fenn.* **166**: 1–48.

7

The demography of European otters *Lutra lutra*

M. L. Gorman, H. Kruuk, C. Jones, G. McLaren and J.W.H. Conroy

Introduction

This paper presents data on the demography of populations of European otters *Lutra lutra* from various parts of the British Isles. The emphasis is on age structure and patterns of mortality.

Otters have declined sharply in Britain over the last few decades, probably largely as a result of contamination by organochlorine pesticides (Mason & Macdonald, 1986; Jefferies, 1989). A similar decline has taken place in many parts of Europe and the species is now extinct in several countries. However, following the withdrawal of a number of pesticides in the UK the otter is now, very slowly, returning to areas from which it had disappeared (Green & Green, 1987; Strachan *et al.*, 1990; Andrews *et al.*, 1992).

A problem that has plagued research on the effects of pollution on populations of otters has been a lack of basic information on demography. In particular, in the absence of animals of known age it has been difficult to answer such pertinent questions as whether or not chemical contaminants accumulate in the bodies of otters as they age. Equally, a knowledge of demographic parameters is important in predicting the rates at which otter populations might recover in the future.

Our aims are:

1 To discuss the methods and associated problems of analysing otter demography.
2 To present data on patterns of mortality and longevity for populations from different areas of the UK.

The sample of otters

Our study is based on 391 otters (Table 7.1) from Shetland, various parts of mainland Scotland, England and Wales. For the purposes of analysis we have dealt with the Scottish mainland otters as one group and we have combined the Welsh and English animals.

Table 7.1. *The sizes of the samples on which the demographic analyses are based*

	Violent	Non-violent	Total
Shetland	83	63	146
Mainland	126	22	148
England & Wales	46	51	97

The otters in the samples are from two very different sources. The first consisted of accumulations of animals that had come to the end of their natural lives and that had died from non-violent causes. The second group contained otters that had met a violent end and that had, therefore, died prematurely. These two groups of animals each require a somewhat different approach for demographic analysis.

It is impossible to assess the relative contribution of violent and non-violent deaths to the overall mortality suffered by a population. This is because animals that die violently, for example as a result of a traffic accident, are much more likely to be found and handed in than are individuals dying naturally, perhaps in remote areas or underground in their holts. It is likely, therefore, that violent deaths, and in particular road deaths, are strongly over-represented in our samples but we cannot quantify this.

We have the most detailed information on causes of death for the samples from Shetland. In the case of the violent deaths, road traffic was the single most important cause of mortality (Fig. 7.1) and probably accounted also for most of the deaths without known cause. The otters killed by dogs were mainly cubs and a few adults were drowned in lobster creels or sewer pipes. For most of the otters dying a non-violent death the exact cause of death was unknown (Fig. 7.2). However, a large proportion had extensive internal bleeding in the gut, a point to which we shall return.

In Shetland, the otters dying violent and non-violent deaths belonged to two demonstrably different samples of individuals. Firstly, there was a marked difference in the times of year at which the two groups of otters were likely to die (Fig. 7.3). Otters were equally likely to die a violent death in any month of the year. In contrast, although otters also died naturally at all times of the year there was a clear peak in natural deaths in May–June. This is the period of lowest food availability in Shetland (Kruuk *et al.*, 1987; Kruuk & Conroy, 1991). Most of the animals in our sample that had died naturally had empty stomachs, often with ulceration and internal bleeding, and were probably starving. Thus, starvation may be an important direct or indirect cause of natural mortality in Shetland otters.

Figure 7.1. The causes of death of
Shetland otters dying violently.
■, traffic; ▨, unknown; ▨, dog;
■, creel; □, sewer.

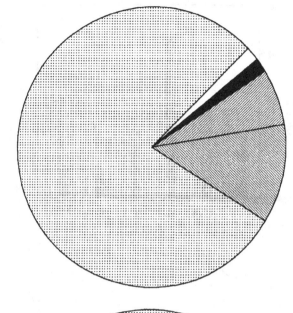

Figure 7.2. The causes of death of
Shetland otters dying non-violent
and natural deaths. ■, unknown;
▨, bleeding; ▨, poisoned;
■, liver neoplasm; □, pneumonia.

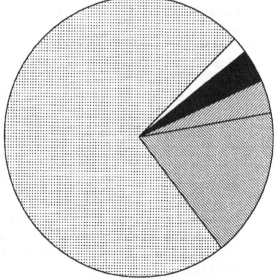

Secondly, the individuals in the two groups of otters were in quite different
states of body condition at the time of their deaths. As a measure of body
condition we calculated the index K (Kruuk *et al.*, 1987), which takes a
relatively low value when animals are in poor condition and a high value when

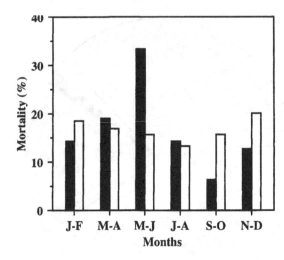

Figure 7.3. The months of death of otters in the samples from Shetland. ■, non-violent; □, violent.

in good condition:

$$K = W/aL^n$$

where: W, weight (kg); L, length (m); $a = 5.02$ (females) or 5.87 (males) and $n = 2.33$ (females) or 2.39 (males).

In Shetland, otters dying a violent death were generally in relatively good condition whereas those dying naturally were in much poorer condition (Fig. 7.4).

Clearly, therefore, the otters dying violent and non-violent deaths represented two very different samples of the population and this must be borne in mind in any analysis of population demography.

Ageing otters

Incremental growth lines are present in the tissues of many animals and sometimes, for example in the case of fish scales and otoliths or the waxy ear plugs of baleen whales, they may be useful for measuring the age of an individual.

In mammals incremental lines are often to be found in the hard parts of the body, particularly in the teeth. Regular variation in environmental conditions appears to give rise to fluctuations in the nature and rates of formation of both cementum and dentine, which assume a layered structure. These incremental growth lines are thought to be present in the teeth of every species of terrestrial mammal with a closed tooth root system (Grue & Jensen, 1979). Detailed

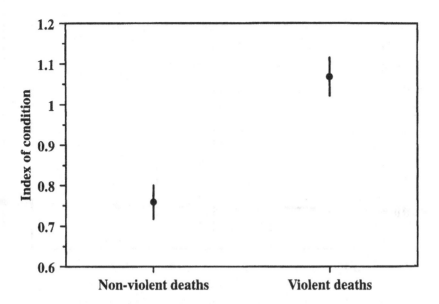

Figure 7.4. Differences in the index of body condition of otters from Shetland which had died violently or non-violently. The values are medians together with their 95% confidence limits.

examination of animals of known age generally shows that one line is formed for each year of life starting at the species-specific age at which the permanent teeth erupt. For example, in a study of the North American river otter, *Lutra canadensis*, Stephenson (1977) recorded a good agreement between the numbers of incremental lines and years in animals of known age.

The physiological factors leading to the formation of incremental lines are not fully understood but are thought to be related to seasonal variation in levels of nutrition (Smith *et al.*, 1994). Thus, Lieberman (1993) was able to produce growth lines in the pre-molars of goats, *Capra hircus*, by experimentally varying the quality of their diet over defined periods.

In this study, growth lines were counted in the lower incisors of otters following decalcification, sectioning and staining. This approach was chosen as it had already been used with success in European otters (Heggberget, 1984) and in sea otters, *Enhydra lutris* (Garshelis, 1984).

The otters were deep-frozen prior to analysis. The incisor teeth were subsequently extracted and placed in 10% formaldehyde in order to fix the tissues. The teeth were then decalcified for 4–5 h in 'Decal Rapid', a proprietary product containing 30.35% HCl. Following decalcification, 15 μm transverse sections were cut on a freezing microtome at −20 °C. Twenty sections were cut from each tooth, immediately in front of the root tip. The sections

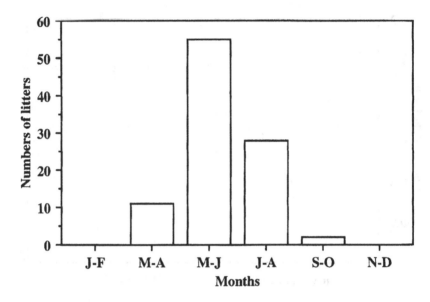

Figure 7.5. The timing of breeding of otters in Shetland.

were stained with 0.1% toluidine blue and mounted in glycerol. The sections were examined under a light microscope, generally at ×10 magnification, and the primary incremental lines were counted recording the maximum present in any of the 20 sections.

When are the growth lines laid down in otters?

Previous work in Denmark, by Grue & Jensen (1979), has shown that growth lines in otters seem to be laid down once per year even in populations where breeding is aseasonal. In Denmark the lines are formed in February–March.

In Shetland, from where we have the best data set, otters are highly seasonal breeders with most litters being born in May–June (Kruuk & Conroy, 1991) (Fig. 7.5). Among the otters from Shetland, individuals with no growth line in their teeth, or just one line, were found to be present in every month of the year, from one breeding season through to the next (Fig. 7.6). This strongly indicates that the first incremental line is deposited around May–June. This is just the time when fish are at their lowest abundance in Shetland (Kruuk *et al.*, 1987), a factor that might result in a seasonal reduction in rates of tooth growth. Thus, the first growth line appears to be deposited as an individual reaches the end of its first year of life.

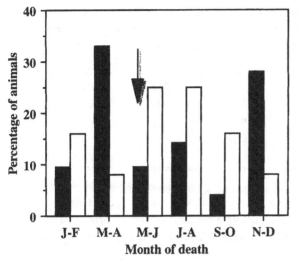

Figure 7.6. The occurrence of individuals with one or no incremental lines in their teeth. The arrow indicates the peak time of breeding. ■, no lines; □, one line.

Demographic analysis

The construction of a life table

Patterns of age-specific mortality are frequently summarized as a life table containing a number of columns, best illustrated with a hypothetical example (Table 7.2).

1   The age interval labelled by the age at the beginning of the interval, x.
2   Survivorship, l_x. The probability at birth of an individual surviving to age x.
3   Mortality d_x. The probability at birth of an individual dying in the interval x to $x+1$.
4   Mortality rate q_x. The probability of an individual aged x dying before it reaches the age $x+1$.

These probabilities are estimated from proportions. The probability of an animal surviving to age x can be estimated, for example, by ringing 1200 fledgling birds and recording the numbers still alive 1 year later, 2 years later and so on (Table 7.2). Survivorship at age 0 is then 1200/1200, i.e. 1.0, by year 1 it has dropped to 500/1200 or 0.42 and by year 2 to 300/1200 or 0.25, and so on.

No further data are needed to complete the life table, since each of the other columns is simply a mathematical manipulation of the l_x column. Mortality, d_x, is calculated as $l_x - l_x + 1$. That is to say $1 - 0.42$, i.e. 0.58 for age $x=0$, and

Table 7.2. *A hypothetical life table showing the various columns*

Age (years) (x)	Frequency (f_x)	Survivorship (l_x)	Mortality (d_x)	Mortality rate (q_x)
0	1200	1.00	0.58	0.58
1	500	0.42	0.17	0.40
2	300	0.25	0.08	0.33
3	200	0.17	—	—
…	…	…	…	…

$0.42 - 0.25$, i.e. 0.17 for age $x = 1$. The mortality rate, q_x, is calculated as d_x/l_x.

In theory, therefore, constructing a life table is straightforward. The difficulty comes in obtaining the necessary data. Ringing 1200 fledglings is easy, estimating the number alive in future years is not. Happily, there are two approximation methods that can be used to construct an approximate life table.

The first involves obtaining an unbiased sample of the ages at which animals die by picking up a collection of animals dying naturally and ageing their remains. This sample can then be treated as a multiple of the d_x column. Our sample of otters dying non-violent deaths can be used for this purpose.

The second approach is to treat the age distribution of a living contemporary population as a surrogate of the survival frequency, (f_x). An unbiased, random sample of the population is required for this. The otters suffering violent deaths probably comes close to such a random sample.

The otter life tables

The appropriate computer programmes listed in Krebs (1989) were used to construct life tables from the samples of otters dying violent and natural deaths.

When using a sample of the living population (the sample of violent deaths) to construct a life table, each age frequency must be lower than the one preceding it. In reality a sampled frequency is likely to be ragged and must be statistically smoothed. This was true for the otter samples (e.g. Fig. 7.7), which were smoothed by a running mean. The zero age frequency, the initial number of otters born, was calculated from the age-specific fecundity rates of the adult females in the sample. To do this we used fecundity data collected from a population in Shetland: i.e. age at first breeding, 2 years; litter size at emergence, 1.64; litter rate, 1 per 1.62 years (Kruuk & Conroy, 1991).

We could detect no significant differences in the survival patterns of male and female otters and therefore we have pooled the two sexes in all analyses.

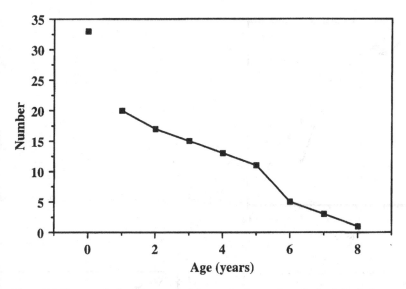

Figure 7.7. The smoothed age frequency of the sample of otters from Shetland that had met a violent death. The zero age frequency was calculated from the fecundity rates of the adults.

Otters from the UK can live beyond 12 years (Fig. 7.8), the oldest animal in our sample being a 16 year old that had died a non-violent death. Most otters, however, die much younger than this. This can best be seen by plotting survivorship curves in which survivors, the l_x column of the life table, are plotted against age (Fig. 7.8(a)–(c)). In such diagrams the survival data are generally plotted on a log scale so that constant age-specific mortality can be readily recognized by the straight-line relationship. Clearly, in the case of otters age-specific survival is not constant. Instead survival is relatively high in the early years of life but decreases with age. This is true for both kinds of life tables, those based on the accumulated remains of animals dying naturally and those based on the violently killed sample of the population age structure.

The median age of death, based on the d_x schedules, was generally around 3–4 years for all samples. The only significant difference in the median age at death for different populations (Mann–Whitney U; $p < 0.02$) was between otters from Shetland and those from England and Wales (Table 7.3). This apparent difference in the longevity of northern and southern populations is reinforced by differences in their rates of age-specific mortality q_x. In all samples the rate of mortality was relatively low early on in life but increased progressively with age (Fig. 7.9). The curves for the populations from Shetland and mainland Scotland were essentially similar but the one for the English and Welsh otter was relatively steep, reflecting the lower median age at death.

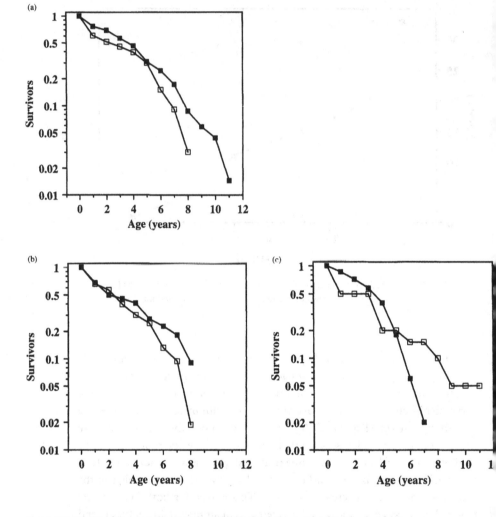

Figure 7.8. Survivorship curves for otters from (a) Shetland, (b) mainland Scotland and (c) England & Wales. ⊟, sample of violent deaths; ■, accumulated non-violent deaths.

Table 7.3. *The median ages of death for different populations of otters*

Shetland	Mainland	England & Wales
4	4	3

  $p = 0.98$
←——————————————→

      $p = 0.27$
   ←————————————————→

   $p = 0.02$
←——————————————————————————→

The statistical probabilities are based on Mann–Whitney *U* tests

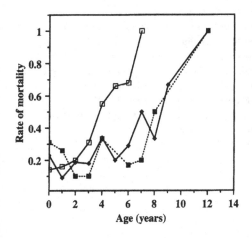

Figure 7.9. Age-specific rates of mortality for otters. ◆, Shetland; - ■ - mainland Scotland; ⊟, England & Wales.

ACKNOWLEDGEMENTS
We are grateful to Don Jefferies for providing material from England and Wales and to Bill Edwards for advice and assistance with histological matters.

References

Andrews, E., Howell, P. &
Johnson, K. (1992). *Otter survey of Wales 1991.* Vincent Wildlife Trust, London.

Garshelis, D. L. (1984). Age estimation of living sea otters. *J.
Wildl. Magmt* **48**: 456–463.

Green, J. & Green, R. (1987).
*Otter survey of Scotland
1984–85.* Vincent Wildlife
Trust, London.

Grue, H. & Jensen, B. (1979). Review of the formation of incremental lines in tooth
cementum of terrestrial mammals. *Dan. Rev. Game Biol.* **11**:
1–48.

Heggberget, T. M. (1984). Age determination in the European
otter *Lutra lutra lutra. Z.
Säugetierk.* **49**: 299–305.

Jefferies, D. J. (1989). The changing otter population of Britain
1700–1989. *Biol. J. Linn. Soc.*
38: 61–69.

Krebs, C. J. (1989). *Ecological
methodology.* Harper & Row,
New York.

Kruuk, H. & Conroy, J. W. H.
(1991). Mortality of otters
(*Lutra lutra*) in Shetland. *J.
appl. Ecol.* **28**: 83–94.

Kruuk, H., Conroy, J. W. H. &
Moorhouse, A. (1987). Sea-
sonal reproduction, mortality
and food of otters (*Lutra lutra*
L.) in Shetland. *Symp. zool. Soc.
Lond.* No. 58: 263–278.

Lieberman, D. E. (1993). Life history variables preserved in den-
tal cementum microstructure.
Science **261**: 1162–1164.

Mason, C. F. & Macdonald, S. M.
(1986). *Otters: ecology and conservation.* Cambridge University Press, Cambridge.

Smith, K. G., Strother, K. A.,
Rose, J. C. & Savelle, J. M.
(1994). Chemical ultrastruc-
ture of cementum growth-
layers of teeth of black bears. *J.
Mammal.* **75**: 406–409.

Stephenson, A. B. (1977). Age determination and morphological variation of Ontario otters.
Can. J. Zool. **55**: 1577–1583.

Strachan, R., Birks, J. D. S.,
Chanin, P. R. F. & Jefferies, D.
J. (1990). *Otter survey of England 1984–1986.* Nature Conservancy Council, Peter-
borough.

8

Habitat use and conservation of otters (*Lutra lutra*) in Britain: a review

H. Kruuk, D.N. Carss, J.W.H. Conroy and M.J. Gaywood

Introduction

This chapter reviews current knowledge and provides previously unpublished data on habitat selection and requirements of the otter (*Lutra lutra*) in Britain. Such information is evaluated in an attempt to provide a basis for habitat management aimed at otter conservation, and to expose areas where further research is needed.

Since the 1950s the otter has declined in many countries in Europe, including Britain (Foster-Turley *et al.*, 1990). In response to this, the British Joint Nature Conservation Committee, following the European Community 'Habitats Directive' (Council Directive 92/43/EEC, 1992), laid down a strategy for otter conservation: 'To maintain existing populations, encourage natural recolonisation, and effectively safeguard viable populations of otters and their habitats throughout their natural range in the United Kingdom' (Anon., 1997). This implies an obligation to conserve those aspects of countryside that are important components of otter habitat. It is these components that the present review attempts to identify.

In principle, conservation management should be directed at environmental factors that limit numbers of the target species. In the case of the otter many of these factors are to be found in the almost linear habitat of the species, following banks and shores. However, although limiting factors should be a primary concern, there are also other aspects of the habitat that are attractive to otters (shown as 'habitat preferences'), but which do not necessarily affect their numbers. Such aspects of habitat could, at least potentially, affect otter numbers at other times and in the absence of any other limiting factors. For instance, if food limited otter numbers in an area, then in the presence of a glut of fish, otter populations could be expected to increase until another factor, such as shelter or predation, might become limiting. It is important, therefore, to identify habitat preferences and to establish from observation or experiment what kind of role the particular factors are likely to play: either as habitat requirements or as mere preferences.

Pollutants are generally acknowledged to be the main cause of the decline of otter populations throughout their former range (Mason & Macdonald, 1986; Foster-Turley *et al.*, 1990), although controversy still surrounds the exact nature of the causal agent (Jefferies, 1989). The question can be asked, therefore, whether habitat conservation is a relevant conservation strategy for this species. We believe it is, for three reasons:

1   Otter populations in some areas of Britain appear to have declined little, if at all, especially in the north and west; their habitat is vital.
2   Recent survey work has suggested that otters are re-colonizing some parts of Britain (Strachan *et al.*, 1990), and this has coincided with a reduction in pollution levels. At least in Scotland, pollution does not appear to be important to populations at present (Kruuk & Conroy, 1996). However, the recovery of otter numbers is relatively slow compared with, for instance, that of piscivorous birds or birds of prey (Newton *et al.*, 1993). Factors other than pollution could be important.
3   Apart from the direct effect of pollution on otters, it may also have acted indirectly through other aspects of the habitat, for instance through its effects on prey populations (Kruuk & Conroy, 1991*a*).

Methods

Most of this review relies on published information, but some unpublished data from our own work in Scotland are also included. There are several problems attached to the relevance of these observations on otter habitat utilization for their use in active conservation measures.

First, there is the general difficulty of deriving 'habitat requirement' from observations on 'habitat preference'. An observed preference may have no population significance: this can only be determined by experiment, or inferred by additional study.

Second, even establishing habitat preference itself is difficult. The usual method (for a review, see Mason & Macdonald, 1986) relies on surveys of faeces ('spraints'); however, spraints have been shown to be unreliable indicators of where otters spend their time (Green, *et al.*, 1984; Kruuk, *et al.*, 1986; Conroy & French, 1987). It has been argued that such surveys should be interpreted with great caution (Kruuk & Conroy, 1987), although this has been challenged by Mason & Macdonald (1987).

An alternative method involves radio-tracking, first carried out for this species by Green *et al.*, (1984), in which the otters' use of different aspects of habitat is compared with availability. This technique is also subject to bias, as such studies can only be carried out where otters are present, and they are undertaken especially in areas where densities are thought to be high. Thus, little information is gained about sub-optimal habitat.

For data on radio-tracking in freshwater areas, as reported in the present study, otters were caught in unbaited wooden box traps of $30 \times 30 \times 200$ cm. They were immobilized and a radio-transmitter was implanted intra-perito-neally (Melquist & Hornocker, 1979, 1983). Radio-tracking was carried out on foot or by vehicle, for whole or part of nights, in the valleys of the Rivers Dee and Don in north-east Scotland (Durbin, 1993; Kruuk et al., 1993). Individual otters were followed for periods of 3–13 months. For further details of capture and radio-tracking see Kruuk (1995).

Habitats along coasts were assessed mainly for depth of foraging areas, the presence on the adjoining land of freshwater pools and streams, height of cliffs, presence of peat, agriculture or human inhabitation (Kruuk et al., 1989). Otter utilization was expressed either by numbers of otter holts, which correlated closely with numbers of otters present (Moorhouse, 1988), or by time spent swimming in 200 m sections of coast, of otters without radio-transmitters (Kruuk et al., 1990). In inland habitats instrumented otters were followed, and they were scored for type of activity and length of time spent in 200 m sections of stream or lake-shore. Sections were characterized in terms of area of water (width: Durbin, 1993; Kruuk et al., 1993). Also, presence and size of reedbeds were scored (Kruuk, 1995), as was presence of islands, trees along banks, marshland (bogs), agriculture and human inhabitation (Durbin, 1993; Kruuk, 1995).

Results

Foraging areas along sea coasts

The diet of otters living along British coasts has been assessed in several studies (Watson, 1978; Mason & Macdonald, 1980, 1986; Herfst, 1984; Kruuk & Moorhouse, 1990; Watt, 1991). Food is obtained almost entirely from the sea, and it consists predominantly of fish, mostly relatively small, benthic species. For instance in Shetland, 75% of prey items ($n = 2031$), and 49% of the estimated annual consumption by weight, consisted of rocklings (mostly *Ciliata mustela*), eelpout (*Zoarces viviparus*), sea-scorpion (*Taurulus bubalis*) and butterfish (*Pholis gunnellus*). The mean mass of individuals of each of these species caught by otters was considerably less than 50 g (Kruuk & Moorhouse,

1990). In some areas crabs were also important, and in winter otters take larger fishes more often than at other times (Kruuk & Moorhouse, 1990; Watt, 1991).

Foraging strategies of coastal otters have been described by Kruuk & Moorhouse (1990), Kruuk *et al.* (1990), Watt (1991), and Nolet *et al.* (1993). Coastal foraging habitat can be defined as areas less than 10 m deep within 100 m of the shore; deep water is avoided. Within that strip almost all food is acquired close to the shore, e.g. 84% of 500 dives occurred within 50 m of the coast (Kruuk & Moorhouse, 1991). Most dives are in water less than 2 m deep (Watt, 1991; Nolet *et al.*, 1993). The preference for foraging in shallow water can be related to energetic costs (with deep dives being more demanding and allowing less time foraging along the bottom; Nolet *et al.*, 1993; Houston & McNamara, 1994). This energetic need for shallow water could possibly explain the low otter densities observed along steep coasts (Kruuk *et al.*, 1989).

Within the shallow zones otters tend to concentrate their foraging effort in particular 'patches'. These are characterized not by greater densities of prey, but apparently by ease of access: patches have more openings in the 'canopy' of phaeophyte algae, enabling otters to get underneath the mass of fronds (Kruuk *et al.*, 1990).

In Shetland different coasts were used seasonally with varying frequency. For instance, the sheltered areas dominated by *Ascophyllum nodosum* and *Fucus vesiculosis* were used especially in summer, when eelpout were abundant. In winter, exposed coasts with *Gigartina stellata* were favoured, and otters caught mostly rockling there (Kruuk *et al.*, 1988; Kruuk & Moorhouse, 1990). Such habitat preferences are well-documented for several of the otters' main prey fish (Gibson, 1972, 1982; Koop & Gibson, 1991).

Prey densities along some Scottish coasts have been estimated using fish traps (Kruuk *et al.*, 1988; Watt, 1991). It has been suggested from estimated rates of consumption by otters, and densities of fish populations, that otters take a high proportion of the population of five-bearded rockling, which is one of the main prey species and occurs only in the shallow zone (Kruuk & Moorhouse, 1990). However, there is no evidence that otters have an effect on the density of any prey species. Depleted foraging patches are recolonized by prey fish within 24 h (Kruuk *et al.*, 1988). Thus, it is possible that otter foraging affected the feeding success of other animals feeding in the same patches during the same day but not over longer periods (Kruuk, 1995), but evidence for this is lacking. Otter mortality is highest during periods of low abundance of fish, and food stress may have played a role in this (Kruuk & Conroy, 1991*b*). Thus, seasonal limitation of otter numbers by prey availability is a possibility, but the mechanism is as yet unclear.

It has been argued that the size of annual ranges utilized by the otters was at least in part determined by the distances between the seasonal feeding areas, i.e. the 'grain size' of the coast (Kruuk, 1995). In Shetland individual range sizes of females varied between 4.7 and 14 km, whilst male ranges have not yet been determined but are much larger. Up to five same-sex otters may share a range, recognizing the same boundaries (Kruuk & Moorhouse, 1991).

The habitat used by males is somewhat different to that of females, as males spend more time on exposed coasts (Kruuk & Moorhouse, 1991). They entered the more sheltered female areas during the spring, which is the mating season along coasts. As a consequence of the difference in habitat use, the diets of males and females varied. Males took more prey from exposed areas, prey which was on average 13% heavier than that of females (Kruuk & Moorhouse, 1990).

Foraging areas may be situated along coasts that are frequently used by people, close to houses and traffic, even next to heavy industry such as at the Sullom Voe oil terminal in Shetland (our unpublished observations).

Washing sites along coasts

When otters forage in seawater they frequently wash in freshwater pools or small streams. This removes salt from the fur and thereby preserves the animals' thermoinsulation; it has been demonstrated that this behaviour is essential for the animals' exploitation of marine resources (Kruuk & Balharry, 1990). Thus, the presence or absence of freshwater along coasts is likely to have a major effect on habitat suitability. These small pools and streams are not used for foraging, as they contain no potential prey.

In Shetland, otters have easy access to subterranean freshwater in areas of peat, where water accumulates in places on top of the impermeable layer of clay or rock, 1–2 m below the peat surface. Consequently, Kruuk *et al.*, (1989) found a strong association between otter density and peat in Shetland, and also with the presence of freshwater pools. In areas of Shetland with apparently high densities of otters, such as the Lunna peninsula with a mean of 1.0 otters/km (Kruuk *et al.*, 1989), there are large numbers of pools and small streams close to the shore. Moorhouse (1988) counted 44 sources of permanent water over a distance of 18 km there (mean 2.4 (±1.3) SD per km), as well as numerous temporary ones.

Well-drained coasts, with little or no freshwater on or close to the surface, have very few otters, although few hard data are available. Moorhouse (1988) compared otter densities along the coasts of Shetland and Orkney and related the much lower numbers in Orkney to the lack of freshwater washing sites. Similarly, it is likely that the distribution of otters along the Scottish west coast and islands, and the virtual absence of coastal otters in east Scotland and the rest of Britain, was at least partly related to the distribution of freshwater streams and pools. This could be a result of differences in geology and rainfall. Similar patterns of otter distribution in relation to coastal freshwater were reported from Portugal (Beja, 1992).

Resting sites along coasts

Otters in Shetland were active in day-time, and in all observations they spent the night underground, inside tunnel systems terms 'holts' (Moorhouse, 1988). Day-time resting sites might also be inside holts but could be anywhere, mostly between boulders or amongst algae near the tide-line. Data from 112 regularly used holts in an intensive study area suggested that they either had a supply of fresh washing water inside or in the entrance (wet holts), or they were situated close to freshwater streams or pools (Moorhouse, 1988). Most holts in Scottish coastal areas were close to the sea (78% were within 100 m in Shetland: Moorhouse, 1988; see also Kruuk & Hewson, 1978), but some were much further inland, sometimes high on flat hill tops, where freshwater gathered underground.

Holt density can reach a mean of 13.3 ($\pm$7.7 SE) per 5 km of coast in peaty areas of Shetland (Kruuk *et al.*, 1989). No holts were found in the well-drained agricultural areas there. There were high densities of holts on small islands, but they were absent from two such islands without freshwater (Moorhouse, 1988).

Otters have been recorded resting inside buildings (such as a pump house at the Shetland oil terminal), close to a quarry (Glensanda, Argyll), under roads and at other sites with frequent human activity (our unpublished observations).

A number of natal holes have been identified in coastal areas, several of them situated far from the sea or other open water, but with their own supply of freshwater underground (Moorhouse, 1988). Such natal holts are difficult to recognize as being used by otters, and they may often be overlooked.

There is a strong correlation between numbers of otters and numbers of used holts along sea coasts (Moorhouse, 1988; Kruuk *et al.*, 1989), but there is no evidence that otter numbers are limited by numbers of holts. *Prima facie* this appears unlikely, given the large numbers of unused holts in the same areas.

Foraging in freshwater areas

The food of otters in inland areas has been particularly well studied, with many of the results summarized by Mason & Macdonald (1986) and Brzezinski *et al.* (1993). The diet consists predominantly of fish, with a minor overall contribution from amphibians, mammals and birds, although in spring and winter some of these components may be very important (e.g. Weber, 1990). The range of fish species and sizes taken is varied but many are small and either truly benthic, or spending at least some time on the bottom whilst inactive. In many northern areas of Britain salmonids and/or eels (*Anguilla anguilla*) are staple foods (Carss *et al.*, 1990).

In streams there is some evidence that large salmon are taken mostly on or near riffles (Carss *et al.*, 1990). Amphibians are taken mostly from the muddy bottom of streams, from ponds or from boggy areas, and may constitute up to half of the prey items in spring (Weber, 1990). The importance of bogs to otters, for catching frogs, is often underestimated because few spraints are deposited there, but observations on animals with radio-transmitters suggest that they may provide a great deal of prey (Green *et al.*, 1984; our unpublished observations).

In many waters in north-east Scotland otters showed distinct preferences for particular widths; there was a strong negative correlation between stream width and time spent per otter per area of water (Kruuk *et al.*, 1993). In the Dinnet Lochs in north-east Scotland 10 otters were radio-tracked for a total of 462 days, of which they spent 39% in small streams and bogs away from the lochs. We estimated that these areas contained less than 5% of the total water surface in the study area.

A likely hypothesis to explain these associations is based on the observation that densities of the main prey fishes vary with stream width. For instance, the biomass of trout (*Salmo trutta*) shows a strong negative correlation with stream width (Durbin, 1993; our unpublished observations), whilst the opposite is true for the less common salmon (*Salmo solar*). Thus, even very narrow tributaries of main rivers (down to 0.5 m in width) may be important foraging sites to otters, even close to human habitation and roads.

There is a highly significant correlation between otter usage and fish biomass in streams in north-east Scotland (Kruuk *et al.*, 1993). Also, otter mortality is highest after months when food is least abundant, and otters take to feeding on prey of inferior quality (Kruuk *et al.*, 1993). These observations suggest that there is a role for prey populations in limiting populations of otters. There is, however, no experimental evidence to support this.

In freshwater, as along coasts, male and female otters tend to use significantly different habitats. For instance, in Deeside in Scotland 62% of adult male otters spent most of their time along the main stem of major rivers ($n = 8$), whereas 87% of females and young males were mostly found in lakes and along tributaries ($n = 15$; Kruuk, 1995).

The length of otter ranges in freshwater can be considerable, each one

Table 8.1. *Resting sites of four otters, followed by radio-tracking along the River Dee and a tributary (the Beltie Burn)*

	No. days in				Total no. of obs.
	Couch		Holt		
River Dee:					
Male N	135	(63.1%)	79	(36.9%)	214
Male R	65	(77.4%)	19	(22.6%)	84
Beltie Burn:					
Female 4	67	(53.6%)	58	(46.4%)	125
Male 2	1209	(48.8%)	126	(51.2%)	246
Total:	387	(57.8%)	282	(42.2%)	669

covering different stream, lake and vegetation types and inhabited by several individuals. Green *et al.*, (1984) estimated ranges of 39 km (male), 16 km and 24 km (females) of river bank in Perthshire, Scotland. In north-east Scotland (the River Dee and River Don areas, over periods of 3–13 months) mean range lengths were estimated at 20 km for females ($n=2$) and 32 km for males ($n=6$) (Kruuk *et al.*, 1993). However, the variation was large, with one male covering 84 km (Durbin, 1993).

Resting sites along lakes and rivers

Radio-tracking otters in freshwater habitats in Scotland showed that most animals spent their resting time above ground in 'couches' (i.e. above-ground resting sites: Hewson, 1969), rather than underground in 'holts'. For instance, Green *et al.*, (1984) identified 24 couches and 21 holts, used by three different animals over a total of 156 days. These authors commented that fewer than 10% of either couches or holts would have been recognized as such without radio-tracking. Along rivers and streams in north-east Scotland we radio-tracked four otters over a total of 669 days; the animals spent 58% of day-time resting periods in couches, 42% in holts (Table 8.1).

Five otters, which we radio-tracked in two locks in the same area for 131 days, slept almost invariably on couches in reed beds or in low shrub; they did not use holts, even in winter. The use and density of these couches was strongly correlated with the size of the area of reed bed, but not with proximity of feeding areas or human disturbance (Table 8.2). Couches in reed beds were often well-constructed, from reeds bitten off nearby. Some were ball-shaped, with a side entrance (Taylor & Kruuk, 1990).

Table 8.2. *Otter resting sites in sections of shore of Loch Davan (Dinnet, Aberdeenshire), in relation to the size of reed beds and distances from feeding areas and potential sources of human disturbance*

Section	(a) % days spent by 5 radio-tagged otters ($n = 131$)	(b) Freq. ranking couches and otter tunnels in veg.	(c) Reed bed size (ha)	(d) Feeding distance index (FDI)	(e) Nearest disturb. distance (in m)
D 1	65	11.0	0.62	345	130
D 2	14	9.0	0.49	291	100
D 3	2	7.0	0.27	426	120
D 4	0	2.5	0.0	261	240
D 5	1	8.0	0.15	328	370
D 6	15	10.0	1.01	480	460
D 7	0	2.5	0.0	690	620
D 8	0	2.5	0.0	484	670
D 9	1	2.5	0.11	451	600
D10	1	5.0	0.18	320	350
D11	2	6.0	0.01	233	330

Observations 1987–1990, from Kruuk (1995).

Feeding distance index (FDI) for each shore (rest) section was calculated as FDI $= \Sigma(D_i \cdot S_i)$, in which D is the distance to any section of (fishing) water of the loch, S the proportion of all otter prey caught in that feeding section. The disturbance distance is the distance to the nearest main road or inhabited house.

Spearman rank correlations:

(a)/(b): $r = 0.91$, $p < 0.001$

(a)/(c): $r = 0.91$, $p < 0.001$

(a)/(d): $r = -0.20$, n.s.

(a)/(e): $r = -0.57$, n.s.

(b)/(c): $r = 0.89$, $p < 0.001$

(b)/(d): $r = -0.19$, n.s.

(b)/(e): $r = 0.55$, n.s.

Along rivers, couches were located in a variety of dense types of cover (Table 8.3), and frequently otters slept on small islands. Couches were not restricted to riparain areas; otters often slept several hundred metres from the river bank. It was difficult to quantify the availability of such sleeping sites because they

Table 8.3. *Sites of otter couches on rivers*

	No. couches (Deeside)		No. days used (Deeside)		No. couches (Perthshire)	
Island	3	(10%)	147	(37%)	5	(21%)
Reeds	3	(10%)	12	(3%)		
Juncus, Deschampsia, etc.	3	(19%)	23	(6%)	7	(29%)
Shrub	9	(31%)	52	(13%)	3	(13%)
Ledge under bank	6	(21%)	62	(15%)		
Under fallen tree	3	(10%)	55	(14%)	6	(25%)
Under boulder	1	(3%)	2	(1%)		
Stick-pile					3	(13%)
Old car	1	(3%)	49	(12%)		
Total	29	(100%)	402	(100%)	24	(100%)

Data from four otters along streams and rivers in Deeside, north-east Scotland (Institute of Terrestrial Ecology, unpublished results), and two otters in Perthshire (Green *et al.*, 1984).

Table 8.4. *Otter holts and their substrate*

	No. holts, Deeside otters		No. days spent by Deeside otters		No. holts, Perthshire otters	
Hole in sand or other soil	2	(22%)	44	(16%)	1	(5%)
Artificial embankment with boulders	4	(44%)	193	(73%)		
Field drain	3	(33%)	30	(11%)		
Tree roots, ±boulders					13	(65%)
Boulders					6	(30%)
Total	9	(99%)	267	(100%)	20	(100%)

Data from four otters in Deeside, north-east Scotland (Institute of Terrestrial Ecology, unpublished results), and from two otters in Perthshire (Green *et al.*, 1984). The substrate 'artificial embankment with boulders' overlaps with or is the same as 'tree roots ± boulders' and 'boulders'. Figures presented are totals for all animals

had few, if any, distinguishing features. Our subjective impression was that they were very abundant, and not likely to be limiting.

Holts used by otters living along rivers occurred in various types of bank

(Table 8.4). Often the animals used man-made structures or artificial embankments, despite the apparent abundance of natural sites. For holts, as for couches, potential sites appear to be abundant everywhere along the streams, therefore not likely to be limiting the numbers of otters.

Natal holts in freshwater areas have been described by Harper (1981). However, in her observations cubs were already several months old when found, and they were therefore likely to have been moved by the mother from the real natal holt elsewhere, as was found in Shetland (Kruuk & Conroy, 1991*b*). One litter was born in a man-made slope of boulders adjacent to a small tributary, away from areas visited regularly by other otters (Durbin, 1993). Cubs may also be born above ground, and a natal couch in a reed bed has been described by Taylor & Kruuk (1990).

Discussion

Data reviewed here are derived mostly from radio-tracking studies or from direct observations where habitat selection could be addressed. This circumvents the problems associated with habitat assessment from 'spraint surveys', where animals may scent-mark some vegetation types, banks or areas, but not others, or they may scent-mark seasonally (Conroy & French, 1987). However, there is also an obvious bias in data from radio-tracking and direct observation (see Methods), as they tend to be collected where otters are plentiful. There are large areas with oligotrophic streams and lakes, with intensive agriculture, industrialization or urbanization, for which no information on otters exists.

With these reservations, the present review shows that in Britain the species occurs along sea coasts as well as in freshwater areas, often in very long and relatively narrow individual ranges. Thus, compared with other native carnivores, the otter's main habitat characteristics are the combination of an aquatic and terrestrial component, and linearity.

Along marine coasts a prominent parameter of the adjacent terrestrial habitat component is access to freshwater, in pools or streams or underground. Coastal otters also usually sleep in underground holts for their main resting period, in contrast to inland otters. Apart from these characteristics otters appear to be very catholic in their choice of areas. The coastal aquatic component includes shallow water, a rocky substrate along shores with sheltered as well as exposed sections, with beds of algae.

In the terrestrial habitat component of freshwater areas otters show preferences for islands and for reed beds, but the animals may rest or sleep almost anywhere with a cover of vegetation or rocks. For their foraging needs the inland otters prefer narrow water bodies that are shallow and with a high fish biomass, and also marshy areas with amphibians. All information from radio-tracking studies, however, is confined to oligotrophic and mesotrophic areas of Scotland, and habitat preferences in more eutrophic regions could be distinctly different: more information is needed.

There are problems with the interpretation of habitat preferences in terms of factors limiting populations (see Introduction). For instance, although otters appear to select small islands as rest sites in catchments in north-east Scotland, there is evidence that fish populations may be limiting otter numbers (see above), at least in some areas (Kruuk *et al.*, 1993). It is possible that under such circumstances a particular habitat preference, for example for rest sites, is of secondary importance, but at least theoretically it could have a limiting function in otter populations under conditions of super-abundant food. Conversely, prey abundance may now be a limiting factor, but only after other habitat requirements, such as appropriate resting sites, have been satisfied.

Thus, although there is some information on otter habitat preferences, additional observations or, ideally, experiments, are needed to confirm our interpretations of those for the purposes of conservation management. For instance, otters were more abundant with an increasing presence of freshwater pools along sea coasts. It has been demonstrated experimentally that for reasons of thermoregulation, when using a marine habitat, the animals frequently need to wash in fresh water, otherwise they would probably not survive (Kruuk & Balharry, 1990). It suggests that this aspect of the habitat is essential for survival. Otters appear to have extra problems with thermoregulation in seawater, and this, coupled with the fact that they almost always sleep in holts in these coastal areas, suggests that the availability of suitable coastal holts is an essential prerequisite for otters, rather than a mere preference. When freshwater and holt sites are abundant, numbers of otters may possibly be limited by fish abundance, by fish biomass or the extent of shallow fishing waters.

Inland there is no evidence that preferences for particular terrestrial habitat components, however obvious, are related to the number of otters using these areas. It has been suggested, from observations on spraint distribution, that cover, trees or other aspects of bank vegetation are vital to otter populations (e.g. Mason & Macdonald, 1986; Bas *et al.*, 1984), and we have seen that radio-tracked otters preferred reed beds and islands. Nevertheless, otters clearly thrive elsewhere without such cover, e.g. on tree-less Shetland. Some terrestrial habitat components, e.g. trees along streams, may affect aquatic prey biomass (Durbin, 1993), and perhaps therefore otter populations; however, there is no direct evidence for this.

Our observations suggest that in inland feeding areas prey productivity is, for otters, the most important aspect of the aquatic habitat component. Prey biomass density may be inversely related to the size of water bodies (Kruuk *et al.*, 1993), and it may be dependent on length of banks and overhanging vegetation (Durbin, 1993). In this case, therefore, the otters' preference for given types of water (i.e. narrow streams) may be a strong indication that the presence of such a habitat feature does, indeed, limit numbers. This aspect of habitat is one that may be affected by environmental management such as canalization of streams (Shackley & Donaghy, 1992).

Clearly, more information on the availability of freshwater habitats is needed, especially from regions where otters are absent or occur in low densities. Despite this, the available information may be used tentatively, as a baseline for habitat conservation policies aimed at the protection of otters in Britain.

When assessing threats to coastal habitats of otters the possibility of land-drainage in areas close to the shore should be evaluated, especially in areas with shallow rocky coasts. Freshwater pools could be created in areas where there are few or no otters (and other habitat aspects are favourable), to encourage recolonization. Other threats, e.g. to the marine inshore fish populations or to holt sites, should also be countered.

In the case of inland otter habitats, bank vegetation may be relatively unimportant in explaining the movements of otters (Durbin, 1993). However, conservation measures should extend over relevant fish populations, especially in small streams. The preference of otters for resting on small islands and in reed beds may imply that perhaps under some conditions, when other limiting factors do not obtain, these habitat elements could be important. They might be considered for special protection. It is clear, however, that further research into the habitat requirements of otters is urgently needed, especially in more eutrophic inland areas.

An important lesson for conservation managers from field studies of otters is that the ranges of individual animals are very long. Thus, management policies must cover many contiguous kilometres of small streams, rivers and coasts, not just small isolated sites. Within such large areas, the species can co-exist with agriculture and many other human activities. There is a clear case for management of large ecosystems on at least a catchment basis, in which otter survival is only one of many priorities.

ACKNOWLEDGEMENTS

We are grateful to Scottish Natural Heritage for funding part of this review. Many people were involved in the field work, including Drs J.-M. Weber, P. Taylor, P. Bacon and L. Durbin, as well as A. Moorhouse, K. Nelson and others. Dr M. Gorman, L. Farrell, D. Howell and Dr P. Boon provided helpful comments on an earlier draft.

References

Anon. (1997). *A framework for otter conservation in the UK: 1995–2000.* Joint Nature Conservation Committee, Peterborough.

Bas, N., Jenkins, D. & Rothery, P. (1984). Ecology of otters in northern Scotland. V. The distribution of otter (*Lutra lutra*) faeces in relation to bankside vegetation on the river Dee in summer 1981. *J. appl. Ecol.* **21**: 507–513.

Beja, P. R. (1992). Effects of freshwater availability on the summer distribution of otters *Lutra lutra* in the southwest coast of Portugal. *Ecography* **15**: 273–278.

Brzezinski, M., Jedrzejewski, W. & Jedrzejewska, B. (1993). Diet of otters (*Lutra lutra*) inhabiting small rivers in Bielowieza National Park, eastern Poland. *J. Zool., Lond.* **230**: 495–501.

Carss, D. N., Krukk, H. & Conroy, J. W. H. (1990). Predation on adult Atlantic salmon, *Salmo salar* L., by otters, *Lutra lutra*(L.), within the River Dee system, Aberdeenshire, Scotland. *J. Fish Biol.* **37**: 935–944.

Conroy, J. W. H. & French, D. D. (1987). The use of spraints to minitor populations of otters (*Lutra lutra* L.). *Symp. zool. Soc. London.* No. 58: 247–262.

Council Directive 92/43/EEC (1992). On the conservation of natural habitats and of wild flora and fauna. *Official J. Eur. Commun.* **L206**: 7–50.

Durbin, L. (1993). *Food and Habitat Utilization of Otters (Lutra lutra L.) in a Riparian Habitat – The River Don in North-east Scotland.* PhD thesis: University of Aberdeen.

Foster-Turley, P., Macdonald, S. M. & Mason, C. F. (Eds) (1990). *Otters: an action plan for their conservation.* IUCN, Gland, Switzerland.

Gibson, R. N. (1972). The vertical distribution and feeding relationships of intertidal fish on the Atlantic coast of France. *J. Anim. Ecol.* **41**: 189–207.

Gibson, R. N. (1982). Recent studies on the biology of intertidal fishes. *Oceanogr. mar. Biol.* **20**: 363–414.

Green, J., Green, R. & Jefferies, D. J. (1984). A radio-tracking survey of otters *Lutra lutra* on a Perthshire river system. *Lutra* **27**: 85–145.

Harper, R. J. (1981). Sites of three otter (*Lutra lutra*) breeding holts in fresh-water habitats. *J. Zool., Lond.* **195**: 554–556.

Herfst, M. (1984). Habitat and food of the otter *Lutra lutra* in Shetland. *Lutra* **27**: 57–70.

Hewson, R. (1969). Couch building by otters *Lutra lutra. J. Zool., Lond.* **159**: 524–527.

Houston, A. I. & McNamara, J. M. (1994). Models of diving and data from otters: comments on Nolet *et al.*, (1993). *J. Anim. Ecol.* **63**: 1004–1006.

Jefferies, D. J. (1989). The changing otter population of Britain 1700–1989. *Biol. J. Linn. Soc.* **38**: 61–69.

Koop, J. H. & Gibson, R. N. (1991). Distribution and movements of intertidal butterfish *Pholis gunnellus. J. mar. biol. Ass. U. K.* **71**: 127–136.

Kruuk, H. (1995). *Wild otters: predation and populations.* Oxford University Press, Oxford.

Kruuk, H. & Balharry, D. (1990). Effects of sea water on thermal insulation of the otter, *Lutra lutra. J. Zool., Lond.* **220**: 405–415.

Kruuk, H. & Conroy, J. W. H. (1987). Surveying otter *Lutra lutra* populations: a discussion of problems with spraints. *Biol. Conserv.* **41**: 179–183.

Kruuk, H. & Conroy, J. W. H. (1991*a*). Mortality of otters (*Lutra lutra*) in Shetland. *J. appl. Ecol.* **28**: 83–94.

Kruuk, H. & Conroy, J. W. H. (1991*b*). Recruitment to a population of otters (*Lutra lutra*) in Shetland, in relation to fish abundance. *J. appl. Ecol.* **28**: 95–101.

Kruuk, H. & Conroy, J. W. H. (1996). Concentrations of some organochlorines in otters (*Lutra lutra* L.) in Scotland: implications for populations. *Envir. Pollut.* **92**: 165–171.

Kruuk, H. & Hewson, R. (1978). Spacing and foraging of otters (*Lutra lutra*) in a marine habitat. *J. Zool., Lond.* **185**: 205–212.

Kruuk, H. & Moorhouse, A. (1990). Seasonal and spatial differences in food selection by otters (*Lutra lutra*) in Shetland. *J. Zool., Lond.* **221**: 621–637.

Kruuk, H. & Moorhouse, A. (1991). The spatial organization of otters (*Lutra lutra*) in Shetland. *J. Zool., Lond.* **224**: 41–57.

Kruuk, H., Conroy, J. W. H., Glimmerveen, U. & Ouwerkerk, E. (1986). The use of spraints to survey populations of otters *Lutra lutra. Biol. Conserv.* **35**: 187–194.

Kruuk, H., Nolet, B. & French, D. (1988). Fluctuations in numbers and activity of inshore demersal fishes in Shetland. *J. mar. biol. Ass. U. K.* **68**: 601–617.

Kruuk, H., Moorhouse, A., Conroy, J. W. H., Durbin, L. & Frears, S. (1989). An estimate of numbers and habitat preference of otters *Lutra lutra* in Shetland, U. K. *Biol. Conserv.* **49**: 241–254.

Kruuk, H., Wansink, D. & Moorhouse, A. (1990) Feeding patches and diving success of otters, *Lutra lutra*, in Shetland. *Oikos* **57**: 68–72.

Kruuk, H., Carss, D. N., Conroy, J. W. H. & Durbin, L. (1993). Otter (*Lutra lutra* L.) numbers and fish productivity in rivers in north-east Scotland. *Symp. zool. Soc. Lond.* No. 65: 171–191.

Mason, C. F. & Macdonald, S. M. (1980). The winter diet of otters (*Lutra lutra*) on a Scottish sea loch. *J. Zool., London.* **192**: 558–561.

Mason, C. F. & Macdonald, S. M. (1986). *Otters: ecology and conservation.* Cambridge University Press, Cambridge.

Mason, C. F. & Macdonald, S. M. (1987). The use of spraints for surveying otter *Lutra lutra* populations: an evaluation. *Biol. Conserv.* **41**: 167–177.

Melquist, W. E. & Hornocker, M. G. (1979). Methods and techniques for studying and censusing river otter populations. *Tech. Rep. Univ. Idaho For. Wildl. Range exp. Stn.* No. 8: 1–127.

Melquist, W. E. & Hornocker, M. G. (1983). Ecology of river otters in west central Idaho. *Wildl. Monogr.* No. 83: 1–60.

Moorhouse, A. (1988). *Distribution of Holts and Their Utilisation by the European Otter (Lutra lutra L.) in a Marine Environment.* MSc thesis: University of Aberdeen.

Newton, I., Wyllie, I. & Asher, A. (1993). Long-term trends in organochlorine and mercury residues in some predatory birds in Britain. *Envir. Pollut.* **79**: 143–151.

Nolet, B. A., Wansink, D. E. H. & Kruuk, H. (1993). Diving of otters (*Lutra lutra*) in a marine habitat: use of depths by a single-prey loader. *J. Anim. Ecol.* **62**: 22–32.

Shackley, P. E. & Donaghy, M. J. (1992). The distribution and growth of juvenile salmon and trout in the major tributaries of the river Dee catchment (Grampian Region). *Scott. Fish. Res. Rep.* No. 51: 1–19.

Strachan, R., Birks, J. D. S., Chanin, P. R. F. & Jefferies, D. J. (1990). *Otter survey of England 1984–1986.* Nature Conservancy Council, Peterborough.

Taylor, P. S. & Kruuk, H. (1990). A record of an otter (*Lutra lutra*) natal den. *J. Zool., Lond.* **222**: 689–692.

Watson, H. C. (1978). *Coastal otters in Shetland.* Vincent Wildlife Trust, London.

Watt, J. P. (1991). *Prey Selection by Coastal Otters (Lutra lutra L.).* PhD thesis: University of Aberdeen.

Weber, J.-M. (1990). Seasonal exploitation of amphibians by otters (*Lutra lutra*) in north-east Scotland. *J. Zool., Lond.* **220**: 641–651.

9

The relationship between riverbank habitat and prey availability and the distribution of otter (*Lutra lutra*) signs: an analysis using a geographical information system

T.J. Thom, C.J. Thomas, N. Dunstone and P.R. Evans

Introduction

This chapter presents preliminary findings of a research programme to determine the factors that influenced the distribution of otter signs in the upper Tyne catchment, Northumberland, England, between 1993 and 1995. The work presented here considers two aspects of the research programme. Firstly, the effect of changing the sampling scale on the pattern of distribution of otter signs in the catchment is assessed. In this context, sampling scale is defined as the size of the sampling unit. Secondly, the relationships between the distribution of otter signs and the availability of different types of riverbank vegetation and the influence of prey availability are investigated.

In the majority of studies of habitat utilization by otters, the distribution of otter spraints has been used as an indicator of the presence or absence of otters (Jenkins & Burrows, 1980; Macdonald & Mason, 1983; Bas *et al.*, 1984). A relationship between the distribution of otter signs and riverbank vegetation, in particular the availability of woodland and scrub habitats, was found in all of these studies. A similar relationship has been demonstrated using radio-telemetry techniques. Jefferies *et al.* (1986) showed that radio-tracked otters spent over 50% of their time in wooded areas, although only a few individuals were tracked. Green *et al.* (1984) showed that one of their radio-tracked otters sprainted more at centres of activity than elsewhere, although analysis of the movements of another individual showed no such relationship. The use of spraint density as an indicator of habitat utilization has been criticized by Kruuk *et al.* (1986) who suggested that, for a population of otters on the coast of Shetland, there was no demonstrable relationship between the time spent in a particular habitat and spraint density.

Kruuk & Conroy (1987) criticized their own methods for the arbitrary choice of sampling scale and discussed the need for the implications of this to be studied in more detail. Jefferies (1986) demonstrated the existence of large-scale relationships between the density of otters and the number of spraints in coastal areas. The effect of scale may also be important when considering the effect of habitat variables on the distribution of otter spraints. For example, it is possible that the pattern of spraint distribution may vary with scale and that this pattern may be associated with different parameters at different scales.

Very few studies of otter ecology have assessed the relationship between otter distribution and prey availability. Kruuk *et al.* (1993) investigated this relationship using radio-tagged individuals. Only one study based on the distribution of spraints has considered the effect of prey availability but prey densities were only estimated as the presence or absence of coarse and game fish (Macdonald & Mason, 1983). In this chapter we present some preliminary results of a comparison between spraint distribution at a variety of scales, riverbank vegetation and prey availability based on the results of an electro-fishing survey to assess prey populations in the upper Tyne catchment (Thom, 1997).

Study area

The Tyne catchment is divided into three main river systems, the main Tyne, the North Tyne and the South Tyne, which form the upland stretches of the catchment and can be classed as fast-flowing upland rivers. This research concentrates on these latter two rivers and associated tributaries (see Fig. 9.1).

The North Tyne rises as a series of streams at an altitude of between 520 and 580 m in the hills forming the border between England and Scotland. Within about 1 m of its source it flows into the upper end of the Kielder Border forest, an extensive area of conifer plantation. Within approximately 10 km of the source the North Tyne's natural flow is interrupted by Kielder reservoir. Below Kielder reservoir the North Tyne flows for a further 50 km through a predominantly rural area, with sheep and cattle pasture being the main form of land use.

A number of small tributaries and streams flow into the reservoir and the North Tyne but only one of these is significant enough to be classed as a river. The Rede is a medium-sized, essentially upland river of 43 km in length and is at the north-eastern extremity of the North Tyne system (see Fig. 9.1). Its source is at an altitude of about 408 m. Like the North Tyne, the river's natural flow is altered after about 7 km by the presence of Catcleugh Reservoir. Redesdale is sparsely populated and much of the valley forms part of the Otterburn military training range. The upper part of the river flows through

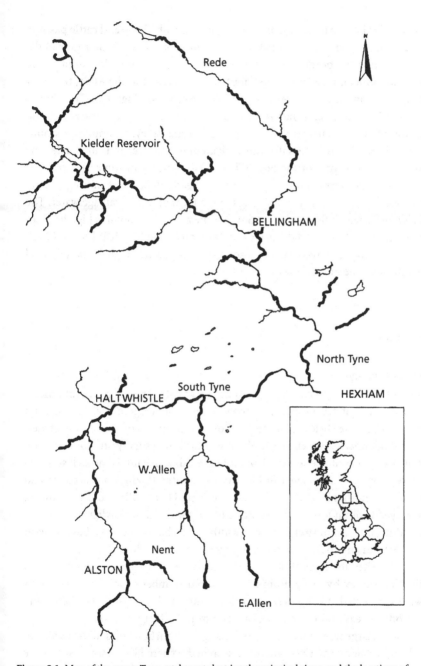

Figure 9.1. Map of the upper Tyne catchment showing the principal rivers and the locations of the 5 km survey stretches (bold lines).

the Kielder Border forest and the rest through mainly sheep and cattle pasture.

The South Tyne rises at an altitude of about 500 m on Alston Moor in the hills forming the border between the counties of Durham, Northumberland and Cumbria and has two main tributaries, the Nent and the Allen, the latter further subdivided into the East and West Allens (see Fig. 9.1). The river is about 65 km in length, fast-flowing and unrestricted by the reservoirs and afforestation so characteristic of the North Tyne valley. Although predominantly rural, the river does flow through an industrial area in Haltwhistle. The predominant feature of the South Tyne is the history of mineral extraction from the Northern Pennines orefield in its headwaters. Most mining activity has ceased but the legacy still remains in the form of mine drainage water. The River Nent, which flows into the South Tyne at Alston, contains high levels of zinc and other heavy metals such as lead and cadmium (Say & Whitton, 1981). Work by Abel & Green (1981) also demonstrated the high levels of metal pollution in the East and West Allens.

Methods

Spraint surveys

Spraint surveys were conducted along 40 5 km stretches of riverbank, measured from digitized Ordnance Survey Maps in a Geographical Information System (GIS) (see below). Figure 9.1 shows the locations of the 5 km stretches.

Five kilometres was chosen as the size of the initial survey unit for two main reasons: (i) it could be surveyed on foot in one day, meaning that each seasonal survey could be completed in less than 2 months (taking into account bad weather conditions), and (ii) this was considered to be a distance that could be surveyed without loss of accuracy due to fatigue or failing daylight.

Along each 5 km stretch a 10 m width of one bank (the same bank in each season) and any rocks or logs in the water up to the middle of the watercourse were searched for signs of otters. The bank to be searched was chosen prior to the first survey by using a binomial random number generator. All surveys were conducted during periods of low water and at least 4 days after any periods of heavy rain or releases of water from Kielder Reservoir.

Rocks, large tree root systems, overhanging branches, holes in the bank and other sites along the riverbank were searched. Where likely spraint sites were inaccessible, for example, on rocks in deep water, binoculars were used to look for otter signs. Tarry secretions and mucus-like deposits varying from white to brown in colour and with the characteristic otter smell were also recorded.

However, the numbers of these were low in all seasons and they were therefore classified as spraints. Paths, flattened areas and above-ground resting sites were also recorded if these could be attributed to the otter (this usually depended on the presence of other signs such as spraints along paths or spread around flattened areas). Spraint heaps, areas of sand or mud scraped into a mound, often with a spraint on top, were also recorded.

Each survey was carried out four times so that the effects of seasonal changes on the distribution and abundance of otter signs could be assessed. The survey periods were Spring 1993 (1 March – 30 April), Summer 1993 (10 June – 10 August), Winter 1994 (5 January – 28 February) and Autumn 1994 (1 October – 20 November).

Riverbank vegetation

Riverbank vegetation was classified into six categories.

1   BARE: No vegetation and close-mown or heavily grazed ground.
2   LOW: Vegetation up to 10 cm in height.
3   MEDIUM: Vegetation greater than 10 cm but less than 1 m in height.
4   DENSE: Herbaceous vegetation greater than 1 m in height with shrubs.
5   OPEN: Open canopy woodland – defined as woodland with the edges of individual tree canopies at least 2 m apart.
6   CLOSED: Closed canopy woodland – defined as woodland with the edges of individual tree canopies touching or overlapping.

The start and end points of each habitat section were recorded onto the 1:10 000 maps and then transferred to the GIS (see below), where the lengths of each section were measured to the nearest 50 m.

Prey availability

Prey availability was determined from the results of an electro-fishing survey conducted during the summer of 1995 at 97 sites throughout the catchment. Two or three electro-fishing sites were chosen within each 5 km stretch from Ordnance Survey maps. It would have been preferable for these sites to have been chosen at random; however, sites had to be accessible by Land Rover because of the amount of equipment involved. Nevertheless, by choosing the sites from the maps prior to the survey, the riverine habitat was unknown and therefore probably sampled at random. The results from each fishing site were pooled for each 5 km stretch and the resultant population estimates assumed to be representative of the fish populations in the whole 5 km stretch (the validity of this assumption is discussed below).

At each site a 100 m^2 area of river or stream was measured as accurately as possible using a tape measure and the downstream and upstream ends were delimited by 9 mm mesh stop-nets, thus restricting fish movements to the fishing area. Where the river or stream was less than 10 m wide the entire width of the stream was enclosed. For those sites that were wider than this, a 10 m width was delimited. Most sites were fished three times with a 30 min break between fishings, although at some sites with low numbers of fish and almost complete depletion on the second fishing only two fishings were carried out. All fish were caught in nets and transferred to holding buckets.

All fish were anaesthetized, identified and then measured (to the nearest millimetre) from tail fork to snout. Fish population estimates were later calculated for each species and size class using the methods described in Bohlin *et al.* (1989). Population estimates were then converted to density (number/ m^2). The following species and size classes were identified and are used in this chapter:

1 Salmon (*Salmo salar*):   (a) less than 80 mm in length (SALLT80)
                                 (b) greater than 80 mm in length (SALGT80)
                                 (c) all salmon (SALTOT)
2 Trout (*Salmo trutta*):   (a) less than 80 mm in length (TRLT80)
                                 (b) greater than 80 mm in length (TRGT80)
                                 (c) all trout (TRTOT)
3 Eel (*Anguilla anguilla*)   (a) less than 30 cm in length (EELLT30)
                                 (b) greater than 30 cm in length (EELGT30)
                                 (c) all eels (EELTOT)
4 Minnow (*Phoxnius phoxinus*) (MIN)
5 Stoneloach (*Neomacheilus barbatulus*) (SL)
6 Stickleback (*Gasterosteus aculeatus*) (SB)
7 Lamprey (*Lampetra* sp.) (LMP)

Thom (1997) provides a full description of methods and detailed analyses of the results of this survey and discusses the relationships between prey availability and otter diet in the upper Tyne catchment.

Data handling
Large quantities of data were collected for a number of environmental variables in addition to the riverbank vegetation and prey availability discussed here. To facilitate storage, manipulation and subsequent analyses, these data were digitized and incorporated into the ARC/INFO GIS (ESRI, 1992). A base GIS for the study area was created by digitizing a number of features such as rivers, streams, lakes, roads and buildings from 1:25 000 Ordnance Survey maps. The majority of the data collected for this study described features of a 10 m wide

strip of riverbank and could, therefore, be considered as descriptions of a linear feature. To handle these kinds of data in the GIS it was necessary to utilize software designed for road traffic management and hydrological applications and ARC/INFO provides a group of commands in Dynamic Segmentation that are capable of storing and manipulating data to describe linear features. Using Dynamic Segmentation routines, linear features such as rivers were converted into a 'route-system' (in this case made up of the 40 5 km stretches of riverbank). Start and end points were associated with each route and attributes (known as 'events' in the Dynamic Segmentation software) such as spraint locations were positioned along the route according to their distance from the start point. Point events, such as spraint location, were described by a single-measure value, and linear events, such as the length of a vegetation type, were described by a start- and end-measure value. Event tables, route-systems and other attributes can then be displayed and queried by using a variety of tools within ARC/INFO and Dynamic Segmentation. Further details on ARC/INFO and Dynamic Segmentation can be found in ESRI (1993). SPSS software (Norusis, 1993*a,b,c*) was used for the analysis of the output from GIS manipulations.

Results

Distribution of otter signs

The number of tarry secretions, mucus-like deposits and other scent marks per 5 km stretch were so low for all surveys that these were combined with the number of spraints per 5 km stretch into one category called SIGN. A second category, SITE, was also established as the number of scent-marking sites per 5 km stretch, each SITE defined as a location containing one or more SIGN separated by at least 1 m from another location containing one or more SIGN.

Figure 9.2 shows the frequency distributions of both SIGN and SITE per 5 km stretch for each seasonal survey. In all seasons, the frequency distributions of both SIGN and SITE per 5 km stretch were strongly negatively skewed, indicating a clumped distribution. To confirm this, the cumulative frequency distributions were compared with the expected cumulative frequency distributions for a Poisson (random) distribution using the Kolmogorov–Smirnov (K–S) goodness-of-fit test. In all seasons the cumulative frequency distributions of both SIGN and SITE differed significantly from a Poisson distribution (Table 9.1), which, given the strong negative skew shown in Fig. 9.2, suggested a clumped distribution.

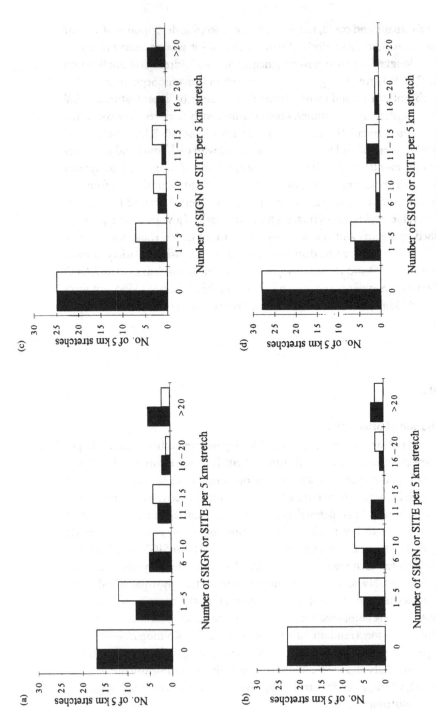

Figure 9.2. Frequency distributions, as the number of the 40 5 km stretches of river surveyed, for the number of SIGNS (■) and SITES (□) found, in four seasonal surveys of the upper Tyne catchment between March 1993 and November 1994. (a) spring; (b) summer; (c) autumn; (d) winter.

Table 9.1. *Kolmogorov–Smirnov (K–S) goodness-of-fit test to determine whether the observed cumulative frequency distributions of the number of SIGN and SITE per 5 km stretch in each season differ significantly from an expected Poisson (random) distribution*

Season	n	K–S$_{(observed)}$ SIGN	SITE	K–S$_{(p = 0.05)}$	K–S$_{(p = 0.01)}$
Spring	40	0.59**	0.52**	0.21	0.25
Summer	40	0.60**	0.62**	0.21	0.25
Winter	40	0.57**	0.56**	0.21	0.25
Autumn	40	0.64**	0.59**	0.21	0.25

* Denotes significance at $p < 0.05$.
** Denotes significance at $p < 0.01$.

Table 9.2. *The total number of 5 km stretches in the upper Tyne catchment with otter signs present or absent in each season*

Season	Present	Absent
Spring	23	17
Summer	15	25
Winter	17	23
Autumn	12	28

($\chi^2 = 6.6$, d.f. $= 3$, NS).

Seasonal differences

Seasonal differences in the distribution of otter signs per 5 km stretch in the catchment were assessed in two ways. Firstly, Table 9.2 shows the number of 5 km stretches classified by presence or absence of otter signs in each seasonal survey. There was no significant difference between seasonal surveys in the proportion of 5 km stretches with otter signs present ($\chi^2 = 6.65$, d.f. $= 3$, NS). Secondly, there were no significant differences between seasons in the median number of SIGN and SITE per 5 km stretch (Kruskal–Wallis One Way ANOVA; SIGN $K_w = 5.8$, d.f. $= 3$, NS; SITE $K_w = 5.6$, d.f. $= 3$, NS). This second analysis was repeated comparing only those stretches that had otter signs present in order to determine whether the high number of negative stretches was biasing the results. No significant difference was found between seasons in the median number of spraints per positive 5 km stretch (Kruskal–Wallis One Way ANOVA; SIGN $K_w = 0.9$, d.f. $= 3$, NS; SITE $K_w = 1.0$, d.f. $= 3$, NS).

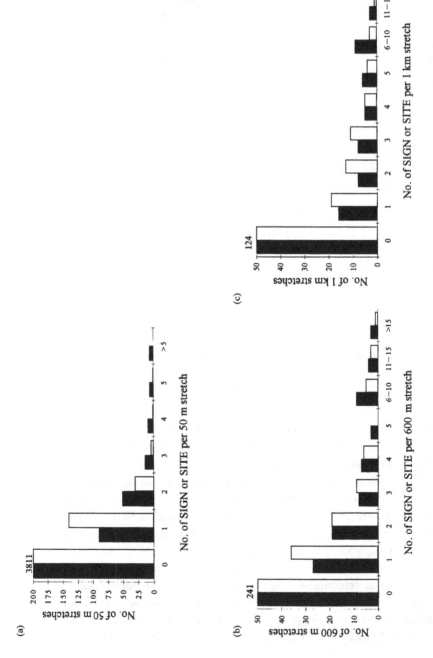

Figure 9.3. Frequency distributions, as the number of 50 m, 600 m and 1 km stretches of river surveyed, for the number of SIGNS (■) and SITES (□) found, in four seasonal surveys of the upper Tyne catchment between March 1993 and November 1994. (a) 50 m units; (b) 600 m units; (c) 1 km units.

Table 9.3. *Results of Kolmogorov–Smirnov (K–S) two-sample test for differences between 600 m (n = 320) and 1 km (n = 183) sampling scales*

Season	K–S$_{(observed)}$	
	SIGN	SITE
Spring	8.8^{NS}	7.6^{NS}
Summer	8.7^{NS}	8.7^{NS}
Winter	5.3^{NS}	5.3^{NS}
Autumn	6.2^{NS}	6.2^{NS}

If K–S$_{(observed)}$ is less than K–S$_{(p = 0.05)}$ (equal to 12.6) then there is no significant difference between sampling scales. NS non-significant.

Effect of sampling scale

Using Dynamic Segmentation in ARC/INFO, each 5 km stretch was divided into smaller units by overlaying onto the SIGN and SITE event data a linear grid divided into different unit lengths. In most recent literature on the use of spraint survey methods the most common sampling scales were 5 km (Macdonald & Mason, 1983), 1 km (Delibes *et al.*, 1991), 600 m (e.g. Strachan *et al.*, 1990) and 50 m (Bas *et al.*, 1984). Each 5 km stretch was subdivided into each of these distances so that revised numbers of SIGN or SITE per sampling unit could be calculated.

Figure 9.3 shows the frequency distributions for the 50 m, 600 m and 1 km sample scales. Only the Spring data are shown but all seasons showed a similar form of the distribution for each sampling scale. The histograms for the 600 m and 1 km sample scales are very similar and no significant difference was found between their cumulative frequency distributions in any season (Table 9.3).

The 1 km sampling scale was also compared with the 5 km sampling scale using a similar test (Table 9.4). These two sampling units did not differ significantly for the Summer and Autumn surveys, which indicates that the distributions of SIGN and SITE were clumped at the 1 km scale during these seasons. A significant difference was found between the two sampling units in both SIGN and SITE in the Spring and in SIGN in the Winter surveys, although comparisons with a Poisson distribution showed that the distributions of both SIGN and SITE were still clumped at the 1 km scale in both of these seasons (Spring SIGN, K–S = 0.5; SITE, K–S = 0.4; Winter SIGN, K–S = 0.4. $n = 83$, $p < 0.01$ in all cases).

Closer inspection of the frequency distribution histograms for the 50 m sampling scale showed that the majority of stretches produced no signs of

Table 9.4. *Results of Kolmogorov–Smirnov (K–S) two-sample test for diferences between 1 km (n = 183) and 5 km (n = 40) sampling scales*

	$K–S_{(observed)}$	
Season	SIGN	SITE
Spring	0.029**	0.25*
Summer	0.18^{NS}	0.17^{NS}
Winter	0.25*	0.23^{NS}
Autumn	0.13^{NS}	0.10^{NS}

If $K–S_{(observed)}$ is less than $K–S_{(p = 0.05)}$ (equal to 0.24) then there is no significant difference between sampling scales. NS, non-significant; *, significant at $p < 0.05$; **, significant at $p < 0.01$.

otters, while those that did produced only small numbers of SIGN or SITE (median = 1) in all seasons. Therefore, the distribution patterns using the 50 m sampling unit were equivalent to the distribution of individual SIGN or SITE. These patterns were assessed using order-neighbour analysis (also known as nearest-neighbour or distance analysis). Distances between the *k*th nearest neighbouring spraints or spraint sites (where *k* = 1st, 2nd or 3rd nearest neighbours) were calculated using Dynamic Segmentation. The cumulative frequency distributions of *k*th nearest neighbour distances for both SIGN and SITE in each season were then compared to a theoretical random distribution of nearest-neighbour distances using the methods described by Aplin (1983). The results of this analysis showed that the distribution of both SIGN and SITE was clumped in the Spring, Summer and Winter surveys, but that the distribution did not differ significantly from random in Autumn (Tables 9.5 and 9.6).

This analysis of the effect of changes in sampling scale showed that the distribution of both SIGN and SITE was clumped at three different sampling scales in at least three of the four seasonal surveys – 50 m (equivalent to individual SIGN or SITE), 600 m/1 km and 5 km.

The relationship between riverbank habitat, prey abundance and the distribution of otter signs
The relationship between environmental variables and the distribution of otter signs was investigated in a number of ways and at two different sampling scales – 1 km and 5 km. Prey density data were only available for 35 of the 40 5 km stretches and at the 1 km scale only 90 stretches coincided with electro-fishing sites. The relationship between prey density and the distribution of otter signs was investigated for the Summer and Autumn surveys only because the

Table 9.5. *Results of nearest-neighbour analysis of the distributions of individual SIGN giving observed and critical values for Kolmogorov–Smirnov (K–S) goodness-of-fit tests for random distributions in the four seasonal surveys at a series of* k *nearest neighbours*

Season	kth neighbour	n	$K-S_{(observed)}$	$K-S_{(p = 0.05)}$	$K-S_{(p = 0.01)}$
Spring	1	323	3.7**	1.1	1.3
	2	303	2.2**	1.1	1.3
	3	283	2.2**	1.1	1.4
Summer	1	216	3.2**	1.3	1.6
	2	197	2.1**	1.4	1.6
	3	187	1.9**	1.4	1.7
Winter	1	219	3.1**	1.3	1.6
	2	201	1.6*	1.4	1.6
	3	188	1.6*	1.4	1.7
Autumn	1	105	2.2*	1.9	2.3
	2	93	1.7^{NS}	2.0	2.4
	3	81	1.4^{NS}	2.1	2.6

NS, non-significant; *, significant at $p < 0.05$; **, significant at $p < 0.01$.

Table 9.6. *Results of nearest-neighbour analysis of the distributions of individual SITE giving observed and critical values for Kolmogorov–Smirnov (K–S) goodness-of-fit tests for random distributions in the four seasonal surveys at a series of* k *nearest neighbours*

Season	kth neighbour	n	$K-S_{(observed)}$	$K-S_{(p = 0.05)}$	$K-S_{(p = 0.01)}$
Spring	1	203	2.3**	1.3	1.6
	2	179	2.0**	1.4	1.7
	3	157	1.6**	1.5	1.8
Summer	1	151	2.0**	1.6	1.9
	2	129	1.7^{NS}	2.4	2.0
	3	109	2.7**	1.8	2.2
Winter	1	158	2.3**	1.5	1.8
	2	147	1.6*	1.6	1.9
	3	127	1.5^{NS}	1.7	2.0
Autumn	1	75	2.2^{NS}	2.2	2.7
	2	58	1.7^{NS}	2.5	3.0
	3	43	0.7^{NS}	2.9	3.5

NS, non-significant; *, significant at $p < 0.05$; **, significant at $p < 0.01$.

electro-fishing survey had been conducted in late summer and it would have been inappropriate to extrapolate to Spring and Winter.

Correlations between the total number of SIGN and SITE, the total length of each habitat type and prey densities for each fish species and size class were investigated using Spearman–Rank correlation coefficients. The large propor-tion of stretches without otter signs at both sampling scales produced too many tied ranks for Spearman–Rank coefficients to be valid, so only those stretches that held signs of otters were used in the analysis. At both the 1 km and 5 km sampling scales there were no significant correlations between the total number of SIGN or SITE (in stretches where these were present) and the total length of any of the vegetation types or the densities of any of the fish species.

The sample units at each scale were then classified into those with and those without otter signs. These two categories were then compared for significant differences in all the riverbank habitat and prey variables using Mann–Whitney *U*-tests. Stepwise logistic regression analysis, with variables excluded on the basis of their Likelihood ratios (Norusis, 1993c), was used to test whether the membership of the two categories was accurately predicted by those environ-mental variables that showed significant differences in the Mann–Whitney *U*-tests (Table 9.7). The resultant logistic model was tested using a jack-knifing procedure where the regression analysis was re-run using $n-1$ of the cases. This was then used to attempt to predict the remaining case. This process was repeated until all cases had been tested and then a re-classification table was produced based on the number of cases that were correctly predicted.

At the 1 km sample unit length there were significant differences between stretches with otter signs and those without in a number of the environmental variables (Table 9.7). In all seasons the median length of vegetation less than 10 cm in height (variable LOW) per 1 km stretch was significantly lower in stretches with otter signs than in those without. The median length of closed canopy woodland (variable CLOSED) per 1 km stretch was significantly high-er in stretches with otter signs in Spring and Summer and the median length of vegetation taller than 1 m with shrubs (variable DENSE) per 1 km stretch was higher in all these stretches in all seasons except Spring. Densities of eels less than 30 cm in length (variable EELLT30), eels greater than 30 cm in length (variable EELGT30), total eels (variable EELTOT), salmon less than 80 mm in length (variable SALLT80), stoneloach (variable SL), minnow (variable MIN) and lamprey (variable LMP) were all significantly higher in stretches with otter signs in Summer, while the density of trout greater than 80 mm in length (variable TRGT80) was lower in these stretches. In the Autumn, densities of stoneloach, minnow and lamprey were all higher in stretches with otter signs

Table 9.7. *Summary of Mann–Whitney* U *tests comparing spraint survey stretches (1 km and 5 km units) with otter signs and those without otter signs using all riverbank habitat and prey variables in all four seasonal surveys of the upper Tyne catchment*

Variable	Significance of Mann–Whitney U for 1 km stretches				Significance of Mann–Whitney U for 5 km stretches			
	Spring	Summer	Winter	Autumn	Spring	Summer	Winter	Autumn
Riverbank vegetation								
BARE	NS	NS	NS	NS	NS	NS	NS	NS
LOW	<0.01	<0.01	<0.01	<0.01	<0.01	<0.01	<0.01	<0.05
MEDIUM	NS	NS	NS	NS	NS	NS	NS	NS
DENSE	NS	<0.05	<0.01	<0.01	<0.05	<0.05	<0.01	<0.05
OPEN	NS	NS	NS	NS	<0.01	NS	<0.01	<0.05
Fish variables								
SALLT80	—	<0.05	—	NS	—	NS	—	NS
SALGT80	—	NS	—	NS	—	NS	—	NS
SALTOT	—	NS	—	NS	—	<0.05	—	NS
TRLT80	—	NS	—	NS	—	NS	—	NS
TRGT80	—	<0.01	—	<0.01	—	NS	—	<0.05
TRTOT	—	NS	—	<0.01	—	NS	—	NS
EELLT30	—	<0.05	—	NS	—	NS	—	NS
EELGT30	—	<0.05	—	NS	—	NS	—	NS
EELTOT	—	<0.01	—	NS	—	NS	—	NS
MIN	—	<0.01	—	<0.01	—	<0.05	—	<0.05
SL	—	<0.01	—	<0.01	—	<0.01	—	<0.01
SB	—	NS	—	NS	—	NS	—	NS
LMP	—	NS	—	<0.05	—	NS	—	NS
FISH	—	NS	—	NS	—	NS	—	NS

NS, no significant difference.
For a description of the variables see Methods.

while the densities of trout greater than 80 mm and total trout were significantly lower in these stretches. Analysis of Spearman–Rank correlation matrices for these variables showed that in the Spring, CLOSED and LOW were negatively correlated. In the Summer all of the eel categories were positively correlated and LOW and DENSE were negatively correlated with CLOSED. In the Autumn trout greater than 80 mm in length (TRGT80) was positively correlated with total trout (TRTOT).

Table 9.8. *Reclassification tables showing the results of jack-knifing tests of the ability of logistic regression models to correctly predict the presence or absence of otter signs in 1 km stretches of river in four seasonal surveys of the upper Tyne catchment*

(a) Spring				(b) Summer			
	No. of correct predictions				No. of correct predictions (%)		
	n	Obs. (%)	Exp.		n	Obs. (%)	Exp.
Absent	61	61(100)	30.5	Absent	62	53(85)	31
Present	29	0(0)	14.5	Present	28	8(29)	14
Overall	90	61(68)	45	Overall	90	61(68)	45

(c) Winter				(d) Autumn			
	No. of correct predictions				No. of correct predictions (%)		
	n	Obs. (%)	Exp.		n	Obs. (%)	Exp.
Absent	57	55(96)	28.5	Absent	70	60(86)	35
Present	33	0(0)	16.5	Present	20	3(15)	11.5
Overall	90	55(61)	45	Overall	90	63(70)	45

Obs., observed predictions of the jack-knifing procedure (percentages in brackets); Exp., expected random prediction.

Table 9.8 shows the jack-knifed logistic regression table for 1 km stretches in each season. In the Spring, variables CLOSED and LOW were used in logistic regression analysis and only CLOSED was retained in the resultant model. This model was tested using the jack-knifing procedure and it was found that the model's predictions were significantly different from the expected random predictions ($\chi^2 = 44.5$, d.f. = 1, $p < 0.001$). All of the stretches with no otter signs were correctly predicted but the model failed to correctly predict those stretches with otter signs. In the Summer, variables LOW, DENSE, CLOSED, SALLT80, EELTOT, MIN, SL and LMP were used in the analysis with EELTOT, SL and LOW being retained in the model. As with Spring the model's predictions were significantly different to random ($\chi^2 = 21.1$, d.f. = 1, $p < 0.001$) but again this was due to the correct prediction of stretches without otter signs but a poor performance in predicting stretches with otter signs. The Winter model was built using variables LOW and DENSE, with only LOW being retained in the model, which was again significantly better than random at successfully predicting stretches without otter signs but completely failed to predict stretches with otter signs ($\chi^2 = 41.1$, d.f. = 1, $p < 0.001$). In the Autumn the model was built with LOW, DENSE, SL, MIN and LMP, with SL and MIN being retained in the model. As with the other seasons the model successfully predicted stretches without otter signs but was no better than random at predicting stretches with otter signs ($\chi^2 = 22.8$, d.f. = 1, $p < 0.001$).

At the 5 km sample unit length the median length of closed canopy woodland (variable CLOSED) per 5 km stretch was significantly higher and the median length of vegetation less than 10 cm in height (variable LOW) significantly lower in those stretches with otter signs than those without, in all four seasons (Table 9.7). In the Spring, Summer and Winter surveys the median length of vegetation taller than 1 m with shrubs (variable DENSE) per 5 km stretch was significantly higher in stretches with otter signs and the median length of open canopy woodland (variable OPEN) was also significantly higher in these stretches but only in Spring. The densities of minnow (MIN) and stoneloach (SL) were significantly higher in stretches with signs of otters in Summer and Autumn whereas the density of salmon was higher in Summer only. The density of trout greater than 80 mm in length (TRGT80) was lower in Autumn in stretches with otter signs than in those without. In the Spring, OPEN was positively correlated with CLOSED. No correlations were found between variables in any other season.

Table 9.9 shows the jack-knifed logistic regression table for 5 km stretches in each season. In the Spring, variables LOW, DENSE, and OPEN and CLOSED combined, were used in the regression analysis but only variable LOW was retained in the model. The model's predictions were significantly better than random, with 82% of stretches with otter signs and 69% of stretches without otter signs being correctly predicted by the model ($\chi^2 = 5.4$, d.f. = 1, $p < 0.05$). In the Summer, variables LOW, DENSE, CLOSED, SAL-TOT, SL and MIN were included in the analysis, but only LOW and SL remained in the model which did not differ significantly from random in correctly predicting stretches with or without otter signs ($\chi^2 = 3.9$, d.f. = 1, NS). In the Winter, variables LOW, DENSE and CLOSED were used in the analysis with only LOW being retained but again the model's predictions did not differ significantly from random ($\chi^2 = 0.3$, d.f. = 1, NS). Finally, in the Autumn, variables LOW, CLOSED, TRGT80, MIN and SL were used in the regression analysis but only SL was retained in the model. The model's predictions differed significantly from random ($\chi^2 = 11.3$, d.f. = 1, $p < 0.001$), but this was due to very poor success at correctly predicting stretches with otter signs.

Table 9.9. *Reclassification tables showing the results of jack-knifing tests of the ability of logistic regression models to correctly predict the presence or absence of otter signs in 5 km stretches of river in four seasonal surveys of the upper Tyne catchment*

(a) Spring		No. of correct predictions		(b) Summer		No. of correct predictions (%)	
	n	Obs. (%)	Exp.		*n*	Obs. (%)	Exp.
Absent	13	9(69)	6.5	Absent	21	17(81)	10.5
Present	22	18(82)	11	Present	14	10(71)	7
Overall	35	27(7)	17.5	Overall	35	27(77)	17.5

(c) Winter		No. of correct predictions		(d) Autumn		No. of correct predictions (%)	
	n	Obs. (%)	Exp.		*n*	Obs. (%)	Exp.
Absent	19	11(58)	9.5	Absent	23	17(74)	11.5
Present	16	9(56)	8	Present	12	1(8)	6
Overall	35	20(57)	17.5	Overall	35	18(51)	17.5

Obs., observed predictions of the jack-knifing procedure (percentages in brackets); Exp., expected random prediction.

Discussion

The importance of scale is very often underestimated or not accounted for in many ecological studies. Wiens (1989), however, provides several examples which demonstrate that the scale of an investigation can have profound effects on the description of the patterns found. Therefore, any investigation that compares two or more parameters such as, in this case, the relationship between spraint density and riverbank vegetation, needs first to determine the appropriate scale at which to make the comparisons. Ideally, this should be done by sampling at the largest scale possible and then taking random subsets at a series of smaller scales, as Kruuk & Conroy (1987) suggested when comparing spraint density with otter activity. However, in many cases the maximum scale is limited by logistical considerations. In this study it was taken as a 5 km length of river as this was the maximum distance that could be covered in a single day in the difficult terrain of the Tyne catchment without reducing sampling accuracy because of fatigue or failing daylight (particularly on short winter days). The results of this study show that the distribution of both spraints and sprainting sites exhibited clumping at this maximum sampling scale and that this pattern was maintained at a series of smaller scales. At the smallest scale, the distribution of individual spraints and spraint sites was investigated using order-neighbour analysis and also revealed a degree of clumping in three out of the four seasons. The low density of otter signs in the autumn survey, however, produced a distribution that did not differ significantly from random. Having established that the distribution of spraints and sites was clumped regardless of the sampling scale, we considered two factors – the type of riverbank vegetation and the abundance of fish prey – that may have caused this clumping.

In the Introduction to this chapter we showed that in most studies that have relied on the distribution of otter signs as an indicator of habitat utilization a relationship has been found between the number of spraints or spraint sites in an area and the availability of riverbank woodland or shrub (Jenkins & Burrows, 1980; Macdonald & Mason, 1983; Bas *et al.*, 1984) and that similar relationships were demonstrated using radio-telemetry techniques (Green *et al.*, 1984; Jefferies *et al.*, 1986). However, Kruuk *et al.* (1986) criticized these findings because they were unable to demonstrate a relationship between a number of measures of otter activity and the number of spraints in a length of coast in Shetland. However, their study was a short stretch of rocky coastline that was known from previous studies to produce the highest number of otters (Kruuk *et al.*, 1989) and where spraints were more common than in other habitats along the Shetland coast (Jenkins & Conroy, 1982). Within this study area 330 m units were then compared for otter activity and level of sprainting. It is possible that the habitat was not significantly different between these sampling units and therefore no differences in sprainting would have been expected if spraint density was an indicator of habitat utilization. In a later study Conroy & French (1987) showed that, using 2 km units of coast over 40 km of coastline in Shetland, there was a relationship between the number of otters seen and the number of spraints found.

In this study we were unable to demonstrate a clear relationship between the number of spraints or sprainting sites in a stretch and any of the riverbank vegetation types studied. Some vegetation types did differ significantly between stretches with otter signs and those without. In all seasons there was significantly more woody vegetation types and significantly less low vegetation (less than 10 cm) in stretches with otter signs than those without in both 1 km and 5 km sampling units. However, these variables were found to be no better than random at predicting the presence of otter signs in logistic regression models at both sampling scales and in three out of four seasons. This lack of statistical significance when assessing the relationship between riverbank vegetation and the distribution of otter signs was found in other studies. Jenkins & Burrows (1980) were able to explain only 10–15% of the variation in spraint numbers by the availability of woodland cover. Macdonald & Mason (1983) showed that the number of mature ash (*Fraxinus excelsior*) trees and otter holts explained only 35% of the variation in sprainting sites and that the number of mature sycamore (*Acer pseudoplatanus*) trees and holts explained only 39% of the variation in spraint numbers. It seems, therefore, that riverbank vegetation may not be a major factor in determining the distribution of otter signs.

Kruuk (1992) suggested that it was prey, rather than habitat availability, which determined the distribution of otters. He demonstrated that Shetland otters sprainted at the beginning of feeding bouts and that the function of this was to signal the use of a prey 'patch' to conspecifics, thus preventing the over-exploitation of resources without the need for aggressive encounters. He also showed that sprainting was seasonal and that this was linked to seasonal changes in prey availability. It is possible, therefore, that areas of high prey availability may also hold higher densities of spraints than other areas. Kruuk *et al.* (1993) showed that the distribution of salmonids had a marked effect on the distribution of otters in the rivers and streams of the Dee catchment in Aberdeenshire, Scotland, because salmonid densities were higher in the narrower streams and tributaries. Durbin (1993) showed that brown trout (*Salmo trutta*) densities were higher in the narrow streams while salmon (*Salmo salar*) were more abundant in wider rivers.

In our study we were unable to demonstrate a relationship between fish prey densities and the number of spraints or spraint sites. However, some prey types did differ between stretches with otter signs and those without at both sampling scales. In the Summer, at the 1 km scale, densities of eels, salmon less than 80 mm in length, stoneloach, minnow and lamprey were all significantly higher in stretches with otter signs, while the densities of trout greater than 80 mm in length were lower in these stretches in both Summer and Autumn. At the 5 km scale, the densities of minnow and stoneloach were significantly higher in stretches with signs of otters in Summer and Autumn. The density of salmon was higher in Summer only while the density of trout greater than 80 mm in length was lower in Autumn in stretches with otter signs than in those without. However, as with riverbank vegetation, when these variables were used in the logistic regression analyses they were found to be poor predictors of the presence of otter signs in either 1 km or 5 km stretches. Three possible reasons for this lack of a relationship are that, firstly, seasonal surveys of spraint numbers were not sensitive to changes in the distribution of otters in relation to changing prey populations (which may occur over shorter temporal scales); and that, secondly, the fish population estimates at each site may not have been representative of the populations in the full 1 km or 5 km stretch. A more intensive electro-fishing study covering more sites in each 1 km or 5 km stretch (which was impractical in this study due to time and financial constraints) may have provided a more detailed assessment of fish populations. Thirdly, the relationship between otter sprainting activity and feeding bouts in a riparian habitat may not be the same as in coastal habitats. Kruuk (1992) showed that, in Shetland, a significant proportion of spraints were short-lived due to the effect of the tide. In freshwater environments, spraints and spraint sites last for much longer and may not therefore be related to periods of feeding activity.

There are a variety of additional possible reasons for the lack of relationships between riverbank vegetation, prey populations and the distribution of otter signs in this study and for the poor statistical performance of riverbank vegetation variables as predictors of the distribution of otter signs in other studies. Firstly, these factors may not be limiting the distribution of otter signs in the Tyne catchment. Other variables, such as physical features of the river (substrate, width, depth, altitude) which determine the distribution of prey, the distribution of otter holts, disturbance by human activities or the effect of heavy metal pollution may be more important and these are currently being investigated (Thom in prep.). Secondly, the Tyne catchment may already provide suitable habitat for otters and the perceived absence of otters from some stretches may be due to a lack of recolonization after the major decline in the 1950s (Chanin & Jefferies, 1978). Thirdly, the spatial scale of investigation may have been too small and inappropriate to the scale of otter movements, which have been shown to be up to 40 km (Durbin, 1993). It may be more appropriate to compare much larger stretches of river such as different watersheds or catchments. Fourthly, sprainting activity may be influenced by population density (Macdonald & Mason, 1983), sex, social and breeding status of individual otters.

Spraint density surveys can provide a simple and cost-effective method of monitoring otter populations. The extension of the use of this method to studies of habitat utilization is more problematic primarily because the function of sprainting is not fully understood. For this reason, proponents of radio-telemetry methods have suggested that more radio-tracking studies of habitat utilization are required. However, many of the organizations involved in otter conservation do not have the resources to conduct these kinds of studies and spraint density surveys provide a useful alternative. We argue, therefore, that organizations with the resources to conduct radio-tracking studies should focus on determining the exact relationship between spraint numbers and otter activity in riparian habitats.

ACKNOWLEDGEMENTS
We would like to thank the members of the Tyne Otter Forum – Chris Spray (Northumbrian Water), David Stewart, Lisa Kerslake, Andrew Bielinski (Northumberland Wildlife Trust), Steve Preston (Northumberland National Park), Colin Blundell (National Rivers Authority), Bill Burlton (Forest Enterprise–Kielder District) and Pippa Merricks (English Nature) for funding and supporting this project. We are indebted to Mick Hanley and Mark Bailey for all their efforts during the electro-fishing survey. We would also like to thank the landowners and managers who gave us permission to cross their land and fish their rivers. Our thanks also to Dr Mark O'Connell for his help in matters statistical. Special thanks to Gill Thom for all her help and support during the project.

References

Abel, P. D. & Green, D. W. J. (1981). Ecological and toxicological studies on invertebrate fauna of two rivers in the northern Pennine orefield. In *Heavy metals in Northern England: environmental and biological aspects*: 109–121. (Eds Say, P. J. & Whitton, B. A.). University of Durham, Durham.

Aplin, G. (1983). *Order-neighbour analysis. Concepts and techniques in modern geography* **36**. Geo Books, Norwich.

Bas, N., Jenkins, D. & Rothery, P. (1984). Ecology of otters in northern Scotland. V. The distribution of otter (*Lutra lutra*) faeces in relation to bankside vegetation on the River Dee in summer 1981. *J. appl. Ecol.* **21**: 507–513.

Bohlin, T., Hamrin, S., Heggberget, T. G., Rasmussen, G. & Saltveit, S. J. (1989). Electrofishing – theory and practice with special emphasis on salmonids. *Hydrobiologia* **173**: 9–43.

Chanin, P. R. F. & Jefferies, D. J. (1978). The decline of the otter *Lutra lutra* L. in Britain: an analysis of hunting records and discussion of causes. *Biol. J. Linn. Soc.* **10**: 305–328.

Conroy, J. W. H. & French, D. D. (1987). The use of spraints to monitor populations of otters (*Lutra lutra* L.). *Symp. zool. Soc. Lond.* No. 58: 247–262.

Delibes, M., Macdonald, S. M. & Mason, C. F. (1991). Seasonal marking, habitat and organochlorine contamination in otters (*Lutra lutra*): a comparison between catchments in Andalucia and Wales. *Mammalia* **55**: 567–578.

Durbin, L. (1993). *Food and Habitat Utilization of Otters (Lutra lutra, L.) in a Riparian Habitat – The River Don in North-east Scotland*. PhD thesis: University of Aberdeen.

ESRI (1992). *Understanding GIS. The ARC/INFO method*. Environmental Sciences Research Institute, Redlands, CA, USA.

ESRI (1993). *ARC/INFO dynamic segmentation manual*. Environmental Sciences Research Institute, Redlands, CA, USA.

Green, J., Green, R. & Jefferies, D. J. (1984). A radio-tracking survey of otters *Lutra lutra* on a Perthshire river system. *Lutra* **27**: 85–145.

Jefferies, D. J. (1986). The value of otter *Lutra lutra* surveying using spraints: an analysis of its successes and problems in Britain. *J. Otter Trust* **1**: 25–32.

Jefferies, D. J., Wayre, P., Jessop, R. M. & Mitchell-Jones, A. J. (1986). Reinforcing the native otter *Lutra lutra* population in East Anglia: an analysis of the behaviour and range development of the first release group. *Mammal Rev.* **16**: 65–79.

Jenkins, D. & Burrows, G. O. (1980). Ecology of otters in northern Scotland. III. The use of faeces as indicators of otter (*Lutra lutra*) density and distribution. *J. Anim. Ecol.* **49**: 755–774.

Jenkins, D. & Conroy, J. W. H. (1982). Methodology for studying habitats used by coastal otters. *Annu. Rep. Inst. terr. Ecol.* **1981**: 19–23.

Kruuk, H. (1992). Scent marking by otters (*Lutras lutra*): signaling the use of resources. *Behav. Ecol.* **3**: 133–140.

Kruuk, H. & Conroy, J. W. H. (1987). Surveying otter *Lutra lutra* populations: a discussion of problems with spraints. *Biol. Conserv.* **41**: 179–183.

Kruuk, H., Conroy, J. W. H., Glimmerveen, U. & Ouwerkerk, E. (1986). The use of spraints to survey populations of otters *Lutra lutra*. *Biol. Conserv.* **35**: 187–194.

Kruuk, H., Moorhouse, A., Conroy, J. W. H., Durbin, L. & Frears, S. (1989). An estimate of numbers and habitat preferences of otters *Lutra lutra* in Shetland, UK. *Biol. Conserv.* **49**: 241–254.

Kruuk, H., Carss, D. N., Conroy, J. W. H. & Durbin, L. (1993). Otter (*Lutra lutra*, L.) numbers and fish productivity in rivers in north-east Scotland. *Symp. zool. Soc. Lond.* No. 65: 171–191.

Macdonald, S. M. & Mason, C. F. (1983). Some factors influencing the distribution of otters (*Lutra lutra*). Mammal Rev. **13**: 1–10.

Norusis, M. J. (1993*a*). *SPSS for Windows. Base system user's guide release 6. 0.* SPSS Inc., Chicago, IL, USA.

Norusis, M. J. (1993*b*). *SPSS for Windows. Professional statistics release 6. 0.* SPSS Inc., Chicago, IL, USA.

Norusis, M. J. (1993*c*). *SPSS for Windows. Advanced statistics statistics release 6. 0.* SPSS Inc., Chicago, IL, USA.

Say, P. J. & Whitton, B. A. (1981). Chemistry and plant ecology of zinc-rich streams in the northern Pennines. In *Heavy metals in northern England: environmental and biological aspects*: 55–63. (Eds Say, P. J. & Whitton, B. A.). University of Durham, Durham.

Strachan, R., Birks, J. D. S., Chanin, P. R. F. & Jefferies, D. J. (1990). *Otter survey of England 1984–1986.* Nature Conservancy Council, Peterborough.

Thom, T. J. (1997). *Factors Affecting the Distribution of Otter (*Lutra lutra* L.) Signs in the Upper Tyne Catchment, N. E. England.* PhD thesis: University of Durham.

Wiens, J. A. (1989). Spatial scaling in ecology. *Funct. Ecol.* **3**: 385–397.

10

Influence of altitude on the distribution, abundance and ecology of the otter (*Lutra lutra*)

J. Ruiz-Olmo

Introduction

The otter (*Lutra lutra*), a semi-aquatic mustelid, has suffered a significant population decline in Europe (Mason & Macdonold, 1986; Foster-Turley *et al.*, 1990). There are several possible causes, although without doubt pollution, the destruction of habitat and the variation of available food have played an important part (Macdonald & Mason, 1992; Kruuk, 1995). The species is one of the most intensively studied of all European mammals, primarily because of its status as a bioindicator species at the top of many freshwater food chains. It has also become an emblem of many conservation projects throughout the world. All these studies have been carried out in coastal environments or in areas of relatively low altitude. This is because few European countries with populations of otters have any significant land that is higher than 1000 m above sea level (a.s.l.) (e.g. Norway, France, Slovakia, Slovenia, Croatia, Bosnia, Yugoslavia, Greece, Bulgaria, Rumania and Spain). In several, the otter has become extinct or scarce, and only two (Greece and Spain) have otter populations in large mountain ranges that exceed 2000 m a.s.l. This means that the effect of altitude, a significantly limiting factor for many species, has not been studied for otters.

Mason & Macdonald (1986) point out that *L. lutra kutab* has been observed at 4120 m in Tibet; others report that it is more abundant in low areas, with fewer tracks and signs with increasing altitude (Green & Green, 1980; Chapman & Chapman, 1982; Delibes, 1990). In Spain, several authors have shown that the density of otter tracks and signs decreases above 800–1000 m (Ruiz-Olmo & Gosálbez, 1988; Lizana & Pérez-Mellado, 1990; Nores *et al.*, 1990; Ruiz-Olmo, 1995). Ruiz-Olmo (1995) showed that when the otter recolonizes an area, it commences in areas of lower altitude and where there is more food. However, the factors that determine this situation have not been studied.

Schmidt-Nielsen (1972) and Hill (1980) analyzed the physiological and adaptive factors of living creatures related to altitude. They emphasize the drop in atmospheric pressure, the climatic factors (principally a drop in temperature) and the reduction of concentration of oxygen with increasing altitude. Changes in oxygen concentration and atmospheric pressure do not seem to be principal limiting factors, since mammals have diverse mechanisms that permit their presence up to 4000–5000 m (Hill, 1980). So climatic factors, and in particular temperature, would seem to be important variables, either directly or indirectly. The first effect of temperature is a factor of great importance for a semi-aquatic carnivore that cannot protect itself by means of a thick layer of fat or fur (Kruuk, 1995): foraging in cold environments may represent a significant loss of heat. In fact, the effect of temperature is even more important in aquatic carnivores and in those semi-aquatic ones with a significantly higher basal metabolism (McNab, 1989). Kruuk (1995) confirmed experimentally that as temperature decreases and approaches 0 °C, the necessity for food per unit foraging time increases; with an ingestion of less than 150 g of food per hour not even a full day's foraging would yield sufficient food to meet such an animal's energy requirements.

The current study sets out to understand how otter populations respond to altitude, considering the influence of food availability – a parameter which affects mortality and reproduction (Kruuk *et al.*, 1987, 1991; Kruuk & Conroy, 1991), determining the distribution and abundance of the otter (Ruiz-Olmo, 1994, 1995).

The study area

The study area occupies some 40 000 km^2 (Fig. 10.1): the whole of Catalunya and a large part of Aragón, with a small portion in the province of Castellón. In this area the otter is only present in some of the Ebro Basin rivers and is absent from areas of low altitude and all the small streams that flow directly to the Mediterranean Sea (Ruiz-Olmo & Gosálbez, 1988; Ruiz-Olmo, 1995).

The topography is complex, with a large range of mountains, the Pyrenees, often exceeding 2000 m in altitude (maximum 3404 m; Aneto). On the marine side the altitudes are lower, although they frequently exceed 500–700 m near the coast (maximum at Montseny, 1717 m; and at Mont Caro, 1447 m). To the south of the study area in the provinces of Teruel, Tarragona and Castellón, land rises to 2019 m (Peñarroya, Sierra de Gúdar). The study area thus provides an ample range of altitudes with which to investigate otter ecology. Furthermore, the physiographical, climatical and biological characteristics are well described in Riba *et al.*, (1979). The climate is very varied, from typical Mediterranean (average temperature in the coldest month is 9–10 °C and is

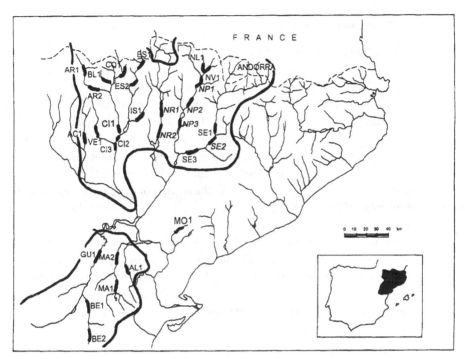

Figure 10.1. Study area. The continuous black line shows otter distribution in 1995. Study stretches are also indicated by thick black lines. They are on the Rivers Arazas (AR1 and AR2), Bellós (BL1), Cinqueta (CQ1), Esera (ES1 and ES2), Isábena (IS1), Cinca (CI1, CI2 and CI3), Veral (VE1), Alcanádre (AC1), Noguere Ribogarçana (NR1 and NR2), Noguere de Lladorre (NL1), Noguere de Vallferrera (NV1), Noguera Pallaresa (NP1, NP2 and NP3), Segre (SE1, SE2 and SE3), Montsant (MO1), Guadalope (GU1), Bergantes (BE1 and BE2), Matarranya (MA1 and MA2) and Algars (AL1).

24–25 °C in the hottest month, 400–500 mm. of rainfall) to alpine tundra environments (average figures of −5 °C and 9–10 °C, respectively; with 1500–2000 mm of rainfall) and includes all the intermediary climates found in this range.

The characteristics of the rivers are very varied according to altitude, orientation, slope and modifications by man (hydroelectric dams are abundant). It is generally possible to distinguish between 'Mediterranean' rivers with a highly fluctuating volume closely linked to rainfall (maximums in spring and autumn) and with frequent floods (Martín-Vide, 1985), and 'Pyrenean' rivers of a nivo-pluvial style, with less variable flow regime (maximum in spring corresponding to the melting of ice).

Materials and methods

Distribution according to altitude

Currently the most common method of studying the distribution of otters is the standard Otter Survey (Mason & Macdonald, 1986). Tracks and spraints (faeces) are recorded along 600 m sampling sites. Even though the validity of this method has been questioned, if it is used solely to estimate distribution at one particular moment, without inferring anything about the abundance of otters or the selection of habitat, it is a useful way of studying the presence–absence of *L. lutra* at regional level. It should not be forgotten that the absence of otter signs does not necessarily imply the absence of the species. Repeated or consecutive absence or presence of these signs in space or in time does however constitute a useful result.

For the current investigation, the entire study area was sampled on three occasions: 1984–85, 1989–90 (also the basins of Guadalope and part of the Cinca in 1987–88: Lacomba & Jiménez, 1988) and 1994–95. The total number of 600 m sections surveyed were 558, 516 and 438, respectively, following the methods described by Delibes (1990) and Ruiz-Olmo (1995).

Fish populations

Electro-fishing was carried out during 1989–90 in a total of 56 stations distributed amongst the sub-basins of the rivers Noguere Ribagorçana, Noguera Pallaresa and the Segre in the Spanish Pyrenees (the Ebro Basin). Here, successive checks were carried out and the method described by Zippin (1958) was applied. The number of fish present and the 'availability' of food (in g/m and g/m^2) of fish with total length (TL) > 5 cm was established. Smaller fish escape the fishing method and are not consumed by *L. lutra* in the study area (Ruiz-Olmo, 1995).

Diet

Twenty-eight separate locations were selected from the whole distribution area of *L. lutra* in the north-east of the Iberian Peninsula (Fig. 10.1). Two of them were studied in two different periods. They are all rivers; carrying out the study in artificial reservoirs was avoided. A total of 3479 spraints were collected between 1984 and 1995. In each place, at least one sample was collected in summer (the period of maximum prey diversity: Ruiz-Olmo *et al.*, 1989) and another in a different season. Wherever possible, a minimum of 50 spraints per site and season was collected. This was not always possible, although in some cases up to 8–10 km were checked; the difficulty was especially evident in areas of greater altitude.

The spraints were examined and prey items determined with the help of a stereomicroscope following the conventional method (Ruiz-Olmo, 1995). For each spraint the minimum number of individuals of each species was established (size and even–odd pieces). The results are presented as relative frequencies (RF) with respect to the total of determined prey items.

$$RF = \frac{\text{number of prey items in a food category}}{\text{total prey items}} \times 100$$

The Hs Index of niche width (Hespenheide, 1974) was also determined from:

$$Hs = (B - 1)/(n - 1)$$

where

$$B = (\Sigma p_i^2)^{-1}$$

and p_i is the prey item found in the diet, and n is the total number of prey items.

Index of abundance and breeding

The abundance of the otter cannot be established directly by counting the number of signs or spraints, given that these are subject to behavioural and temporal variations (Kruuk, 1995). Because otters are frequently nocturnal, a method of establishing an approximate index of otter abundance (besides radio-tracking, e.g. Kruuk et al., 1993) is to use tracks (e.g. Erlinge, 1968; Ruiz-Olmo, 1985; Reid et al., 1987; Sidorovich, 1991). These studies are much easier to carry out in places where there is plenty of snow (although in these periods the ecology of the otters is profoundly modified and the results may not be representative of the whole year; Reid et al., 1994a) or mud. In places without snow or mud, results may be invalid because of the non-detection of some otters. This may not happen in areas of low otter density, but problems can appear in places where there are many otters, several of them of the same dimensions. In several of our study areas this was the problem; we were able to find hundreds of tracks (often similarly sized) in relatively short distances (Ruiz-Olmo, 1995).

However, direct observation of individuals, whenever possible, is a method for establishing the density of otters (Kruuk et al., 1989; Lejeune & Frank, 1990; Udevitz et al., 1995). The elevated rate of observations of active otters at dusk and dawn (for periods of 1–2 h) permitted the development of a method for estimating the abundance of otters (Ruiz-Olmo, 1996). This method was based on observations on the riverbank during the summer (May–July), from 2 h before until 10–20 min after dark in the evening and from 30–60 min before until 2 h after the first light at dawn. Otters are more active in this period, because of the short nights (Green *et al.*, 1984; Jefferies *et al.*, 1986; Ruiz-Olmo *et al.*, 1995; Rosoux, 1996; A. Kranz personal communication). Also, a few of the otters can be observed during the waiting period in the dark, which increases the probability of observation. Studies of otter activity carried out in Spain and France by means of radio-tracking showed that practically 100% of otters were active outside the resting site between May and early July, during daylight, at both dusk and dawn (Ruiz-Olmo *et al.*, 1995; Rosoux, 1996). This method should thus ensure that most of the otters are detected.

Each census was carried out along stretches of river 8–12 km long, placing an observer every 500 m (they can be placed within a range of ±100 m) at a favourable place (repeated during dawn and dusk). The otters were distinguished by group type (i.e. female with young, single, or other group size), relative position (simultaneous observation in different places), and time of observation; to distinguish individuals, we considered the data on average speed (0.8–2.0 km/h) and maximum speed (3.0–6.0 km/h) (Green *et al.*, 1984; Jefferies *et al.*, 1986; Kruuk, 1995; Ruiz-Olmo *et al.*, 1995).

Although in fact data are an index of abundance, based on the probability of observing an otter, data are presented in terms of observed otters per hour of vigil, and individual otters observed per kilometre; the latter number is presumed to be close to the actual density. Following Kruuk (1995), data on otters observed per hectare are also presented. For this calculation, the average width of the studied stretch was taken and 20 m added, 10 m to each bank (the range within which 100% of the otters have moved about or been tracked by our observers or been followed using radio-tracking; it did not extend into the areas where almost no tracks or spraints had been found).

A total of 64 censuses were carried out in seven stretches of river in the Spanish Pyrenees (Fig. 10.1), between 1990 and 1995. In total, 1183 individual vigils were carried out, amounting to 2160 h of vigil and 167 individual otters were observed. The characteristics of those stretches (vegetation, bankside structure and size) make it easy to watch the otters when they are present. Visibility is even better in the upstream stretches.

Throughout the present study special attention was paid to the detection of otter litters or pregnant females. It was thus possible to establish the altitudes in which reproduction of *L. lutra* has been verified. 'Breeding' was defined as the presence of females seen with young, young lost or abandoned by mother, or young found dead, captured females or nursing mothers discovered dead or with embryos, and clear tracks of the young (e.g. Sidorovich, 1991). Where possible, the number of young accompanying the female was established (not taking nursing mothers into account).

Results

Distribution
Throughout the 11 years (1984–1995) the distribution of otters has apparently expanded significantly in the study area. In 1984–85 only 13.3% of the stations were positive, while in 1989–90 24.4% were positive and in 1994–95 37.0% were positive. Throughout this time the otter remained in the southern tributaries of the River Ebro but has apparently disappeared from the small, far populations of the Rivers Muga and Montsant. In the Pyrenean mountains a spectacular recovery has been witnessed in all the sub-basins. No significant differences were observed between the different otter surveys in the altitude of positive sites (Fig. 10.2). The otter is absent in the lowest areas and its tracks and signs have not been found above 1800 m.

The structure of fish populations
The abundance of fish (almost the exclusive food sources of the otter in the area studied) was influenced by altitude in a very important way. Below 500–800 m altitude, the abundance of fish was elevated, and increased rapidly (Fig. 10.3(a)). Between 300 m (minimum altitude studied) and 500 m, estimated biomass was between 20 and 280 g/m^2 or 200 and 2700 g/m of stream length. However, above 800 m, these values were much lower and decreased progressively, varying between 2 and 30 g/m^2 or between 16 and 170 g/m, respectively. It should be noted that above this altitude 48.1% of the places that were fished had less than 10 g/m^2 of fish and 85.2% has less than 15 g/m^2.

The biomass of fish present in the high river areas decreased dramatically, as did the variety and diversity of species (Fig. 10.3(b)). Between 200 and 400 m the fish population was composed of 6–10 species, whilst above 1000 m only *S. trutta* was found. Between 700 and 1000 m, only two species of fish were found, normally the brown trout and *Barbus haasi* or *Phoxinus phoxinus*.

Diet
In almost all study locations, fish (cyprinids and salmonids) represented 85–100% of the prey items recorded ($n = 6766$). From a trophic point of view, there was a strong dependency on altitude; the proportion of fish in the diet increased with altitude (Fig. 10.4(a)), above 800 m fish were almost exclusively taken (85–100%). Thus the relative frequency of salmonids (almost exclusively *Salmo trutta*) and cyprinids (mainly *Barbus graellsi*, *B. haasi* and *Chondrostoma toxostoma*) varied with altitude (Fig. 10.4(b)). *Salmo trutta* increases in importance above 500 m and is the only species above 800 m. Conversely,

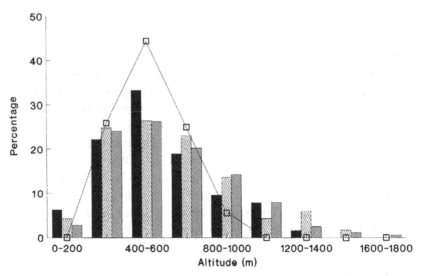

Figure 10.2. Distribution of the altitude of the positive sites in the Otter Surveys and of breeding sites. The number of positive stations increased significantly between otter surveys (1984–85/1989–90: $\chi^2 = 24.09$; d.f. = 1, $p < 0.0001$; 1989–90/1994–95: $\chi^2 = 22.76$; d.f. = 1, $p < 0.0001$). No significant differences were found between the different otter surveys in the altitudes of positive sites (1984–85/1989–90: $\chi^2 = 5.861$, d.f. = 7, $p = 0.556$; 1985–85/1994–95: $\chi^2 = 4.856$, d.f. = 8, $p = 0.773$; 1989–90/1994–95: $\chi^2 = 5.228$, d.f. = 8, $p = 0.733$). The distribution of altitudes of breeding events was practically significantly different to the distribution of altitudes of the positive stations ($\chi^2 = 10.551$; d.f. = 5, $p = 0.061$). ■, 1984–85, $n = 63$; ▩, 1989–90, $n = 117$; ▦, 1994–95, $n = 162$; ⊟, litters, $n = 36$.

cyprinids are dominant at altitudes below 500–600 m, but absent above 800 m.

The fish component of the diet becomes less diverse with increased altitude, particularly above 800 m, where the remains of only one species were recorded (Fig. 10.4(c)) and the breadth of the trophic niche is at a minimum (Hs < 0.1) (Fig. 10.4(d)). The number of fish species eaten (Fig. 10.4(c)) is lower than the number recorded as present because some species are very rare in the different stretches (Fig. 10.3(b)) (detected only by electro-fishing and then in very low numbers). Except in places above 500 m the number of species of fish increases greatly (5–7) in those diets that include crayfish (*Austrapotamobius pallipes* and *Procambarus clarki*), amphibians (*Rana perezi*) and reptiles (*Natrix maura*).

Index of abundance and breeding

In the different stretches of river (for key see legend to Fig. 10.1), an interannual variation in the index of the abundance of otters was observed:

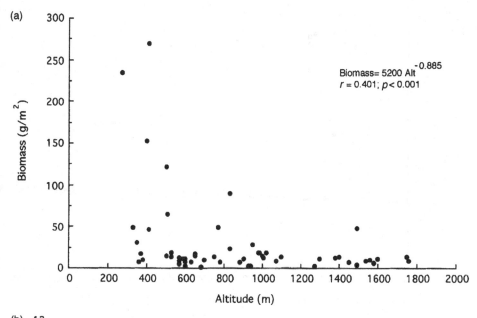

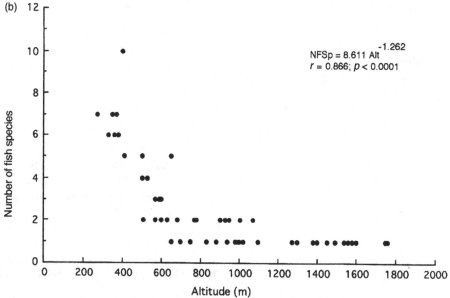

Figure 10.3. Structure of the fish community with attitude: (a) fish biomass (g/m^2) (b) number of fish species (NFSp).

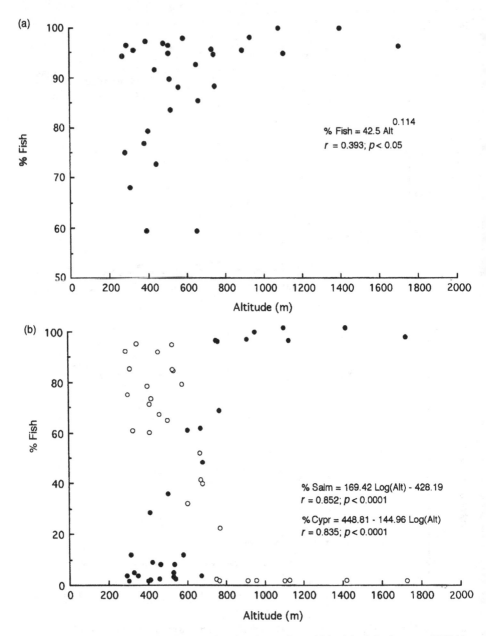

Figure 10.4. Relationships between altitude and some diet variables: (a) relative frequency (% Fish) of fish, (b) frequency of salmonids (●) and cyprinids (○), (c) number of fish species (NFSp) eaten, (d) Hs niche width index.

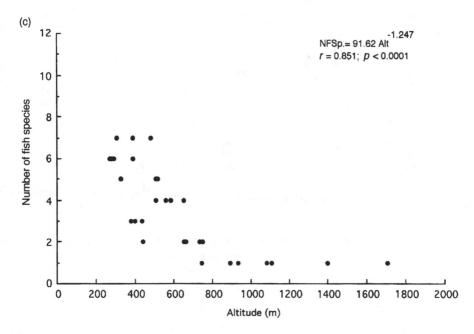

(c)

$$NFSp.= 91.62 \, Alt^{-1.247}$$
$$r = 0.851; \; p < 0.0001$$

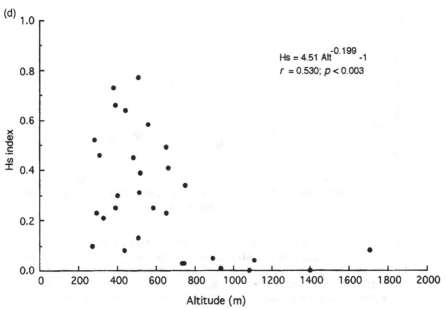

(d)

$$Hs = 4.51 \, Alt^{-0.199} -1$$
$$r = 0.530; \; p < 0.003$$

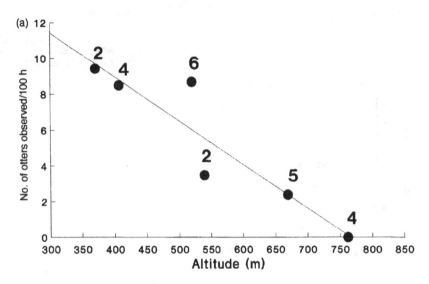

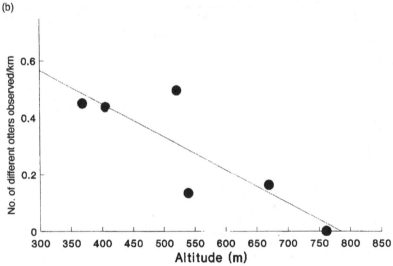

Figure 10.5. Relationships between altitude and average estimated otter abundance. Numerals indicate how many censuses were carried out at each site; some censuses carried out in 1990 (years developing the method) are not included in the Figure. (a) Number of otters observed/h; (otters observed/100 h = −0.011 altitude + 0.917; $r = 0.925$; $p = 0.008$), (b) number of different otters observed/km (different otters observed/km = −0.024 altitude + 18.76; $r = 0.85$; $p = 0.033$).

1   (Otters observed per 100 h): 0–7 (NR1), 3–16 (NR2), 0 (NP1), 1–6 (NP2),
     3–19 (NP3) and 6–13 (SE1).
2   (Otters observed per kilometre): 0–0.35 (NR1), 0.24–0.82 (NR2), 0 (NP1),
     0.10–0.17 (NP2), 0.24–0.70 (NP3) and 0.35–0.55 (SE2).
3   (Otters observed per hectare): 0–013 (NR1), 0.09–0.23 (NR2), 0 (NP1),
     0.12–0.04 (NP2), 0.09–0.25 (NP3), 0.07–0.11 (SE2).

The number of otters observed per hour and per kilometre were inversely correlated with altitude, reaching a minimum at ca. 800 m (Fig. 10.5). No significant correlation was found in terms of number of otters observed per hectare.

A total of 36 different reproductive events were detected in the study area, at an average altitude of 534.6 m ($\text{SD} = 129.96$). The minimum altitude was 390 m and the maxima were 835 and 820 m (Fig. 10.2). The altitudinal distribution of these events was similar to that of the stations that were positive in at least one of the three Otter Surveys (i.e. 500–700 m in altitude). However, there was a significant difference attributable to the fact that the reproduction of otters was not recorded in some of the places in which the species regularly occurs (i.e. 200–300 m and 900–1800 m). The breeding events tended to be also negatively correlated with altitude (but without a significant correlation; $r = 0.771$; $p = 0.14$); no litters of four young were recorded above 500 m altitude.

Discussion

Our results show that altitude acts as a significant limiting factor for the distribution and abundance of otters. Their continued presence in the Iberian Peninsula above 2000 m thus seems difficult to maintain. Water conditions and ambient temperature must have some influence, but it seems that these may not be the most important factors, because the otter can live in many other European countries where, at lower altitudes, climatic conditions similar to those in the mountainous zones of Spain are found. Ice cover may be an important factor. In the mountainous zones in our study area, this appears annually in lakes, dams and rivers above 1700–2000 m. At these heights the ice can remain for periods of a few weeks to several months, putting food resources out of the reach of *L. lutra*. This fact has been shown for *L. lutra* in other parts of Europe (V. Sidorovich, personal communication; A Kranz, personal communication) and also in *L. canadensis* (Reid *et al.*, 1994a,b). This also seems to be one of the main factors that prevent these two species living in arctic and subarctic environments (Sage, 1986). However, outside the period when there is ice, lakes and rivers at altitude remain exposed and food sources remain available. In fact, Reid *et al.*, (1994*a*) demonstrated that otters were capable of adapting themselves to conditions of extreme winter harshness, modifying their use of space throughout the year. So, if otters do not exceed certain altitudes in summer or become scarce, other limiting factors must exist. Lizana & Pérez-Mellado's (1990) results from mountains in the centre of Spain suggested that in these zones otters can ascend to over 2000 m during certain periods in spring, and spend several weeks there. At this time of year amphibians, principally *Bufo bufo*, reproduce in large numbers, providing the otters with sufficient food for a short time.

Our results concur with those of Kruuk (1995). Thus the decrease in the abundance and presence of *L. lutra* with altitude is a product of the combination of two of the factors that seem most to affect their ecology in natural conditions: the availability and diversity of food, and water and air temperature. The lower these are, the greater the need to food and energy for thermoregulation. Kruuk *et al.*, (1993) found that otters tend to use areas with a greater availability of food more frequently – especially those where salmonids exceed 8–9 g/m^2 (although otters also consume a small proportion of eels). Nores *et al.* (1990) did not find otters in mountainous places with 1.8–3.3 g of fish/m^2, but did find them in those areas that have 18–60 g of fish m^2. Our data (Ruiz-Olmo, 1994, 1995) coincide with this information, the otter being more widespread and apparently more abundant as the availability of food increases; in our study area, the otter was present in places with more than 10–20 g of fish m^2. Above an altitude of 800–1000 m, the abundance of fish lies within, or below, this range, which places this mustelid very close to this range of food availability, especially if it is considered that important intra- and interannual fluctuations occur.

Bearing in mind that water temperatures fluctuate between 0 and 10 °C for a great part of the year (and little more in summer), conditions are those which Kruuk (1995) proposed as the least favourable for foraging. According to this author's study, the otter needs to obtain 180–200 g of food/h in these conditions (investing 5–7 h of activity); if the animal cannot do this, the point is reached where it is forced to spend 24 h a day in foraging activity, and therefore it cannot survive.

In the present study it is suggested that lower temperature and insufficient availability of food are responsible for the progressive decline in otter presence with altitude. We then have to ask ourselves, what are the response mechanisms of this animal? To some extent physiological mechanisms and the increase of insulation (greater fur covering) can help to mitigate the effect of adverse conditions, but these mechanisms are not well known. Without doubt, the principal mechanism is a decrease in the number of individuals in proportion to a decrease in food. This mechanism has already been proposed by Kruuk *et al.* (1993), Kruuk (1995) and Ruiz-Olmo (1994, 1995). However, the regulation of the number of individuals (apart from migration) is determined by mortality and reproduction (Kruuk & Conroy, 1991; Kruuk *et al.*, 1987, 1991; Heggberget, 1993). In our study area, we have only detected otter litters up to 800–900 m in altitude; they appear much more frequent at lower altitudes. Even though it is possible that some otters reproduce in higher areas, everything seems to indicate that when altitude increases, reproduction becomes more difficult. In our study area, the average number of young per litter also seems to decrease with altitude, which would support this. Otters present between 1000 and 2000 m altitude would be males, subadults, dispersing young or non-reproductive females, and they would use the range temporarily or permanently in relation to the availability of food.

But this species' great necessity for space, up to tens of kilometres, has to be remembered (Green, *et al.*, 1984; Jefferies *et al.*, 1986; Kruuk 1995; Ruiz-Olmo *et al.*, 1995; A. Kranz, personal communication). Depending on the area, an individual home range that extends over ±1000 m in altitude will offer significant variations in food resources which the otter can exploit by using different food patches at different times.

With increased altitude, there is less food, a less ample trophic niche (diversity of food) and a reduction in the number of species on which the otter's diet is based. This is in fact a reflection of the impoverishment of the available food supply. Above 800 m the otter feeds almost exclusively on one species, the brown trout. This extreme specialization makes populations of otters very vulnerable and very dependent on the evolution and fluctuations of the populations of this salmonid. In this sense, catastrophic floods (frequent in the Mediterranean area; Martín-Vide, 1985), fish management, epizootic disease or competition with other piscivorous predators could have a significant influence on the populations of otter. In fact, there are several facts that show this. The floods in 1982 gravely affected the trout populations in the whole of the Pyrenees, which forced the otters to remain below 1200 m (Ruiz-Olmo, 1995). Subsequently, a progressive recuperation of the species with altitude has been observed. Other rivers, e.g. the Noguera de Tor or the Noguera de Lladorre, have seen the otters leave in some years because of the effect of floods or excess fishing (our unpublished data).

However, in our study area, the otter is absent in areas of low altitude below 250 m, precisely the range where *L. lutra* is most frequent in the rest of Europe. This phenomenon is common throughout the Mediterranean and in the

Iberian Peninsula (Delibes, 1990), being an effect of contamination through agriculture, industry, cities and tourism (Ruiz-Olmo, 1995; Chapter 18). In these low-altitude areas, where the levels of contamination of the last 30–40 years have brought about this significant regression (Ruiz-Olmo & Gosálbez, 1988), the otter has not yet recolonized them. In spite of the fact that *L. lutra* at altitudes as low as 200 m, few litters are recorded below 400 m and none below 300 m. These results show that this area does not provide the conditions for *L. lutra* to manage to reproduce easily. However, sufficient food is available, which suggests that something else is the cause of this situation. Everything seems to indicate that the levels of contamination are not sufficient to cause the death of individuals, because they are still living there, but are sufficient to inhibit or hinder reproduction. This effect is consistent with the presence of compounds that can progressively accumulate in the trophic network, such as organochloride compounds, heavy metals and others (Mason, 1989; Chapter 18). This situation helps us to understand the relationship between contamination and the otter, because we found areas where otters are still present, but no breeding events were detected.

ACKNOWLEDGEMENTS
Many people have participated in these studies. Special thanks must go to the following: Santiago Palazón, José María López-Martín, José Antonio Muñoz, Josep Jordana, Antonio Berenjeno, 'Cinto' Medina, Antoni Margalida, Marc Alonso, Jordi Canut, Vittorio Pedrocci, Daniel Oro, Oscar Arribas and Diego García. Emma O'Dowd translated the text.

References

Chapman, P. J. & Chapman, L. L. (1982). *Otter survey of Ireland, 1981–82.* The Vincent Wildlife Trust, London.

Delibes, M. (1990). *La nutria (Lutra lutra) en España.* Serie Tècnica. ICONA, Madrid.

Erlinge, S. (1968). Territoriality of the otter *Lutra lutra* L. *Oikos* **19**: 81–98.

Foster-Turley, P., Macdonald, S. M. & Mason, C. F. (eds) (1990). *Otters. An action plan for their conservation.* IUCN/

SSC Otter Specialist Group, Gland.

Green, J. & Green, R. (1980). *Otter survey of Scotland, 1977–79.* The Vincent Wildlife Trust, London.

Green, J., Green, R. & Jefferies, D. J. (1984). A radio-tracking survey of otters *Lutra lutra* on a Perthshire river system. *Lutra* **27**: 85–145.

Heggberget, T. M. (1993). *Reproductive Strategy and Feeding Ecology of the Eurasian Otter*

Lutra lutra. DSc thesis: University of Trondheim.

Hespenheide, M. A. (1974). Prey characteristics and predation niche width. In *Ecology and evolution of communities*: 158–180. (Eds Cody, M. L. & Diamond, J. M.). Harvard University Press, Cambridge, MA.

Hill, R. W. (1980). *Fisiologia animal comparada.* Ed. Reverté, S. A., Barcelona.

Jefferies, D. J., Wayre, P., Jessop, R. M. & Mitchell-Jones, A. J.

(1986). Reinforcing the native otter *Lutra lutra* population in East Anglia: an analysis of the behaviour and range development of the first release group. *Mammal Rev.* **16**: 65–79.

Kruuk, H. (1995). *Wild otters: predation and populations.* Oxford University Press, Oxford.

Kruuk, H. & Conroy, J. W. H. (1991). Mortality of otters (*Lutra lutra*) in Shetland. *J. appl. Ecol.* **28**: 83–94.

Kruuk, H., Conroy, J. W. H. & Moorhouse, A. (1987). Seasonal reproduction, mortality and food of otters (*Lutra lutra* L.) in Shetland. *Symp. zool. Soc. London.* No. 58: 263–278.

Kruuk, H., Moorhouse, A., Conroy, J. W. H., Durbin, L. & Frears, S. (1989). An esimate of numbers and habitat preferences of otters *Lutra lutra* in Shetland, UK. *Biol. Conserv.* **49**: 241–254.

Kruuk, H., Conroy, J. W. H. & Moorhouse, A. (1991). Recruitment to a population of otters (*Lutra lutra*) in Shetland, in relation to fish abundance. *J. appl. Ecol.* **28**: 95–101.

Kruuk, H., Carss, D. N., Conroy, J. W. H. & Durbin, L. (1993). Otter (*Lutra lutra* L.) numbers and fish productivity in rivers in north-east Scotland. *Symp. Zool. Soc. London.* No. 65: 171–191.

Lacomba, J. I. & Jiménez, J. (1988). *La nutria* (Lutra lutra) *y su conservación en Aragón.* Departmento de Agricultura, Ganadería y Montes, Diputación General de Aragon.

Lejeune, A. & Frank, V. (1990). Distribution of *Lutra*

maculicollis in Rwanda: ecological constraints. *IUCN Otter Spec. Group Bull.* No. 5: 8–16.

Lizana, M. & Pérez-Mellado, V. (1990). Depredación por la nutria (*Lutra lutra*) del sapo de la sierra de Gredos (*Bufo bufo gredosicola*). *Doñana Acta Vert.* **17**: 109–112.

Macdonald, S. M. & Mason, C. F. (1992). *Status and conservation needs of the otter* (Lutra lutra) *in the Western Palearctic.* Council of Europe, Strasbourg.

Martín-Vide, X. (1985). *Pluges I inundacions a la Mediterrània.* Collecció Ventall, Ketres Ed., Barcelona.

Mason, C. F. (1989). Water pollution and otter distribution: a review. *Lutra* **32**: 97–131.

Mason, C. F. & Macdonald, S. M. (1986). *Otters: ecology and conservation.* Cambridge University Press, Cambridge.

McNab, B. K. (1989). Basal rate of metabolism, body size and food habits in the order Carnivora. In *Carnivore behavior, ecology, and evolution:* 335–354. (Ed. Gittleman, J. L.) Chapman and Hall, London.

Nores, C., Hernández-Palacios, O., García-Gaona, J. F. & Naves, J. (1990). Distribución de señales de nutria (*Lutra lutra*) en el medio ribereño cantábrico en relación con los factores ambientales. *Revta Biol. Univ. Oviedo* **8**: 107–117.

Reid, D. G., Bayer, M. B., Code, T. E. & McLean, B. (1987). A possible method for estimating river otter, *Lutra canadensis,* populations using snow tracks. *Can. Fld-Nat.* **101**: 576–580.

Reid, D. G., Code, T. E., Reid, A. C. H. & Herrero, S. M. (1994a). Food habits of the river otter in a boreal ecosystem. *Can. J. Zool.* **72**: 1306–1313.

Reid, D. G., Code, T. E., Reid, A. C. H. & Herrero, S. M. (1994b). Spacing, movements, and habitat selection of the river otter in boreal Alberta. *Can. J. Zool.* **72**: 1314–1324.

Riba, O., de Bolòs, O., Panareda, J. M., Nuet, J. & Gosálbez, J. (1979). *Geografia física dels Països Catalans.* Ketres Ed., Barcelona.

Rosoux, R. (1996). Cycle journalier d'activités et utilisation des domaines vitaus chez la loutre d'Europe (*Lutra lutra* L.) dans le Marais Poitevin (France). *Cah. Ethol.* **15**: 283–305.

Ruiz-Olmo, J. (1985). *Distribución, Requerimientos Ecológicos y alimentación de la Outria* (Lutra lutra) *L., 1758) en el N. E. de la Península Ibérica.* Licenciature thesis: University of Barcelona.

Ruiz-Olmo, J. (1994). Influence of food availability on otter distribution and abundance. In *Seminar on the conservation of the European otter* (Lutra lutra), *Leeuvarden, The Netherlands, 7–11 June 1994:* 114–116. Council of Europe, Strasbourg.

Ruiz-Olmo, J. (1995). *Estudio bionómico sobre la nutria (*Lutra lutra L., 1758) *en aquas continentales de la Península Ibérica.* PhD thesis: University of Barcelona.

Ruiz-Olmo, J. (1996). Visual census of Eurasian otter (*Lutra lutra*): a new method. *Habitat* **11**: 125–130.

176   *Altitude and European otter ecology*

Ruiz-Olmo, J. & Gosálbez, J.
(1988). Distribution of the ot-
ter, *Lutra lutra* L. 1758 in the
NE of the Iberian Peninsula.
Publines Dep. Zool. Barcelona
14: 121–132.
Ruiz-Olmo, J., Jordán, G. &
Gosálbez, J. (1989). Alimen-
tación de la nutria (*Lutra lutra*
L., 1758), en al NE de la
Península Ibérica. *Doñana Acta
vert.* 16: 227–237.

Ruiz-Olmo, J., Jiménez, J. &
López-Martin, J. M. (1995).
Radio-tracking of otters *Lutra
lutra* in north-eastern Spain.
Lutra 38: 11–21.
Sage, B. (Ed.) (1986). *The Arctic
and its wildlife*. Croom Helm,
London.
Schmidt-Nielsen, K. (1972) *How
animals work?*. Cambridge
University Press, Cambridge.

Sidorovich, V. E. (1991). Structure, reproductive status and
dynamics of the otter population in Byelorussia. *Acta
theriol.* 36: 152–161.
Udevitz, M. S., Bodkin, J. L. &
Costa, D. P. (1995). Detection
of sea otters in boat-based surveys of Prince William Sound,
Alaska. *Mar. Mamm. Sci.* 11:
59–71.
Zippin, C. (1958). The removal
method of population estimation. *J. Wildl. Mgmt* 22: 82–90.

11

Diets of semi-aquatic carnivores in northern Belarus, with implications for population changes

V. Sidorovich, H. Kruuk, D. W. Macdonald and T. Maran

Introduction

In this chapter we present data on the diet of the guild of semi-aquatic carnivores, the European mink (*Mustela lutreola*), the American mink (*M. vison*), the polecat (*M. putorius*) and the otter (*Lutra lutra*). These species share habitats in rivers, streams and lakes in northern Belarus. The data are used to test predictions from the hypothesis that the decline of European mink is caused by changes in prey availability, or by competition for food with other carnivores within the guild.

There are a number of carnivore species in Europe that may be termed 'semi-aquatic', species that live close to water, and which capture at least part of their food by swimming and diving. These include the otter, which feeds mostly on fish and some amphibians (for summaries, see Mason & Macdonald, 1986; Kruuk, 1995), the European mink with a diet of amphibians, small mammals, fish and crayfish (Sidorovich, 1992a), the American mink, which has been introduced in many areas since the 1930s and which feeds on small mammals, fish, amphibians and crayfish (for review, see Dunstone, 1993), and the polecat with a diet of mostly small mammals and amphibians (Sidorovich, 1992a). There have been no studies of this whole complement of semi-aquatic predators in any one area, however. Such an approach is necessary in order to assess possible competition for food, to draw comparisons between the effects of prey species on different predators, and to study the effects of these species on each other.

The European mink has disappeared from large parts of its former range in Europe (Sidorovich 1992b; Chapter 17). It is present now only in small areas of France and Spain, and in areas east of Estonia, in north-eastern parts of Belarus and in restricted parts of Russia. Other semi-aquatic mammals have either held their own (polecat), or decreased far less dramatically (otter: Foster-Turley *et al.*, 1990), or have spread and substantially increased in numbers (American mink: Chapter 19).

A number of hypotheses have been put forward to explain the decline of *M.*

lutreola; here we test the idea that a change in food availability is the cause. In north-eastern Belarus there is an opportunity to investigate this possibility, in an area where the European mink is still present but declining (Sidorovich, 1992*b*), and where we can compare its feeding ecology with that of taxonomically close and ecologically similar species in the same area. We think that the possibility of changes in available prey biomass as a cause of population changes in such predators should be taken seriously. There are strong suggestions, at least in some areas, that prey populations may be limiting numbers of semi-aquatic predators such as the otter (Kruuk *et al.*, 1993; Kruuk, 1995).

Our hypothesis states that the decline in numbers of the European mink is caused at least partly by changes in food availability, because of either (i) declining prey populations or (ii) increased competition for food with other predators. Some predictions from (i), which we test here, are that:

P1   The European mink is a more highly specialized predator than the American mink, otter or polecat.

P2   The European mink is more dependent on prey that has declined and is declining that the other predators are.

Some predictions from (ii) are:

P3   There is a large overlap in diet with other predators that have increased in numbers.

P4   Such overlap occurs especially over prey species that are scarce.

In this chapter we will be concerned only with diet, and not with foraging behaviour; thus, we will not address the possibility of direct, aggressive competition over food between species.

Study areas

The observations were made in various water bodies in an area of about 20 km by 40 km, at the head of the River Lovat in Belarus (Vitebsk region, Gorodok district; 56 °N 32 °E). The area is wooded, with little agriculture and sparse human habitation; the dominant vegetation consists of alder (*Alnus glutinosa* and *A. incana*), birch (*Betula pubescens* and *B. pendula*), spruce (*Picea abies*), aspen (*Populus tremula*) and oak (*Quercus robur*). Preliminary data suggest that there is no significant pollution (V. Sidorovich, unpublished observations).

In this area the European mink was common in all aquatic habitats, but it has sharply declined in the few years leading up to 1995. The American mink is now common, after its arrival in the study area in 1988 (Sidorovich, 1992*b*). Polecats are relatively common everywhere, but they are not confined to riparian strips, as are the other three species studied. The otter is common in all waters. Many of the water bodies are inhabited (and have been modified) by beaver (*Castor fiber*). Other common large predators include the wolf (*Canis lupus*), bear (*Ursus arctos*) and lynx (*Lynx lynx*).

The major aquatic habitats consisted of two glacial lakes, Lake Zavesno and Lake Zadrach, about 20 km apart, and various feeding streams and outflows. The river Lovat flows into Lake Zavesno, then on to Lake Zadrach, and later through a wide flood plain to Lake Mezha (outside our study area). As habitat categories we distinguished (i) fast-flowing streams and rivers, (ii) slow-flowing rivers (including drainage canals), and (iii) lakes.

1   Fast-flowing rivers and small streams. The flow rate is between 0.3 and 1.0 m/s, there is no flood plain, and the banks are mostly wooded. These include the River Lovat above Lake Zadrach, and below Lake Zadrach to the Ljahovsky drainage systems; the Servajka, Uzhovsky, Bibinsky, Borkovsky, Rudnjansky, Trubachovsky, Mahalovsky and Skljanka streams. These streams are shallow, may dry up in summer and they may be up to 8 m wide. The smaller ones may contain fish during only part of the year. The species diversity of fish is low (there are no salmonids), Crayfish (*Astacus astacus*) may be moderately abundant or absent; there are few water voles (*Arvicola terrestris*) and birds, but many common frogs (*Rana temporaria*).

2   Slow-flowing rivers and drainage canals. The rivers are up to 25 m wide and up to 2.5 m deep, with a flow rate less than 0.3 m/s. The flood plain along the margins tends to be covered in swamp dominated by bullrush (*Typha latifolia* and *T. angustifolia*) and reeds (*Phragmites communis*). Large-scale flooding occurs in spring. There is a considerable diversity of fish species (but this is low in the artificial drainage canals). Crayfish abundance is variable; water voles, birds and common frogs are common on the flood plains. In winter access by the animals to the rivers and drainage canals themselves may be limited because of ice.

3   Lakes Zavesno and Zadrach are about 40 ha and 100 ha in area, respectively, and up to about 5 m deep. They are surrounded by forest and also some agriculture. Both lakes have marginal reed beds and swamps. There are high densities and diversity of fish species. Crayfish fluctuate between very abundant and absent; water voles, birds and common frogs are very common. Often the animals have no access to water in winter, because the lakes freeze over.

Methods

We collected faeces at various dates between April 1988 and May 1995, mostly from dens or holts. Faeces were collected only if we had a positive identification of the species; this was made from the appearance of the faeces for otter (Mason & Macdonald, 1986). Scats of the two mink species and polecat were usually collected near dens, and the occupant was identified where possible from tracks (Sidorovich, 1994) or, more often, after capture with a box-trap at the entrance, before releasing the animal again.

In the laboratory, faeces were dissected dry, or washed with detergent. The contents were identified microscopically, using published keys of mammalian hair (Day, 1966; Teerink, 1970; Debrot *et al.*, 1982), fish scales and other bones (Galkin, 1953; Pucek, 1981) or pharyngeal teeth (Zhukov, 1965, 1988), amphibian bones (Bohme, 1977) and by comparisons with our own reference collection. It was decided to concentrate on aquatic and semi-aquatic prey, therefore most mammalian and bird remains in the faeces were identified to Class only. Hair of the carnivores themselves was ignored.

For statistical treatment results were expressed as the percentage of faeces containing a given prey category, and for a general overview prey occurrence as a percentage of all occurrences was calculated. For calculations of statistical significance the data were analysed using SAS (1991), using Kruskal–Wallis χ^2 approximation. Levels of significance are shown.

Results

A total of 4312 faeces were collected and analysed, 1930 from American mink, 1474 from European mink, 641 from otter and 267 from polecat. The prey categories identified are listed in the Appendix.

The results from all areas and seasons are summarized in Table 11.1. as the percentage of faeces containing a given prey category. Proportions of faeces containing each prey category were significantly different between predators, for all prey categories except 'other'. In American mink scats small mammals, amphibians and fish dominated and occurred about equally often; in European mink scats amphibians were by far the most important, with fish also present in large numbers; for otters fish dominated all other kinds of prey, closely followed by amphibians; polecat scats were usually full of small mammal remains, again followed in importance by amphibians. The prey category that was strikingly important for all four predators was amphibians.

In general, these same trends and differences were also present when the

Table 11.1. *Percentages of faeces of American and European mink, otter and polecat, containing various prey categories*

Species	No. of faeces	Small mammals	Amphib- ians	Fish	Crust- aceans	Bird/ Reptiles	Other
American mink	1930	32.0	26.0	32.4	6.7	8.2	6.5
European mink	1474	14.5	56.5	26.6	10.9	3.5	7.6
Otter	641	0.3	45.1	61.2	14.5	1.7	1.2
Polecat	267	64.4	30.0	1.9	0.7	9.0	4.5
Significance:		***	***	***	**	*	n.s.

All habitats and seasons combined. The statistical significance is shown for differences between predator species for each prey category (Kruskal–Wallis χ^2 approximation, d.f. = 3). n.s., not-significant; *, $p < 0.05$; **, $p < 0.01$; ***, $p < 0.001$.

four predators were compared for different habitats: fast-flowing streams, slow-flowing rivers and lakes (Table 11.2). However, amphibians were less important for both mink species when feeding in lakes compared with rivers ($p < 0.05$), whereas crayfish were important to them especially in lakes ($p < 0.05$ and < 0.01). Most other differences in prey between habitats were not significant. When comparing the prey of the four predators within each habitat type, it was noticeable that differences were especially large in slow-flowing rivers.

We obtained few samples during the autumn. Although there was considerable variation in diet between the seasons, little of that was statistically significant (Table 11.3). It was striking that, in spring, European mink faeces contained more amphibians than did the scats of any of the other predators ($p < 0.05$). American mink scats showed many more small mammal remains in winter ($p < 0.05$).

Amongst small mammal prey, the species identified most often for both American and European mink was the water vole (*Arvicola terrestris*). Polecats took many bank voles (*Clethrionomys glareolus*), and to a lesser extent so did American mink (see Appendix). Amongst amphibian prey, the common frog (*Rana temporaria*) was by far the most important for all four carnivores. *Rana arvalis* could not be distinguished from *R. temporaria* in the faeces, but was uncommon in the study area, where *R. temporaria* was abundant. Very few toads (*Bufo bufo*) were taken, despite their great abundance in the area. Fish

Table 11.2. *Prey remains in faeces from different habitats in north-east Belarus*

Species	No. of faeces	Prey categories					
		Small mammals	Amphib- ians	Fish	Crust- aceans	Bird/ Reptiles	Other
A. Fast-flowing streams							
American mink	1200	29.2	34.2	30.3	4.9	9.6	6.2
European mink	992	16.2	64.2	21.9	8.0	3.5	9.8
Otter	143	0.0	53.8	76.9	2.1	0.0	1.4
Polecat	53	64.2	34.0	3.8	0.0	5.7	3.8
Significance:		*	*	n.s.	n.s.	*	n.s.
B. Slow-flowing rivers							
American mink	466	35.5	51.2	14.1	1.9	6.2	6.0
European mink	337	16.0	68.9	14.3	10.3	3.1	11.5
Otter	399	0.4	36.1	57.1	26.7	1.2	0.2
Polecat	156	75.7	20.6	0.6	0.0	5.2	3.2
Significance:		***	*	***	**	n.s.	n.s.
C. Lakes							
American mink	264	36.3	12.8	29.5	14.5	5.4	11.5
European mink	145	19.9	42.4	29.1	24.5	6.6	2.8
Otter	99	0.0	30.3	52.5	38.4	1.0	4.0
Polecat	58	55.2	31.0	3.4	3.4	19.0	6.9
Significance:		n.s.	*	n.s.	n.s.	n.s.	n.s.

Table shows the number of faeces examined and the percentage of faeces containing prey re♦

prey was very varied for the three more aquatic predators; in otter spraints pike (*Esox lucius*), roach (*Rutilus rutilus*) and perch (*Perca fluviatilis*) predominated, and although these species were also important to the two minks, in their scats the fish diversity appeared to be greater than for otters.

To compare the overall diet diversity in the four carnivores, we calculated

Table 11.3. *The effects of seasonality: percentage of faeces containing prey categories*

	Spring	Summer	Autumn	Winter
A. American mink				
No. sampling periods	8	9	1	2
No. faeces	454	1210	48	218
Small mammal	23.8	25.8	35.4	83.0
Amphibians	53.7	17.1	50.0	11.9
Fish	20.5	41.9	27.1	6.0
Crustaceans	5.1	7.9	0.0	5.0
Birds/Reptiles	3.7	11.2	2.1	2.3
Others	5.7	7.4	4.2	3.7
B. European mink				
No. sampling periods	7	9	1	5
No. faeces	369	849	30	226
Small mammal	15.7	11.2	20.0	24.3
Amphibians	70.7	45.2	73.3	73.5
Fish	14.4	35.2	10.0	16.4
Crustaceans	9.2	12.5	0.0	8.8
Birds/Reptiles	4.9	3.5	3.3	1.3
Others	6.8	8.7	20.0	3.1
C. Otter				
No. sampling periods	4	1	1	1
No. faeces	357	99	143	42
Small mammal	0.6	0.0	0.0	0.0
Amphibians	46.2	30.3	53.8	40.5
Fish	55.2	52.5	76.9	78.6
Crustaceans	13.2	38.4	2.1	11.9
Birds/Reptiles	2.8	1.0	0.0	0.0
Others	0.6	4.0	1.4	0.0
D. Polecat				
No. sampling periods	3	2	1	3
No. faeces	61	89	53	64
Small mammal	34.4	61.8	64.2	96.9
Amphibians	60.7	28.1	34.0	0.0
Fish	1.6	2.2	3.8	0.0
Crustaceans	0.0	2.2	0.0	0.0
Birds/Reptiles	3.3	16.9	5.7	6.3
Others	9.8	4.5	3.8	0.0

Table 11.4. *Overlap of food niche (Pianka index), based on six prey categories*

	European mink	Otter	Polecat
American mink	0.83	0.79	0.78
European mink		0.86	0.58
Otter			0.27

Levin's index of niche breadth (Levin, 1968; Ciampalini & Lovari, 1985) for the six food categories. Thus, the index may vary between 1 and 6; it was 4.31 for American mink, 3.33 for European mink, 2.57 for otter, and 2.37 for polecat. This shows that the American mink had the most varied diet, and the polecat was the most specialized; both minks are considerably less specialized than either otter or polecat.

In order to evaluate the overlap in diet between the species we used Pianka's index (Pianka, 1973; Ciampalini & Lovari, 1985). Comparisons between the four carnivores are shown in Table 11.4. At the broad level of our prey categories there was a large overlap of prey selection by European mink with both American mink and otter. The lowest coincidence of prey species occurred between polecat and otter, and between polecat and European mink.

Discussion

The observations reported here are based on faecal analyses, and we do not have the necessary information to convert these into estimates of diet (as e.g. Lockie, 1959; Carss & Parkinson, 1996). Thus, in our comparisons between species we will assume that such conversion factors are substantially the same for all of them. It has been demonstrated for otters that when estimating diet from faecal analyses, even from very large samples, the actual percentage occurrence in the diet may be substantially different from that in the spraints (Carss & Parkinson, 1996).

Frequently taken prey and large items tend to be underestimated, rare prey and small items overestimated. Nevertheless, in these studies with captive animals the rank-order of importance of similar-sized prey was the same in diet and spraints. Clearly, we have to keep these reservations in mind when drawing conclusions from our analyses.

The observations in the Belarus study area suggest that the two mink species both occupy a food niche that is very wide, with a diet covering a spectrum of small mammals, amphibians, fish and various other prey. They are much more general predators than either otter or polecat, therefore. Our first prediction (P1) that European mink are more specialized than the others in this complement of semi-aquatic carnivores, is not borne out.

These trends appeared to hold, with small modifications, throughout the seasons and in the different habitats that we considered.

A second, important, prediction (P2) from our starting hypothesis was that European mink are more dependent on prey categories that are declining than the other predators are. Our data suggest that the most important single prey species for European mink is the common frog, occurring in almost half of the European mink's scats. We have no information on population trends for this species, but in the study area it was very abundant indeed, with numbers being relatively stable, and it has been so for a least the previous 10 years (densities of up to 12 frogs/10 m^2 of stream and riverbank: M. Pikulik, unpublished observations).

The common frog was also the single most important prey species in the diet of other predators that appear to be thriving, and frogs are extremely abundant in north-east Belarus. Thus, although there have been suggestions that many amphibians have declined everywhere (Pechmann *et al.*, 1991; Wake 1991), this was not likely to be the reason why this one predator declined and is still declining in our study areas.

The single prey species that is next in importance for the European mink, though far less important than frogs, is the crayfish, which is abundant in many places in our study area. There have been large fluctuations in crayfish populations in Eastern Europe, and the species has disappeared from many areas in Europe due to 'crayfish plague' (Maran & Henttonen, 1995). However, in our study area it is common, especially in lakes (our unpublished observations). It also appears to be at least as important to otters as it is to European mink, if not more so (see Appendix), and the evidence is that otter numbers are being maintained.

Thus, there is no evidence to link the observed decline of European mink in Belarus (Sidorovich, 1992*b*) to changes in biomass of its available main prey species, and our second prediction is not supported.

European mink are slightly more specialized than American mink, but between the two species there is a very large overlap in dietary interest, as there is also between European mink and otters. In the case of the otter, we are concerned with a species that has been present in the same areas as the European mink for a long time in evolutionary terms. The American mink, however, is a recent arrival (see Chapter 17). The coincidence of dietary interest between the two minks confirms our third prediction (P3), based on competition for food between the established European mink and the newly arrived American species.

Thus, the possibility of competition for food between these two in our study sites should be considered. The prey species that figured especially strongly in the zone of overlap were rodents and frogs. Our preliminary and unpublished observations suggest that both these categories were abundant. We were not able to quantify this, but competition between the two mink species for this resource did appear to be unlikely.

In many areas the decline in European mink predated the arrival of the American mink by many decades (see Chapter 17). On present evidence, this decline cannot be explained by competition for resources with the other semi-aquatic predators. Nor can we suggest some species-specific response to a declining resource as a cause, a response by only the European mink and different from that by the otter and polecat in the past, and different from those two as well as from the American mink more recently. Generally, the European mink has a very catholic diet compared with the otter and polecat, though perhaps slightly less so than the American mink.

In conclusion, our data do not provide support for the hypothesis that changes in availability of prey, or competition for prey, are a main cause for either the gradual long term or the accelerated recent decline of the European mink.

ACKNOWLEDGEMENTS
We are grateful to Professor M. Pikulik for unpublished data and assistance in the field, and to the British Government Darwin Initiative for Financial support.

References

Bohme, G. (1977). Zur Bestimmung quartärer Anuren Europas an Hand von Skelettelementen. *Wiss. Z. Humboldt-Univ. Berlin, math.-nat. Reshe* **26**: 283–300.

Carss, D. N. & Parkinson, S. G. (1996). Errors associated with otter *Lutra lutra* faecal analysis. I. Assessing general diet from spraints. *J. Zool., London.* **238**: 301–317.

Ciampalini, B. & Lovari, S. (1985). Food habits and trophic niche overlap of the badger (*Meles meles* L.) and the red fox (*Vulpes vulpes* L.) in a Mediterranean coastal area. *Z. Säugetierk.* **50**: 226–234.

Day, M. G. (1966). Identification of hair and feather remains in the gut and faeces of stoats and weasels. *J. Zool., London.* **148**: 201–217.

Debrot, S., Fivaz, G., Mern & Weber, J.-M. (1982). *des poils de mammifères* pe. Inst. Zool., Univ. Neuchatel, Neuchtel.

Dunstone, N. (1993). *The* T. & A. D. Poyser, Lond

Foster-Turley, P., Macdon M. & Mason, C. F. (eds) (1990). *Otters, an action for their conservation.* IU Gland.

Galkin, G. G. (1953). *Atlas of scales of freshwater fishes.* Publ. Res. Inst. River and Lake Econ., Moscow.

Kruuk, H. (1995). *Wild otters: predation and populations.* Oxford University Press, Oxford.

Kruuk, H., Carss, D. N., Conroy, J. W. H. & Durbin, L. (1993). Otter (*Lutra lutra* L.) numbers and fish productivity in rivers in north-east Scotland. *Symp. zool. Soc., Lond.* No. 65: 171–191.

Levin, R. (1968). *Evolution in changing environments.* Princeton University Press, Princeton, NJ.

Lockie, J. D. (1959). The estimation of the food of foxes. *J. Wildl. Mgnt.* **23**: 224–227.

Maran, T. & Henttonen, H. (1995). Why is the European mink (*Mustela lutreola*) disappearing? – a review of the process and hypotheses. *Annls. zool. Fenn.* **32**: 47–54.

Mason, C. F. & Macdonald, S. M. (1986). *Otters: ecology and conservation.* Cambridge University Press, Cambridge.

Pechmann, J. H. K., Scott, D. E. Semlitsch, R. D., Caldwell, J. P., Vitt, L. J. & Whitfield Gibbons, J. (1991). Declining amphibian populations: the problem of separating human impacts from natural fluctuations. *Science* **253**: 892–895.

Pianka, E. R. (1973). The structure of lizard communities. *Annu. Rev. Ecol. Syst.* **4**: 53–74.

Pucek, Z. (1981). *Keys to vertebrates of Poland Mammals.* PWN, Warsaw.

SAS (1991). *Software systems.* SAS Institute Inc., Cary, NC, USA.

Sidorovich, V. E. (1992*a*) Comparative analysis of the diets of European mink (*Mustela lutreola*), American mink (*M. vison*) and polecat (*M. putorius*) in Byelorussia. *Small Carniv. Conserv.* No. 6: 2–4.

Sidorovich, V. E. (1992*b*). Gegenwärtige Situation des Europäischen Nerzes (*Mustela lutreola*)

in Belorusland: In *Semiaquatische Säugetiere:* 316–328. (Eds Schröpfer, R., Stubbe, M. & Heidecke, D.). Hypothese seines Verschwindens. Martin-Luther-Univ. Halle-Wittenberg.

Sidorovich, V. E. (1994). How to identify the tracks of the European mink (*Mustela lutreola*), the American mink (*M. vison*) and the polecat (*M. putorius*) on waterbodies. *Small Carniv. Conserv.* No. 10: 8–9.

Teerink, B. J. (1970). *Hair of Western European mammals.* Cambridge University Press, Cambridge.

Wake, D. B. (1991). Declining amphibian populations. *Science* **253**: 860.

Zhukov, P. I. (1965). [*Fishes in Byelorussia.*] Nauka and Tehnika Publ., Minsk. [In Russian.]

Zhukov, P. I. (1988).]*Ecology of freshwater fishes.*] Nauka and Tehnika Publ., Minsk. [In Russian.]

Appendix

Percentages of the total number of occurrences of different prey, in faeces of European mink, American mink, otter and polecat. All habitats, all seasons combined. One occurrence is the observation of one prey (species, or category) in a scat.

American mink:　$n = 2160$ occurrences (in 1930 faeces).
European mink:　$n = 1763$ occurrences (in 1474 faeces).
Otter:　　　　　$n = $　795 occurrences (in　641 faeces).
Polecat:　　　　$n = $　295 occurrences (in　267 faeces).

	American mink	European mink	Otter	Polecat
A. Mammals				
Total	28. 4	11. 9	0. 0	58. 7
Arvicola terrestris	2. 2	1. 8	0. 3	3. 1
Clethrionomys glareolus	2. 1	0. 7	0. 0	5. 8
Microtus agrestis	0. 4	0. 3	0. 0	2. 7
M. arvalis	0. 3	0. 1	0. 0	2. 4
M. oeconomus	0. 4	0. 1	0. 0	0. 0
Apodemus sylvaticus	0. 5	0. 0	0. 0	0. 0
Ondatra zibethica	0. 0	0. 0	0. 0	1. 0
Unidentified rodent	12. 3	4. 8	0. 0	38. 3
Neomys fodiens	0. 5	0. 3	0. 0	0. 0
Sorex spp.	3. 7	1. 4	0. 0	1. 7
Talpa europaea	0. 6	0. 1	0. 0	0. 0
Unidentified small mammal	5. 4	2. 3	0. 0	3. 7
B. Birds				
Total	5. 8	2. 7	1. 3	8. 1
Unidentified	5. 8	2. 7	1. 3	8. 1
C. Reptiles				
Total	1. 4	0. 3	0. 0	0. 0
Lacerta vivipara	0. 9	0. 1	0. 0	0. 0
Natrix natrix	0. 5	0. 2	0. 0	0. 0
D. Amphibians				
Total	22. 7	47. 4	35. 6	27. 2
Rana temporaria/arvalis	20. 5	46. 7	35. 3	24. 1
Rana esculenta complex	1. 7	0. 5	0. 3	0. 0
Bufo bufo	0. 5	0. 2	0. 0	3. 1

E. Fishes

Total	31. 1	22. 8	49. 9	2. 0
Esox lucius	3. 0	1. 0	12. 0	1. 0
Abramis spp.	0. 4	0. 1	2. 3	0. 0
Alburnoides bipunctatus	1. 2	1. 1	0. 4	0. 0
Alburnus alburnus	2. 3	1. 1	2. 0	0. 0
Blicca bjoerkna	0. 4	0. 4	1. 1	0. 0
Leuciscus cephalus	1. 0	0. 5	0. 0	0. 0
L. idus	0. 4	0. 2	0. 6	0. 0
L. leuciscus	0. 8	0. 7	0. 0	0. 0
Rutilus rutilus	7. 4	4. 7	12. 4	0. 0
Scardinius erythrophthalmus	0. 5	0. 3	0. 8	0. 0
Unidentified cyprinid	0. 7	0. 1	0. 4	0. 0
Cobitis taenia	1. 8	2. 8	1. 1	0. 0
Misgurnus fossilis	0. 2	0. 3	1. 1	0. 0
Nemacheilus barbatulus	0. 2	0. 6	0. 0	0. 0
Lota lota	1. 1	0. 9	0. 0	0. 0
Gasterosteus aculeatus	1. 5	3. 5	1. 5	0. 0
Perca fluviatilis	5. 8	3. 4	9. 3	0. 0
Gymnocephalus cernus	1. 1	0. 6	2. 2	0. 0
Stizostedion lucioperca	0. 0	0. 0	1. 4	0. 0
Unidentified fish	0. 2	0. 5	0. 3	1. 0
F. Crustaceans				
Total	5. 9	9. 1	11. 9	0. 7
Astacus astacus	5. 9	9. 1	11. 9	0. 7
G. Insects				
Total	5. 0	5. 2	1. 0	3. 4
Dytiscus spp.	2. 4	4. 2	1. 0	1. 4
Unidentified insect	2. 6	1. 0	0. 0	2. 0
H. Molluscs				
Total	0. 3	1. 4	0. 0	0. 0
Unidentified mollusc	0.3	1.4	0.0	0.0
I. Vegetable matter				
Total	0.6	0.0	0.0	0.0
Bilberry	0.6	0.0	0.0	0.0

12

Otter (*Lutra lutra*) prey selection in relation to fish abundance and community structure in two different freshwater habitats

D.N. Carss, K.C. Nelson, P.J. Bacon and H. Kruuk

Introduction

The otter, *Lutra lutra*, was once a top piscivorous predator in coastal and riparian habitats throughout much of Britain, continental Europe, the former USSR, Asia and Japan (Mason & Macdonald, 1986). However, otter populations have declined severely over much of this range during the last century, particularly in freshwater habitats, and the species is now rare or absent in many countries (Mason & Macdonald, 1986; Foster-Turley et al., 1990). The species is listed on the EU Habitats Directive, the Bern Convention and the Convention on International Trade in Endangered Species (CITES), and is classified by the International Union for Conservation of Nature and Natural Resources (IUCN) as 'vulnerable' as a result of the declining or endangered status of many of its populations (for full legislative details, see JNCC, 1996). Several countries, including the UK (see JNCC, 1996), have conservation strategies for the species that include such things as population monitoring, enhancement of populations through habitat management, the promotion of natural recolonization, and improved knowledge through appropriate research.

Several of these factors involve knowledge of the relationships between otters and their food supply. Two approaches have been adopted to investigate the use of food resources by animals. Firstly, descriptive studies have quantified the diets of particular populations of many species and, for otters, descriptions of diet in various habitats are available from either faecal (spraint) analysis (see Mason & Macdonald, 1986; and also Adrian & Delibes, 1987; Carss et al., 1990; Kemenes & Nechay, 1990; Beja, 1991; Heggberget & Moseid, 1994) or direct observations (e.g. Kruuk, 1995). Secondly, the preferences shown by predators for particular prey can be interpreted in terms of optimal foraging theory (see Krebs & Kacelnik, 1991) in which foraging behaviour is modelled under the constraints of fixed hunting strategies and homogeneous food supply. Such situations are, however, rare in nature and so optimal foraging theories may be inappropriate (Stephens & Krebs, 1986) where prey availability varies significantly in time and/or space. A complementary modelling approach involves 'prey switching' as the availability and profitabilities of prey items change (see Begon *et al.*, 1996) due to predation or environmental factors (e.g. tidal phase, time of day, etc). Prey switching often occurs when different prey occupy different habitats (Begon *et al.*, 1996). Many aspects of otter feeding ecology, in particular prey selection, are currently unresolved (Carss, 1996).

Many predators, including otters, are assumed to take prey species in proportion to their 'availability' in the environment (e.g. for otters, see Mason & Macdonald, 1986; Kruuk, 1995). The accessibility of prey to a predator may be highly variable, changing with such things as time of day, season, habitat, prey type or the feeding motivation of the predator. This makes prey availability difficult, if not impossible, to measure. As a consequence, many predator–prey studies rely on measures of prey abundance as an index of prey availability, assuming a close correlation between the two (for a review, see Hutto, 1990). For otters, very few studies have attempted to study diet and fish abundance simultaneously (see Kruuk & Moorhouse, 1990; Heggberget & Moseid, 1994, for coastal habitats) and, in freshwater habitats, these have been limited in their value because of methodological problems with both spraint analysis and fish sampling.

For instance, Erlinge (1967) investigated the feeding habits of otters in lakes, streams and marshes in southern Sweden, determining diet by spraint analysis and comparing this with concurrent measurements of prey populations (fishes, amphibians, crustaceans, birds). Fish populations (total number and/ or number per session) were determined by electro-fishing but statistical comparisons were not made with the results of spraint analysis. Wise *et al.* (1981) attempted to determine 'the effects of prey availability' on the diet of otters (and mink, *Mustela vison*), as determined by spraint analysis, in a eutrophic lake and a moorland river in south-west England. However, most of the information presented on fish 'availability' (in reality fish abundance) was anecdotal and rigorous comparisons between assessments of otter diet and fish abundance were not made.

The present study tests the hypothesis that otters take prey fishes in proportion to their abundance in the environment. We measured the abundance (density and biomass) of the main prey fishes, brown trout (*Salmo trutta*), Atlantic salmon (*S. salar*), eel (*Anguilla anguilla*), perch (*Perca fluviatilis*) and pike (*Esox lucius*) in two different freshwater habitats (streams and lochs) in north-east Scotland. We also assessed otter diet in the same habitats from

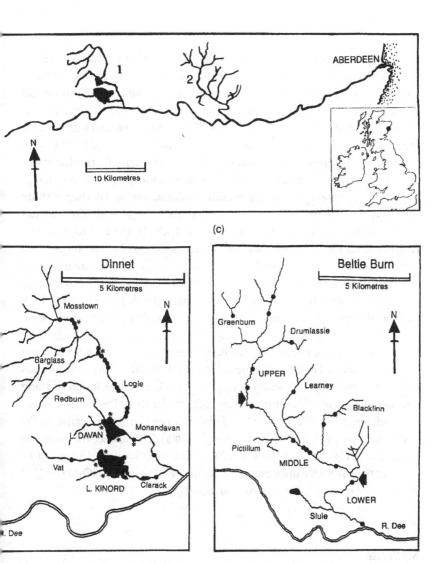

Figure 12.1. (a) The lower 80 km of the River Dee, showing the locations of study areas; (b) the Dinnet catchment; (c) the Beltie Burn catchment. In (b) and (c) (●) shows the electro-fishing sections, and (*) the spraint collection sites. Arrows in (c) show the upper limits of the lower and middle reaches of the Beltie Burn catchment.

spraint analysis by frequency of occurrence, incorporating recent findings which quantify some major errors associated with this method (Carss & Parkinson, 1996). Annual, seasonal and spatial variations in fish abundance and otter diet were assessed and methodological limitations are discussed.

Study areas

The work was undertaken on two major tributaries, and their associated sub-catchments, in the River Dee system in north-east Scotland (Fig. 12.1(a)). These tributaries flow through regions of agriculture, with mixed pasture and cereal farming, natural and commercial coniferous forests and other natural woods including birch (*Betula* spp.), alder (*Alnus glutinosa*) and willow (*Salix* spp.). The River Dee is of national conservation importance and has a unique assemblage of aquatic plants because, unlike most rivers, its trophic status changes little throughout the catchment, beginning as oligotrophic and changing to oligo-mesotrophic at the mouth (Holmes, 1985). The larger study tributary, the 'Dinner catchment', comprises the Logie and Monandavan Burns and their associated streams, as well as Lochs Davan and Kinord (Fig. 12.1(b)). Loch Kinord (Ordnance Survey national grid reference NO 442 995, surface area 82 ha, mean depth 1.52 m) is fed from a small, granite-based moorland sub-catchment to the west and is oligotrophic. Loch Davan (grid reference NJ 442 007, surface area 42 ha, mean depth 1.22 m) is fed by a number of streams, the largest of which, the Logie Burn (ca. 10 km long, max. width 5 m), rises at 570 m above sea level (a.s.l.), drains a relatively large agricultural sub-catchment, carries treated effluent from a nearby village and enters L. Davan at 160 m a.s.l. (grid reference NJ 442 010). The Monandavan Burn (ca. 4 km long, max. width 4 m) flows from L. Davan at 160 m a.s.l. and downstream carries effluent from both lochs to the River Dee at 150 m a.s.l. (grid reference NO 472 989) and is slightly nutrient-enriched relative to the main river and its upper tributaries (Pugh, 1985). The smaller tributary, the Beltie Burn (Fig. 12.1(c)), rises at 320 m a.s.l. and flows ca. 16 km before entering the River Dee upstream of Banchory (grid reference NO 671 964) at 90 m a.s.l. (see description in Kruuk *et al.*, 1993).

Methods

Fish
Species composition and numbers
Fish in streams and lochs were sampled by electro-fishing between June and September (Table 12.1(a)). In the streams, each sampling unit was a measured section, electro-fished either two or three times on any one occasion with a break of approximately 30 min between fish removals (Bohlin *et al.*, 1989). In the lochs each sampling unit was a known area of water bounded by a

Table 12.1.

(a) The location and number of electro-fishing sites sampled during each year of the study

Location	Year						Total
	1989	1990	1992	1993	1994	1995	
A. Dinnet catchment							
Mosstown	—	—	—	—	4	—	4
Logie	—	—	—	—	12	—	12
Monandavan	—	—	—	—	2	—	2
Clarack	—	—	—	—	1	—	1
L. Davan (N)	—	—	7	6	—	3	16
L. Davan (W)	—	—	2	2	—	4	8
L. Davan (S)	—	—	1	2	—	2	5
L. Kinord (W)	—	—	6	5	—	—	11
L. Kinord (N & S)	—	—	1	6	—	—	7
B. Beltie catchment							
Upper	3	3	—	—	—	—	6
Middle	3	3	—	—	—	—	6
Lower	2	2	—	—	—	—	4
C. Totals	8	8	17	21	19	9	82

(b) The location (site or reach) and number of otter spraints collected during each year of the study

Location	Year					Total
	1989	1990	1992	1993	1994	
A. Dinnet catchment						
Mosstown	—	—	0	73	11	84
Logie	—	—	100	127	68	295
Monandavan	—	—	8	35	13	56
Clarack	—	—	46	57	27	130
L. Davan (N)	—	—	100	90	118	308
L. Davan (W)	—	—	47	56	79	182
L. Davan (S)	—	—	88	109	206	403
L. Kinord (W)	—	—	158	112	49	319
L. Kinord (N & S)	—	—	43	38	16	97
B. Beltie catchment						
Upper	26	5	—	—	—	31
Middle	26	21	—	—	—	47
Lower	24	38	—	—	—	62
C. Totals	76	64	590	697	587	2014

rectangular enclosure, made by extending a 60 m seine net (10 mm square mesh) from the shore into the loch and back to the shore, and fished in the same manner as stream sections. Benthic fishes may be under-represented in such catches (Bohlin *et al.*, 1989) and so careful searches for immobilized eels were made during fishing sessions, in an attempt to reduce this potential bias. Both lochs were shallow, particularly in summer, and water depths here were similar to the deeper sections of some of the streams. Based on catches at each site, total fish numbers were calculated by the weighted maximum likelihood model described by Carle & Strub (1978). Sometimes fish catches were very small or the number caught increased between successive removals. On such occasions the actual number of fish caught was taken as the minimum population estimate. The fork lengths (FL) of all fish caught were measured (to the nearest 1 mm) and a subsample of salmonids was weighed (to the nearest 1 g) to produce a series of length : weight relationships (see below). After processing, all fish were returned alive to the water.

Biomass

For each electro-fished area, fish biomass was calculated by multiplying the estimated total number of fish by the average weight of those caught. For some fishes, length : weight relationships were taken from the literature: eel (Carss & Elston, 1996), perch (Craig, 1974), pike (Frost & Kipling, 1967). For salmonids, weight was estimated by the following equations, based on fish caught in the study area:

June	brown trout	$\text{Log } W = -5.05 + 3.07 \text{ Log FL}$	$n = 56,\ r^2 = 0.99$
	Atlantic salmon	$\text{Log } W = -4.97 + 3.04 \text{ Log FL}$	$n = 48,\ r^2 = 0.99$
September	both species	$\text{Log } W = -5.00 + 3.02 \text{ Log FL}$	$n = 142,\ r^2 = 0.99$

where W is weight (g) and FL is fork length (mm). Catches of salmonids could be separated into age classes from the modes of length–frequency histograms. Kruuk *et al.* (1993) found that otters on the Beltie Burn ate salmonids mostly longer than 50 mm, appearing to ignore the smaller (0+ age class) fry. Moreover, Carss & Elston (1996) demonstrated that the remains of some of the smallest (i.e. < 40 mm) salmonids were not recovered in otter spraints. Thus in the present study, values for 0+ fry were excluded from calculations of small salmonid density and biomass. Eels and perch could not be separated easily like this and so all age classes were combined. A total of 10 'large' pike (FL range: 315–755 mm) were caught within enclosures in the lochs, either singly (eight occasions) or two together (one occasion). These fish were scarce (6% of all pike caught) but, perhaps more importantly, their inclusion in biomass calculations had a large influence. They had an average weight of 1396 g (range

224–3333 g) compared to the average weight of 45 g for the remaining ($n = 152$) pike. We therefore decided to exclude them from calculations of prey biomass. This is an example of the patchily-distributed prey problem that we highlight in the discussion.

At the end of this process a biomass value was determined for each fish species caught at each of 82 sample locations. The total biomass, of fishes within the size-range of otter prey, at each location was then calculated by summing these species-specific biomass estimates. Some samples (e.g. Moss-town Burn, $n = 4$ locations) were combined to produce average biomass values for a particular 'site', others, on the Beltie Burn (e.g. upper, $n = 6$), were combined in the same way for a particular 'reach' i.e. upper, middle and lower (Fig. 12.1(c)).

Otter diet

The composition of otter diet was assessed from spraints collected monthly in both catchments (Table 12.1(b)). Within the Dinnet catchment, spraint collection was restricted to 10 defined areas (Fig. 12.1(b)), samples from two of these (Loch Kinord north and south, both rocky shores) were combined for all subsequent analyses, to give samples from a total of nine separate sites (the same as those for electro-fishing samples). Collection sites in the Beltie Burn catchment were not restricted to particular areas and spraints were collected (from stretches close to roads, bridges and the confluence of streams) at a total of 43 sites. Spraints were later combined, according to location, into samples from the three reaches (see above).

Spraints were frozen immediately after collection and later thawed for preparation and analysis (see Carss & Parkinson, 1996). They were soaked in a saturated solution of biological washing powder for approximately 48 h at room temperature, then rinsed through two sieves (1.0 and 0.5 mm) and dried at room temperature for 24–48 h before examination.

Fish remains were identified to species by examining undigested vertebrae and comparing them with a reference collection of fish caught locally. Further-more, for spraints collected along the Beltie Burn, which has an annual run of spawning salmon and sea trout, it was possible to categorize salmonid remains into those from 'large' (> 30 cm) and 'small' (i.e. 4–30 cm, see above) fish (see Carss et al., 1990). Other prey groups (amphibian, bird and mammal) were easily identified from bones, feathers and fur, respectively. All identifiable parts in the spraints were registered (i.e. presence or absence), and results expressed in terms of percentage frequency (the percentage of spraints con-taining a particular item). Carss & Parkinson (1996) tested the validity of this technique by feeding trials involving captive, tame otters and computer simulation of various spraint sub-sampling regimes. They concluded that it was not possible to quantify otter diet accurately by frequency-of-occurrence methods but that the resulting estimated mean values could be used to determine the rank order of prey groups consumed. These authors also demonstrated that perch scales continued to be recorded in spraints up to 10 days after ingestion while most, if not all, other hard parts were expelled within 24 h. Subsequent spraint analysis thus resulted in a large over-estimation of the amount of perch consumed. In an attempt to reduce such over-estimation, those records of perch from spraints containing only their scales, but no other hard parts, were excluded from the frequency-of-occurrence analyses.

Carss & Parkinson (1996) also found that for the most commonly recorded prey groups in spraints (> 10% mean frequency), the coefficients of variation (standard deviation as a percentage of the mean) around the estimated means were less than 50%, even for relatively small samples of spraints (e.g. < 25). In the present study some spraint samples were quite small, but scarcer items were excluded from analyses that concentrated on the most commonly recorded fishes in the diet of otters. Thus it is likely that our analyses gave an accurate determination of the rank order of the main prey consumed.

Statistical analyses
Otter diet
Occurrence of prey remains in spraints was analysed by fitting a generalized linear model, with binomial (i.e. presence/absence of prey species, see above) errors, to the data (see McCullagh & Nedler, 1989; Trexler & Travis, 1993). Several factors were used to explain the variation in probabilities of remains from particular prey being present. These were year (1992, 1993, 1994), season (winter: December–February; spring: March–May; summer: June–August; autumn: September–November), and site (see Table 12.1(b)) for the Dinnet catchment, and season and reach (upper, middle and lower) for the Beltie catchment. The importance of each of these factors was determined by the corresponding analysis of deviance table. These results are presented by expressing the deviance explained by each factor relative to the total deviance of the data. The significance of explanatory factors was assessed by taking the ratio of their mean deviance to the residual mean deviance (RMD), and comparing this with the appropriate *F* distribution value. All calculations were carried out using the statistical package Genstat 5.

Assessments of otter diet in relation to those of fish abundance
Assessments of otter diet (i.e. percentage occurrence values for each fish species) were compared with the percentage of the total estimated biomass

values calculated from the analysis of electro-fishing catches. These measures were chosen because Carss & Parkinson (1996) found that percentage occurrence estimates from spraint analysis were best fitted to the biomass, rather than the number, of prey ingested.

Results

Fish communities

Within the Dinnet catchment, the fish communities of both lochs comprised mostly eel, perch and pike, although brook lamprey (*Lampetra planeri*) was occasionally caught in Loch Davan. In the streams the fish community was dominated by salmonids (brown trout and a few salmon) and eel, although brook lamprey, perch and pike were present in small numbers. In the Beltie Burn catchment, the fish community comprised trout, salmon and eel, and small numbers of brook lamprey, three-spined stickleback (*Gasterosteus aculeatus*), and a single perch were also caught. At both lochs, most of the estimated fish biomass comprised eel, pike and perch, and most in the Dinnet and Beltie burns consisted of salmonids and eel (Fig. 12.2(a)). Fishes that contributed < 0.2% to overall biomass were excluded.

Assessments of otter diet

Analyses were based on 2014 spraints (Table 12.2). Those from both lochs contained mostly eel, followed by perch, pike and small salmonids. Small salmonids were the most frequently recorded items in spraints from the Dinnet burns, followed equally by eel and perch, then pike. Spraints from the Beltie Burn were dominated by small salmonids, followed by large salmonids and eel, while a few also contained perch (see Fig. 12.2(b)). However, assessments of otter diet varied with year, season and location (site or reach).

Within the Dinnet catchment there was evidence of significant year effects for both salmonid and pike remains in spraints and also significant interactions between year and season for all four prey items (Table 12.3(a)). However, these effects (year, plus interactions with season and site) were relatively small and explained a total of no more than 15% of the total deviance, whereas site alone explained between 47% and 81% of total deviance. The site factors explained the largest proportion of the deviance for both eel and salmonids (Fig. 12.3(a) and 12.3(b): 81% and 67%, respectively), with season explaining relatively little (4% and 3%, respectively). In general, eel remains predominated in spraints from the lochs and salmonid remains predominated in those

(a)

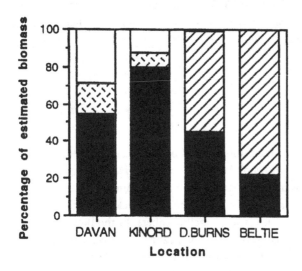

(b)

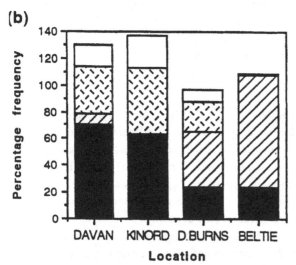

Figure 12.2. Fish communities (i.e. the proportion of estimated total biomass for each species caught) in lochs and streams, as determined by (a) electro-fishing and (b) analysis of spraints from the same areas by percentage frequency; values are independent and thus totals do not constitute 100%. □, pike; ⊠, perch; ▨, salmonids; ■, eel.

from the streams, throughout the year. For perch and pike, although the largest proportion of deviance was again explained by site (Fig. 12.3(c) and 12.3(d): 47% and 48%, respectively) with most remains occurring in spraints from the lochs, there was a larger seasonal effect with proportions generally being lowest in summer and highest in winter; season explained 25% and 19% of the deviances, respectively (Table 12.3).

Table 12.2. *Overall assessment of otter diet from spraint analysis*

Prey	L. Davan	L. Kinord	Dinnet burns	Beltie Burn
Large (> 30 cm) salmonids	—	—	—	32.5
Small (< 30 cm) salmonids	9.0	0.9	41.1	83.5
Eel	67.9	62.3	23.9	23.5
Perch	35.0	49.8	23.4	0.7
Pike	16.1	24.5	8.6	—
No. spraints (= 100%)	3779	1582	2367	425

Figures for each prey type are percentage frequencies (no. spraints = 100%); these are independent (i.e. several prey types may occur in the same spraint) and so do not add up to 100%.

Spraint samples from the Beltie Burn catchment were too small for monthly comparisons and so they were categorized by season and reach (see Fig. 12.1(c)). For small salmonids, analysis of deviance showed that reach explained more deviations than season (40% compared to 15%, respectively) but neither was significant (Table 12.3(b)). Most of the deviance for large salmonids and eel was explained by season, while reach also had a significant effect for eel. Thus, most spraints from each reach in every season contained the remains of small salmonids (Fig. 12.4(a)), proportions of spraints containing the remains of these fishes being unaffected by either season or location. Remains of large salmonids and eel were highly seasonal in otter spraints from this area, large salmonids occurring mostly in autumn and winter (Fig. 12.4(b)) and eels in spring and summer (Fig. 12.4(c)). The proportion of spraints containing eel remains was also influenced by location within the catchment, most being recorded from the lower reaches in every season (Fig. 12.4(c)).

Assessments of otter diet in relation to those of fish abundance

For eel, there was a strong positive correlation ($r^2 = 0.66$, $t = 6.22$, d.f. = 18, $p < 0.001$) between the percentage of estimated biomass in the environment that comprised eel, and the frequency of eel remains in spraints (Fig. 12.5(a)). The relationship was apparently a straight line but it did not pass through the origin. Similar relationships also appeared to hold within stream habitats (Fig. 12.5(a), $r^2 = 0.65$, $t = 4.16$, d.f. = 8, $p = 0.003$) while that for loch habitats was close to significance ($r^2 = 0.32$, $t = 2.27$, d.f. = 8, $p = 0.053$). There was also a strong positive correlation ($r^2 = 0.87$, $t = 11.23$, d.f. = 18, $p < 0.001$) between the percentage of estimated biomass in the environment that comprised

Table 12.3. *Spraint analysis for the Dinnet and Beltie Burn catchments, showing the relative importance of various factors in determining the presence of particular prey remains in a spraint*

(a) Dinnet

Factor	d.f.	Prey			
		Eel	Salmonids	Perch	Pike
Season	3	4.5	3.6	25.4	19.2
Site	8	81.0	67.4	46.8	47.8
Combined	—	84.5	71.0	72.2	67.0
Season × Site	24	(4.3)	(5.3)	(6.0)	12.7
Year	2.0	(0)	2.2	(0.8)	3.0
Year × Season	6.0	2.7	7.6	4.5	3.6
Year × Site	16.0	(2.2)	5.2	(4.7)	5.6
Combined	—	4.9	15.0	10.0	12.2
Residual	105.0	4.9	8.6	11.8	7.9

(b) Beltie Burn

Factor	d.f.	Prey		
		Salmonids		Eel
		Large	Small	
Season	3	82.8	(15.0)	77.0
Reach	2	7.8	(39.5)	17.8
Residual	6	9.3	45.2	5.1

Figures are percentages of the total deviance of the data that are explained by each factor; those in parenthesis are not significant at $p = 0.05$.

salmonids, and the frequency of small salmonid remains in spraints (Fig. 12.5(b)). Again, this was best summarized as a straight-line relationship that did not pass through the origin but the data set was essentially two clusters of points representing low and high salmonid biomass, with only one intermediate point. Salmonids were not caught by electro-fishing in the lochs but the relationship still held within the stream habitat alone (Fig. 12.5(b), $r^2 = 0.65$, $t = 4.2$, d.f. $= 8$, $p = 0.03$), after exclusion of the loch data.

For perch and pike, which were only caught by electro-fishing in the lochs, there were no relationships between percentage of estimated biomass and percentage frequency of remains in spraints (Fig. 12.5(c), $t = 0.44$, d.f. = 38, NS; Fig. 12.5(d), $t = 1.14$, d.f. = 38, NS, respectively). However, the range of values for these fishes was smaller than those for eel and salmonid, with no more than 25% frequency in spraints and few biomass estimates over 50%.

Discussion

The present study was an investigation into the relationship between electro-fishing samples and the contents of otter spraints from the same area. There were almost certainly biases in both sets of data (demonstrably so for the latter, see Carss & Parkinson, 1996) but the techniques used were the most rigorous available and were unlikely to have affected the overall conclusions of the study.

Fish communities

Population estimates for salmonids in small streams are generally thought to be accurate, particularly if, as in the present study, standard electro-fishing techniques are used (Bohlin *et al.*, 1989). These authors also concluded that electro-fishing could be used to sample fish in the littoral zone of lakes, although in open water areas of depths greater than ca. 1 m it was generally unsuccessful, probably because of the flight reactions of the fishes and particularly when water transparency was poor. In the present study, littoral sampling was sometimes undertaken in water deeper than 1 m (max. 1.5 m) but it was conducted within a fine-meshed net enclosure, thus eliminating the problem of fish evading capture by flight. The probability of capture here was also increased because the water was exceptionally clear and the bottom substrate could be seen at all times.

Electro-fishing catches showed the fish communities in streams and lochs to comprise mostly salmonids/eels, and eels/perch/pike, respectively. In general this is typical of the distribution of these fishes in Britain, since here, and more so in Scotland, the freshwater fish fauna is impoverished compared with that of continental Europe (for a major review see Maitland & Campbell, 1992). Fast-flowing moorland streams in southern Britain, and most riverine habitats in Scotland, are cool and nutrient-poor and are dominated by salmonids. Lowland water bodies, particularly in southern Britain, tend to be eutrophic and have relatively high summer water temperatures, and are often dominated by cyprinid fishes (see also Varley, 1967; Winfield & Nelson, 1991). Cyprinids

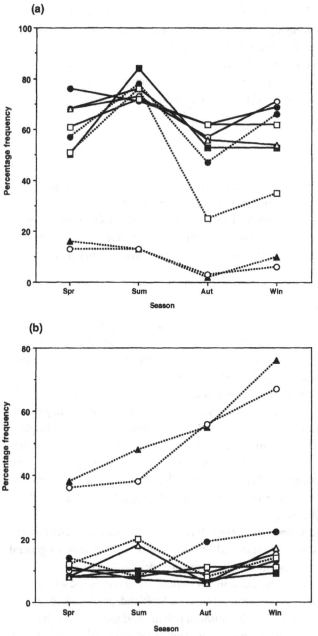

Figure 12.3. The percentage frequency of spraints containing remains of (a) eel, (b) small salmonids, (c) perch, and (d) pike in relation to seasons and site within the Dinnet catchment. Key to Lochs: -○-, Davan, north; -●-, Davan, west; -□-, Davan, south; -■-, Kinord, north & south; -△-, Kinord, west. Key to streams: -▲-, Mosstown; -○--, Logie; --●--, Monandavan; --□--, Clarack.

(c)

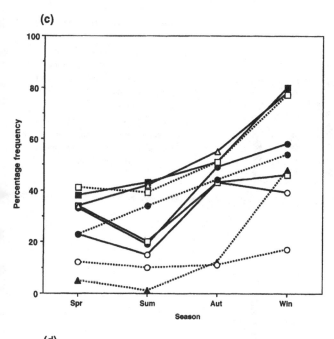

(d)

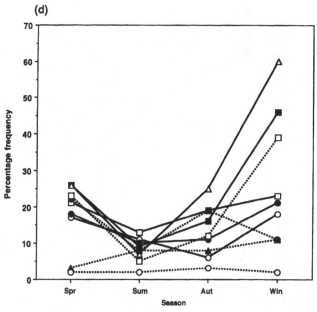

(a)

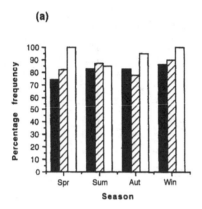

(b)

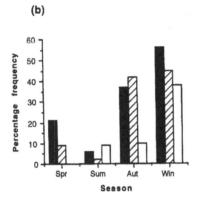

(c)

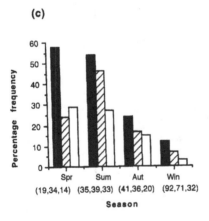

Figure 12.4. The percentage frequency of spraints containing remains of (a) small salmonids (< 30 cm); (b) large salmonids (> 30 cm) and (c) eel in relation to season and reach within the Beltie Burn catchment. Figures in parenthesis are total numbers of spraints analysed in each season from the (■) lower, (▨) middle and (□) upper reach, respectively

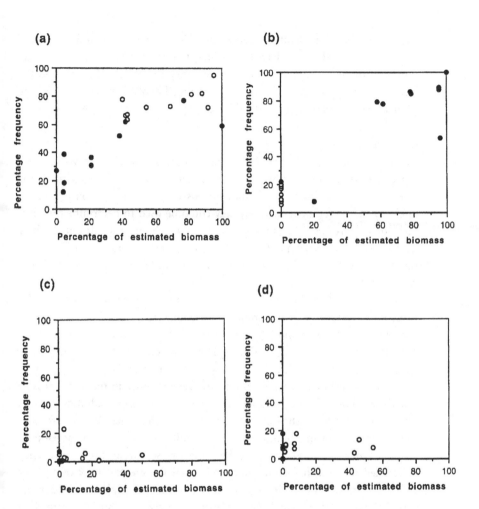

Figure 12.5. Relationships between the percentage of estimated fish biomass and frequency of occurrence of remains in spraints for (a) eel, (b) salmonids, (c) perch and (d) pike. Data from specific sites or reaches within the Dinnet and Beltie Burn catchments, respectively. ○, lochs; ●, streams.

are scarce in Scotland, particularly north of the Forth–Clyde canal (Mills, 1969; Maitland, 1972) and only the minnow (*Phoxinus phoxinus*) is widely distributed. Eels are widespread in all freshwater habitats throughout Britain, including all of Scotland. Little is known about their habitat preferences here, but on the basis of commercial harvesting (see Maitland, 1994), it appears that they are more abundant in lochs than in rivers and streams. Apart from salmonids, minnow and three-spined stickleback, perch and pike are the most widely distributed fishes in Scottish freshwaters. These species are indigenous to south-eastern England but have been widely redistributed by man throughout much of Britain, including Scotland, though they do not occur in the far north. Both fishes occur in lakes and slow-flowing stretches of streams and rivers, but as such rivers are scarce in Scotland, most populations are in lochs.

Otter diet

There are a number of potential errors associated with frequency-of-occurrence analyses of otter spraints (see discussion in Carss & Parkinson, 1996). These authors concluded that it was not possible to quantify otter diet accurately by these methods; however, they could be used to determine the rank order of prey groups. Thus, in the present study, although the calculated values for frequency-of-occurrence for items in otter spraints almost certainly do not represent the true proportions eaten, the rank order of items was probably close to reality.

In the present study, site was the most important factor influencing the presence of specific prey in spraints. Those collected around the lochs contained more eel, perch and pike remains than those from the streams, which contained mostly salmonids. Although in some cases differences in spraint analysis methods preclude direct comparisons, this finding endorses those from similar habitats elsewhere and from previous work in the same areas in north-east Scotland. For example, Wise *et al.* (1981) examined otter spraints from Slapton Ley, a shallow eutrophic lagoon, and the Rivers Dart and Webburn, fast-flowing moorland rivers, in south-west England by developing a bulk estimate method involving the subjective scoring of prey items. They concluded that roach (*Rutilus rutilus*) formed the greatest bulk of all prey remains from spraints at the lagoon (45.4% of the diet [sic]), followed by eel (26.6%), perch (10.6%) and pike (9.2%); in the rivers, salmonids comprised 59.1% of the overall diet, eel 16.2%, and bullhead (*Cottus gobio*) 5.8%. In another study in south-west England, Chanin (1981) examined spraints by relative frequency (no. items = 100%) and concluded that on the River Teign, salmonids formed the staple diet (60.1% of 353 items) with eel an important secondary item (29.5%), and at Slapton Ley (cf. Wise *et al.* (1981) above) cyprinid remains were most common (32.1% of 607 items), followed by perch (28.2%), eel (23.2%), and pike (6.3%).

In north-east Scotland, Jenkins *et al.* (1979) analysed (by percentage frequency: i.e. no. spraints = 100%) spraints collected from the Dinnet lochs in 1975 and 1976. They found that the main item in spraints was eel (almost always > 85% in each month), followed by perch (< 50% of spraints each month), pike (7%), and salmonids (3%). This study was extended in 1976–

1978 to include a stretch of the main stem of the River Dee and three tributaries in mid-Deeside as well as the Dinnet lochs (Jenkins & Harper, 1980). These authors found that eel was again the main prey of otters at the lochs, remains being found in over 85% of spraints (and always > 43%) in most months, followed by perch, pike and salmonids. In the River Dee, however, salmonids predominated in spraints (usually > 65%), followed by eel (0–70% depending on season, see later). Smaller samples were available from the tributaries but they showed that salmonids were most commonly recorded, followed by eel.

There were also seasonal patterns in the occurrence of items in spraints collected during the present study. On average, proportions of spraints containing eel remains were highest in the summer and often lowest in the autumn, whilst those containing salmonids increased throughout the year from a low in spring to a maximum in winter. These findings were similar to those of previous studies (e.g. Jenkins *et al.*, 1979; Jenkins & Harper, 1980; Chanin, 1981; Wise *et al.*, 1981). There were also significant seasonal patterns for perch and pike remains in otter spraints. Similarities between the various studies, both spatially (i.e. south-west England and north-east Scotland) and temporally (i.e. north-east Scotland 1975–1978 and 1992–1994) suggest that otter diet does indeed change seasonally, presumably as a response to changes in fish availability. This must remain speculative, however, because year-round fish abundance data were not available from the present study. It was, however, possible to compare fish abundance and otter diet from samples collected concurrently during the summer.

Otter prey selection

In order to investigate prey selection, one must be confident that estimates of predator 'diet' and prey 'availability' are robust. It did seem likely that this was the case for assessments of otter diet from spraints (see above). There was, however, some discrepancy between the prey anticipated from certain habitats and the remains recorded in spraints collected there. This was presumably a result of otters eating an item caught in one habitat and then sprainting in another. The scope for such 'transfer' is obvious, as otters are known to travel considerable distances, often within short periods. For instance Kruuk *et al.* (1993) calculated that the mean length of river or stream used by otters in north-east Scotland was 34.8 km for males ($n = 6$) and 20.0 km for females ($n = 2$). Moreover, the three otters radio-tracked by Durbin (1993) travelled nightly, on average, 8–13 km (SD = 3–5 km, respectively) of stream within the catchment of the River Don in Aberdeenshire. If remains of prey caught and eaten in one habitat appear in spraints left in a different habitat, as demonstrated in the present study, we would not expect graphs of 'diet against prey biomass' to pass through the origin. It would have thus been erroneous and misleading to force such regressions through the origin, and this may often be the case. It is also not clear whether these relationships are truly linear: it is possible that they are asymptotic curves, not necessarily reaching maximum values of 100% frequency in spraints because of the 'transfer' process discussed above. Unfortunately, within site types, our data are insufficient to clearly distinguish these alternatives.

Electro-fishing catches showed that the fish communities in lochs and streams were different. Salmonids, and to a lesser extent eels, predominated in the streams, whereas eels, followed by perch and pike, did so in the lochs. Similar trends were apparent in the remains of prey fishes recovered in spraints, supporting the hypothesis that otters take the most abundant fish species in a particular habitat. However, care is needed in interpreting electro-fishing data. Here it was assumed that prey fishes were distributed evenly within habitats (i.e. lochs and/or streams), or that they were essentially a patchless resource. This appeared to be a valid assumption for salmonids in streams and for eels in the lochs and streams, because these fishes occurred in almost every electro-fishing sample. The positive correlations found between the contribution a particular fish species (salmonids or eel) made to the estimated total fish biomass and the proportion of spraints containing its remains showed that, within these habitats, otters did take fishes in proportion to their estimated abundance there.

Problems could arise in interpreting electro-fishing data if prey were patchily distributed, for instance if they occurred in shoals or were confined to specific microhabitats. In such cases, estimates of 'availability' would be dependent on the siting and timing of electro-fishing samples, which by chance may hit (or miss) these patches/shoals. The 'patchless resource hypothesis' may have been invalid for perch and pike as there were many records of either no fish or 'presence' (i.e. very low catches) from electro-fishing catches. Most of the perch and pike caught by electro-fishing were small (D.N. Carss & K.C. Nelson, unpublished data) and such young perch may be expected to form shoals, with older fish becoming more solitary (Bruylants et al., 1986), at least during the summer (cf. Johnson & Evans, 1991). Similarly, small pike may aggregate in dense aquatic vegetation (e.g. Wright, 1990), larger individuals tending to be solitary and found in more open water (Chapman & MacKay, 1984). The presence of aggregations of small perch and pike could explain why no relationship was found between the proportion of estimated biomass and the frequency of occurrence in spraints for these fishes, especially if otters are readily able to find such aggregations. An alternative explanation could be that these fishes are taken less frequently during the summer (the season of comparison) than at other times (e.g. winter) and that summer may not have been the best season within which to investigate otter diet/prey availability relationships for perch and pike.

Having found a relationship between otter diet and a rigorous index of prey availability, at least for the most commonly-taken fishes during the summer, it is tempting to claim that otters show no selection for prey species. Indeed, data from the present study support the hypothesis that otters take fish in accordance with their abundance *in any one habitat*, and any possible selection effects are so small that they do not affect the rank order of prey remains in spraints. However, it is possible that otters do select for prey species, by electing to forage in habitats with particular, predictable, fish communities. For salmonid (brown trout and salmon) communities it is possible to search for direct evidence of species-selection because biasses in population sampling and spraint analysis are exactly the same for each species (see Bohlin *et al.*, 1989; Carss & Elston, 1996). Moreover, rigorous techniques are now available to investigate possible size-selection by otters for trout, salmon and eels (Carss & Elston, 1986).

ACKNOWLEDGEMENTS

We are grateful to the many riparian owners who kindly allowed us access to their land, particularly Scottish Natural Heritage for permission to work at Dinnet NNR. Our special thanks to Jim Parkin, Warden at Dinnet, for his great interest, help and encouragement throughout our work there. Intensive field and laboratory work was made possible by the cheerful hard work of many people. L. A. Carss, K. Duncan, G. Evans, M. McCann, R. McDonald, G. Olsthoorn, S. Racey, D. Riddell, J. Rook, A. Webb and Stuart Bell assisted with electro-fishing, E. Bacon with spraint collection and analysis and S. Black, S. From, L. Hannaford-Hill, K. Marshall and G. Olsthoorn with spraint analysis. Particular thanks must go to H. Visser for much field, laboratory and computing help and to D. A. Elston (Biomathematics and Statistics Scotland at the Macaulay Land Use Research Institute, Aberdeen) who provided valuable statistical and computing advice. Robert Moss and Lorna Brown made many valuable comments on the manuscript.

References

Adrian, M. I. & Delibes, M. (1987). Food habits of the otter (*Lutra lutra*) in two habitats of the Doñana National Park, SW Spain. *J. Zool., Lond.* **212**: 399–406.

Begon, M., Harper, J. L. & Townsend, C. R. (1996). *Ecology*. (3rd edn). Blackwell Scientific, Oxford.

Beja, P. R. (1991). Diet of otters (*Lutra lutra*) in closely associated freshwater, brackish and marine habitats in south-west Portugal. *J. Zool., Lond.* **225**: 141–152.

Bohlin, T., Hamrin, S., Heggberget, T. G., Rasmussen, G. & Saltveit, S. J. (1989). Electrofishing – theory and practice with special emphasis on salmonids. *Hydrobiologia* **173**: 9–43.

Bruylants, B., Vandelannoote, A. & Verheyen, R. (1986). The movement pattern and density distribution of perch, *Perca fluviatilis* L., in a channelized lowland river. *Aquacult. Fish. Mgmt* **17**: 49–57.

Carle, F. L. & Strub, M. R. (1978). A new method for estimating population size from removal data. *Biometrics* **34**: 621–630.

Carss, D. N. (1996). Foraging behaviour and feeding ecology of the otter *Lutra lutra*: a selective review. *Hystrix* **7**: 179–194.

Carss, D. N. & Elston, D. A. (1996). Errors associated with otter *Lutra lutra* faecal analysis. II. Estimating prey size distribution from bones recovered in spraints. *J. Zool., Lond.* **238**: 319–332.

Carss, D. N. & Parkinson, S. G. (1996). Errors associated with otter *Lutra lutra* faecal analysis. I. Assessing general diet from spraints. *J. Zool., Lond.* **238**: 301–317.

Carss, D. N., Kruuk, H. & Conroy, J. W. H. (1990). Predation on adult Atlantic salmon, *Salmo salar* L., by otters, *Lutra lutra* (L.), within the River Dee system, Aberdeenshire, Scotland. *J. Fish Biol.* **37**: 935–944.

Chanin, P. R. F. (1981). The diet of the otter and its relations with the feral mink in two areas of south-west England. *Acta theriol.* **26**: 83–95.

Chapman, C. A. & MacKay, W. C. (1984). Versatility in habitat use by a top aquatic predator, *Esox lucius* L. *J. Fish Biol.* **25**: 109–115.

Craig, J. F. (1974). Population dynamics of perch, *Perca fluviatilis* L., in Slapton Ley, Devon. I. Trapping behaviour, reproduction, migration, population estimates, mortality and food. *Freshwat. Biol.* **4**: 417–431.

Durbin, L. S. (1993). *Food and Habitat Utilization of Otters (Lutra lutra L.) in a Riparian Habitat.* PhD thesis: University of Aberdeen.

Erlinge, S. (1967). Food habits of the fish-otter, *Lutra lutra* L., in south Swedish habitats. *Viltrevy, Stokh.* **4**: 371–443.

Foster-Turley, P., Macdonald, S. M. & Mason, C. F. (Eds) (1990). *Otters: an action plan for their conservation.* IUCN/ SSC Otter Specialist Group, Gland.

Frost, W. E. & Kipling, C. (1967). A study of reproduction, early life, weight-length relationship and growth of pike, *Esox lucius* L., in Windermere. *J. Anim. Ecol.* **36**: 651–693.

Heggberget, T. M. & Moseid, K.-E. (1994). Prey selection in coastal Eurasian otters *Lutra lutra*. *Ecography* **17**: 331–338.

Holmes, N. T. H. (1985). Vegetation of the River Dee. In *The biology and management of the River Dee*: 42–55. (Ed. Jenkins, D.). ITE/NERC Publication, HMSO, London.

Hutto, R. L. (1990). Measuring the availability of food resources. *Stud. avian Biol. No.* **13**: 20–28.

Jenkins, D. & Harper, R. J. (1980). Ecology of otters in northern Scotland. II. Analyses of otter (*Lutra lutra*) and mink (*Mustela vison*) faeces from Deeside, Scotland in 1977–78. *J. Anim. Ecol.* **49**: 737–754.

Jenkins, D., Walker, J. G. K. & McCowan, D. (1979). Analyses of otter (*Lutra lutra*) faeces from Deeside, N. E. Scotland. *J. Zool., Lond.* **187**: 235–244.

JNCC (1996). *A framework for otter conservation in the UK: 1995–2000.* Joint Nature Conservation Committee, Peterborough.

Johnson, T. B. & Evans, D. O. (1991). Behaviour, energetics, and associated mortality of young-of-the-year white perch (*Morone americana*) and yellow perch (*Perca flavescens*) under simulated winter conditions. *Can. J. Fish. aquat. Sci.* **48**: 672–680.

Kemenes, I. & Nechay, G. (1990). The food of otters *Lutra lutra* in different habitats in Hungary. *Acta theriol.* **35**: 17–24.

Krebs, J. R. & Kacelnik, A. (1991). Decision-making. In *Behavioural ecology: an evolutionary approach* (3rd edn): 105–136. (Eds Krebs, J. R. & Davies, N. B.). Blackwell Scientific, Oxford.

Kruuk, H. (1995). *Wild otters: predation and populations.* Oxford University Press, Oxford.

Kruuk, H. & Moorhouse, A. (1990). Seasonal and spatial differences in food selection by otters (*Lutra lutra*) in Shetland. *J. Zool., Lond.* **221**: 621–637.

Kruuk, H., Carss, D. N., Conroy, J. W. H. & Durbin, L. (1993). Otter (*Lutra lutra* L.) numbers and fish productivity in rivers in north-east Scotland. *Symp. zool. Soc. Lond.* No. 65: 171–191.

Maitland, P. S. (1972). A key to the freshwater fishes of the British Isles. *Scient. Publs. freshwat. biol. Ass.* No. 27: 1–139.

Maitland, P. S. (1994). Fish. In *The fresh waters of Scotland*: 191–208. (Eds Maitland, P. S., Boon, P. J. & McLusky, D. S.). Wiley, Chichester.

Maitland, P. S. & Campbell, R. N. (1992). *Freshwater fishes of the British Isles.* Harper Collins, London.

Mason, C. F. & Macdonald, S. M. (1986). *Otters: ecology and conservation.* Cambridge University Press, Cambridge.

McCullagh, P. & Nedler, J. A. (1989). *Generalized linear models.* Chapman & Hall, London.

Mills, D. H. (1969). The growth and population densities of roach in some Scottish waters. *Proc. Br. coarse Fish Conf.* **4**: 50–57.

Pugh, K. B. (1985). The chemistry of the river system. In *The biology and management of the River Dee*: 34–41. (Ed. Jenkins,

D.). ITE/NERC Publication, HMSO, London.

Stephens, D. W. & Krebs, J. R. (1986). *Foraging theory.* Princeton University Press, Princeton.

Trexler, J. C. & Travis, J. (1993). Nontraditional regression analyses. *Ecology* **74**: 1629–1637.

Varley, M. E. (1967). *British freshwater fishes: factors affecting their distribution.* Fishing News (Books), London.

Winfield, I. J. Nelson, J. S. (1991). *Cyprinid fishes: systematics, biology and exploitation.* Chapman & Hall, London.

Wise, M. H., Linn, I. J. & Kennedy, C. R. (1981). A comparison of the feeding biology of mink *Mustela vison* and otter *Lutra lutra. J. Zool., Lond.* **195**: 181–213.

Wright, R. M. (1990). The population biology of pike, *Esox lucius* L., in two gravel pit lakes, with special reference to early life history. *J. Fish Biol.* **36**: 215–229.

13

Diet, foraging behaviour and coexistence of African otters and the water mongoose

D. T. Rowe-Rowe and M. J. Somers

Introduction

In Africa the Eurasian otter, *Lutra lutra*, occurs in streams flowing from the Atlas Mountains in the three north African countries of Morocco, Algeria, and Tunisia. There are no otters in the arid Sahara region. Three endemic otters occur in the areas of sub-Saharan Africa that receive an annual rainfall in excess of about 500 mm. They are the Cape clawless otter, *Aonyx capensis*, the Congo clawless otter, *Aonyx congica*, and the spotted-necked otter, *Lutra maculicollis*. Occurring in all of the habitats in which the otters have been recorded is a fourth endemic amphibious carnivore, the water mongoose, *Atilax paludinosus* (Fig. 13.1). In this chapter we will not deal with the Eurasian otter. In order to synthesize much recent work on African otters and water mongooses, particularly in southern Africa, a review concerning diet, foraging behaviour and coexistence of African otters and the water mongoose is presented in this chapter.

Very little information is available on *Aonyx congica*. It occurs in rain forests and lowland swamp forests of the Congo River Basin, as well as the forests and wetland areas of Rwanda, Burundi and south-western Uganda (Rowe-Rowe, 1990). Pygmies in the Ituri Forest of Zaire stated that *A. congica* lives on fish and crabs (Capaneto & Germi, 1989), while Baranga (1995) associated the distribution of this otter with habitats in which there was an abundance of freshwater crabs, giant earthworms, mudfish and clawed toads (*Xenopus* sp.). Kingdon (1977) suggested that it is dependent on 'worms, insects, molluscs, crustaceans, and amphibians'. Owing to the paucity of data on *A. congica*, we will deal with it only superficially, and will concentrate on the two other otters, *Aonyx capensis* and *Lutra maculicollis*, as well as the water mongoose, in freshwater habitats. While reviewing this information we test three hypotheses: (i) the species differ in their prey selection; (ii) they have different foraging behaviours; and (iii) they select different macrohabitats.

Both the Cape clawless otter and the water mongoose also forage in marine or estuarine habitats where their diets have been studied (Whitfield & Blaber,

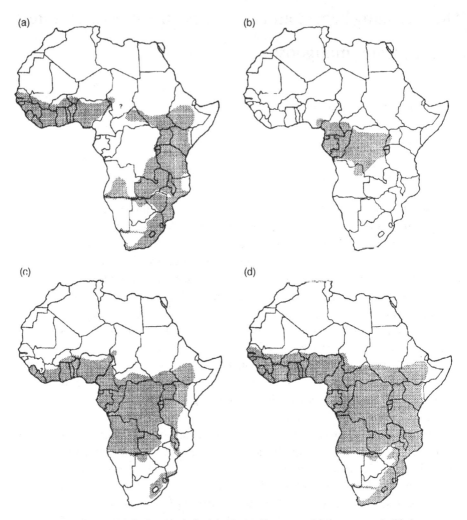

Figure 13.1. Distribution (shaded) of the three African otters (a) *Aonyx capensis*, (b) *Aonyx congica* and (c) *Lutra maculicollis* (from Rowe-Rowe, 1990), and (d) the water mongoose, *Atilax paludinosus* (based on Skinner & Smithers, 1990).

1980; Van der Zee, 1981; Arden-Clarke, 1983; Louw & Nel, 1986; MacDonald & Nel, 1986; Verwoerd, 1987).

Diet

Data were obtained from the results of studies in South Africa (in the provinces of KwaZulu-Natal, Eastern Cape, Free State and Western Cape), Zimbabwe,

Table 13.1. *Sources of information of data used to compile Tables 13.2 to 13.4 on the diets of* Aonyx capensis, Lutra maculicollis *and* Atilax paludinosus *at various freshwater localties*

Species	Locality		n	Reference
A. capensis	KwaZulu-Natal		1361	Rowe-Rowe (1977a)
	Free State		34	Purves & Sachse (1998)
	Eastern Cape		66	Somers & Purves (1966)
	Western Cape:	Clanwilliam	132	Purves et al. (1994)
		Montagu	105	Ligthart et al. (1994)
	Zimbabwe		255	Butler & du Toit (1994)
L. maculicollis	KwaZulu-Natal		294	Rowe-Rowe (1977a)
	Eastern Cape		79	Somers & Purves (1996)
	Rwanda:	Lake Muhazi	150	Lejeune (1990)
	Tanzania:	Lake Victoria	61	Kruuk & Goudswaard (1990)
At. paludinosus	KwaZulu-Natal: highland		210	Rowe-Rowe (1977a)
		lowland	349	Maddock (1988)
	Free State		103	Purves & Sachse (1998)
	Eastern Cape		31	Somers & Purves (1996)
	Western Cape:	Vanrynsdorp	57	Louw & Nel (1986)
		Clanwilliam	70	Purves et al. (1994)
	Zimbabwe		29[†]	Smithers & Wilson (1979)

The sample size (n) reflects the number of scats examined.
[†] Stomachs, not scats.

Tanzania and Rwanda (Table 13.1). In all of these studies, with the exception of the information on the water mongoose from Zimbabwe, information was obtained from the results of faecal analyses. Relative percentage occurrence was used to express the proportions of items, calculated by totalling all occurrences (i.e. presence or absence in a scat) and calculating actual occurrence of each item as a percentage of the total. (See Rowe-Rowe (1977a) or Butler & du Toit (1994) for more detail on the method.)

Aonyx capensis

Items recorded in the scats of the Cape clawless otter from four South African provinces and from Zimbabwe are indicated in Table 13.2. Results from the two separate studies done in Western Cape (Table 13.1) were combined, as the

Table 13.2. *Items recorded in* Aonyx capensis *scats at five freshwater localities in sourthern Africa, expressed as relative percentage occurrence*

	Locality and sample size				
Item	KwaZulu-Natal (1361)	Free State (34)	Eastern Cape (66)	Western Cape (237)	Zimbabwe (255)
Crab	65	57	51	44	42
Frog	23	15	9	11	11
Fish	4	5	18	15	23
Insect	2	8	19	15	17
Bird	1	2	0	1	< 1
Reptile	1	3	0	0	< 1
Mammal	< 1	0	0	8	1
Mollusc	< 1	0	0	2	0
Other	3	10	3	4	5

Samples sizes (number of scats examined) are indicated in brackets. Sources of information are as listed in Table 13.1.

study areas were considered similar in terms of aquatic conditions, prey species, vegetation and climate.

Freshwater crabs (*Potamonautes* spp.) were most important in the diet of *A. capensis* at all of the localities, supplemented by either frogs, fish, or aquatic insects (mainly Odonata larvae). The importance of different secondary items is probably related to local prey availability. For example, the highest percentage of crab and the lowest percentage of fish were recorded from KwaZulu-Natal. This study was done at midland to highland elevations (1060–1650 m above sea level) where fish faunas are poor and crabs abundant (Rowe-Rowe, 1977a). It was also at this wetland-rich locality that the frog component was highest and constituted the most important supplement to crabs. In Zimbabwe, where mountain catfish and eels were abundant, the percentage of fish was highest (Butler & du Toit, 1994).

The relative percentage occurrence of purely aquatic prey was 87–97% at all localities.

Lutra maculicollis

The results of four studies are reflected in Table 13.3. In the study done in the high-rainfall, wetland-rich highland area of KwaZulu-Natal, where fish faunas were poor, almost equal percentages of crab and fish were the main items in the

Table 13.3. *Items recorded in* Lutra maculicollis *scats from four freshwater localities in Africa, expressed as relative percentage occurrence*

	Locality and sample size			
Item	KwaZulu-Natal (294)	Eastern Cape (79)	Rwanda (154)	Lake Victoria (61)
Fish	35	47	80	99
Crab	37	38	0	1
Frog	21	8	<1	0
Insect	4	2	11	0
Bird	2	1	1	0
Mollusc	0	0	3	0
Other	1	3	5	0

Sample sizes (number of scats examined) are indicated in brackets. Sources of information are as listed in Table 13.1.

diet, supplemented by frogs (Rowe-Rowe, 1977a). In the more arid Eastern Cape study area, fish was the main item (47%) in the diet, supplemented by a high (38%) crab component and a low frog component (Somers & Purves, 1996). In the fish-rich lakes of central Africa (Lake Muhazi, Rwanda) and east Africa (Lake Victoria, Tanzania), fish was the main item in the diet, with little else being eaten (Lejeune, 1990; Kruuk & Goudswaard, 1990).

The relative percentage occurrence of purely aquatic prey was 97–100% at the four localities.

Atilax paludinosus

Items recorded in water mongoose scats at five South African localities, as well as in 29 stomachs from Zimbabwe, are listed in Table 13.4. Data for KwaZulu-Natal were obtained from a highland study area, the same as that in which the otters were studied, and a very different lowland study area (Table 13.1). These two localities were treated separately. Results from the two studies done in Western Cape (Table 13.1), however, were combined as the study areas were similar.

Freshwater aquatic prey, mainly crabs, frogs, some aquatic insects and fish, made up between 33 and 59% of the diet (relative percentages) in the studies based on faecal analysis. The highest percentage (59%) of aquatic prey was recorded in the highland KwaZulu-Natal area, where there were rivers, many minor streams and extensive marshy habitats.

Table 13.4. *Items recorded in* Atilax paludinosus *scats at six freshwater localities in sourthern Africa, expressed as relative percentage occurrence*

	Locality and sample size					
	KwaZulu-Natal		Free State	Eastern Cape	Western Cape	Zimbabwe
Item	Highland (210)	Lowland (349)	(100)	(31)	(127)	(29)[†]
Crab	43	21	25	26	31	24
Frog	14	17	8	14	7	29
Fish	2	0	0	3	11	5
Mammal	14	15	25	15	9	24
Bird	14	2	5	7	6	0
Reptile	1	2	5	5	1	0
Insect	2	19	23	28	21	19
Arachnid	0	6	<1	0	0	0
Millipede	0	8	<1	0	0	0
Plant	2	9	8	0	10	0
Carrion	5	0	0	0	0	0
Other	3	2	<1	2	4	0

Sample sizes (number of scats examined) are indicated in brackets. Sources of information are contained in Table 13.1.
[†] Stomachs, not scats.

Terrestrial prey comprised mainly mammals, birds, insects and some reptiles. At some of the localities fruits and seeds were eaten.

Foraging behaviour and adaptations

The three carnivores differ in terms of body size and social organization. *Aonyx capensis* is the largest, its body weight being 11–16 kg; that of *Lutra maculicollis* is 4–6 kg, and that of *Atilax paludinosus* is 3–5 kg. In both of the otters male weights are about 35% greater than female. Cape clawless otters occur in groups of up to four, very occasionally five, in KwaZulu-Natal (Rowe-Rowe, 1978, 1992*a*), where spotted necked otters have been recorded in similar-sized groups (Rowe-Rowe, 1978, 1992*a*; Carugati, 1995; D'Inzillo Carranza, 1997). In Lake Victoria groups of up to 10 or 20 spotted-necked otters have been recorded (Proctor, 1963; Kruuk & Goudeswaard, 1990). Water mongooses are generally solitary (Rowe-Rowe, 1978; Maddock & Perrin, 1993).

Aonyx capensis

Cape clawless otters are crepuscular, usually starting to forage about 1–2 h before sunset (Rowe-Rowe, 1977b, 1978). Although they may occur in groups, they forage singly. In shallow, stony-bottomed streams, the Cape clawless otter walks in the water, submerging the head occasionally, feeling with the forefeet under and between stones for prey (Rowe-Rowe, 1977b). In deeper water the otter dives to the bottom and forages on the substratum, once again using the forefeet (Rowe-Rowe, 1977b; Purves et al., 1994). In water 1.5 m deep, dives had a mean duration of 17.4 s (range 8–26 s). Total foraging time at a single session was between 35 and 39 min for three otters (Rowe-Rowe, 1977b). Almost all prey is captured using the forefeet, then bitten. Where larger prey are involved, e.g. large fish and frogs, the killing bite is directed at the head. Prey is fed into the mouth using the forefeet. Frogs and fish, except large Clarias spp., are eaten head first. Crabs of all sizes are consumed entirely (Rowe-Rowe, 1977b). In experiments (Rowe-Rowe, 1977c) it was found that slow-swimming fish were more easily captured than fast-swimming species, and small fish more easily than larger specimens of the same species.

Aonyx capensis has broad, bunodont molars (Roberts, 1951; Skinner & Smithers, 1990), highly suited for crushing crustaceans. The clawless, unwebbed, manually dextrous forefeet enable it to locate and grasp slow-moving prey, such as crabs, by feeling for them among stones or in holes (Rowe-Rowe, 1977b). It is also hypothesized that the Cape clawless otter's many long vibrissae possibly assist it in locating prey such as crabs (Rowe-Rowe, 1977b). In an investigation on relative brain size (cranial volume in relation to body weight) among 30 South African carnivores, Sheppey & Bernard (1984) found that Aonyx capensis had the largest brain: 2.08 times the expected size. They hypothesized that this accounted for the Cape clawless otter's advanced dexterity.

Lutra maculicollis

Spotted-necked otters have been recorded foraging during daylight in the lakes of east Africa and central Africa (Proctor, 1963; Lejeune, 1989; Kruuk & Goudswaard, 1990). In the streams and impoundments of highland KwaZulu-Natal most foraging is done during the early morning and late afternoon, sometimes continuing until after dark (D'Inzillo Carranza, 1997). Although spotted-necked otters in the large African lakes move in groups of up to 10–20 individuals (Proctor, 1963; Lejeune, 1989; Kruuk & Goudsward, 1990), they forage individually, usually within 2–10 m of the shore. In Lake Muhazi, Lejeune (1989) recorded dives of between 5 and 40 s, most (51%) having a duration of 16–25 s. In a shallow (1.5 m deep) oxbow lake in KwaZulu-Natal, Rowe-Rowe (1977*b*) recorded dives of between 5 and 20 s. Periods at the surface between unsuccessful dives are brief (mainly 6–10 s) as are those when the otters surface to consume small fish they have captured (mainly 11–15 s) (Lejeune, 1989).

All prey is captured in the mouth (Rowe-Rowe, 1977*b*). Small fish (<60 mm) are eaten from the head, but all larger fish and frogs are eaten tail first. Most of the fish captured are small: Rowe-Rowe (1977*a*) and Lejeune (1990) recorded 77% and 81%, respectively, that were <100 mm long (fork length), while Kruuk & Goudswaard (1990) estimated that almost 90% were <150 mm long. In observations on a captive spotted-necked otter, Rowe-Rowe (1977*b*) noted that large crabs, carapace width >50 mm, were avoided while those of <30 mm width were readily captured and eaten.

The dentition of *L. maculicollis* differs from that of *A. capensis* in that the molars and premolars are not as robust (Roberts, 1951). The upper PM4 and lower M1 form a carnassial shear (Skinner & Smithers, 1990). It is less suited for feeding on crustaceans, having smaller crushing teeth than *A. capensis* has, but is better adapted for eating fleshy prey. The spotted-necked otter differs too in that the feet are fully webbed, thus making it better suited for pursuing fast-moving prey, such as fish (Rowe-Rowe, 1977*a,b*).

Atilax paludinosus

Adult water mongooses are generally solitary (Rowe-Rowe, 1978; Baker, 1989; Maddock & Perrin, 1993). Occasionally two are seen together. According to Rowe-Rowe (1978), water mongooses forage mainly during early morning and late afternoon. Maddock & Perrin (1993), however, using telemetry, concluded that this mongoose is nocturnal. The water mongoose does not swim readily, but does most of its aquatic hunting while walking in shallow water or along the shore (Rowe-Rowe, 1978). Sometimes when pursuing prey in shallow water, it runs after it with its head submerged. The hand-like forefeet are used to locate and capture most aquatic prey, feeling for it in the water, under stones, or in holes and crevices (Rowe-Rowe, 1978; Baker, 1989). Hard-shelled prey (mussels, snails, eggs) are picked up in the forefeet and thrown directly downwards while the mongoose stands upright on its hind legs. If this method does not succeed (e.g. with mussels) the mongoose carries the prey to the nearest large stone or other hard object which it uses as an anvil (Rowe-Rowe, 1978). When killing crabs, snakes, or large rodents, the mongoose seizes the prey in its mouth and flicks it sideways, apparently to stun it before delivering a killing bite (Rowe-Rowe, 1978; Baker, 1989). Insects in flight are captured in the air by grasping them in the mouth or with the forefeet.

The water mongoose is behaviourally a very adaptable predator, being able to vary its hunting and killing patterns to deal with a wide variety of prey types, including both aquatic and terrestrial prey. The molars and premolars are not as robust as those of the otters (Roberts, 1951), but at the same time the carnassial shear is not as well-developed as that in more predacious viverrids (Skinner & Smithers, 1990). The degree of manual dexterity places it in a category different from other viverrids, but similar to *Aonyx capensis*.

Coexistence

In the overall distribution of the otters there is very little overlap between *Aonyx capensis* and *A. congica* (Fig. 13.1). The distribution of *Lutra maculicollis* completely overlaps that of *A. congica* and there is about 70% overlap with *A. capensis*. Exclusive to *A. capensis* are the more arid areas into which *L. maculicollis* does not extend. The overall distribution of *Atilax paludinosus* overlaps the ranges of all three otter species.

Two of the studies on diet dealt with in this chapter included three sympatric amphibious carnivores: Cape clawless otter, spotted-necked otter, and water mongoose (Rowe-Rowe, 1977a; Somers & Purves, 1996). A further two (Purves *et al.*, 1994; Purves & Sachse, 1998) dealt with two sympatric carnivores: Cape clawless otter and water mongoose. In all of these studies the diets of the two otters reflect a much greater dependence on aquatic prey than does that of the water mongoose (Table 13.2–13.4).

Comparison of the diets of the two otters in South African studies (Rowe-Rowe, 1977a; Somers & Purves, 1996) reflected a greater dependence on crabs by *A. capensis*, and a higher amount of fish in the diet of *L. maculicollis*. Assessments of the status of otters in Africa (Rowe-Rowe, 1990, 1991) revealed that in countries that are poor in fish faunas, such as South Africa (Skelton, 1994), *A. capensis* is fairly common and *L. maculicollis* is rare. On the other hand, in the African floodplain rivers and large lakes of central and east Africa, which are particularly rich in fish (Fryer & Iles, 1972; Welcomme, 1979), *L. maculicollis* is common and *A. capensis* is rare. It is in such areas that very little overlap in the diets of these two species is likely to occur, with *L. maculicollis* subsisting almost entirely on fish and *A. capensis* subsisting mainly on crabs, as suggested by Kruuk & Goudswaard (1990).

Freshwater crabs featured prominently as shared food in the diets of sympatric carnivores in the South African studies. In four studies, investiga-

Table 13.5. *Ranges of carapace widths (in mm) of freshwater crabs eaten by three amphibous carnivores at four South African localities*

| Locality | Carapace widths (mm) of crabs eaten by | | | Source |
	A. capensis	*L. maculicollis*	*At. paludinosus*	
Western Cape	4–61	–	9–48	Purves *et al.* (1994)
Free State	8–58	–	7–41	Purves & Sachse (1998)
Eastern Cape	3–66	9–44	10–41	Somers & Purves (1996)
KwaZulu-Natal	16–51	14–43	–	Carrugati (1995)

tions were done to determine whether the carnivores fed on crabs of different sizes (Purves *et al.*, 1994; Carugati, 1995; Somers & Purves, 1996; Purves & Sachse, 1998). Carapace widths of crabs eaten were determined from lengths of eyestalks found in scats, since a numerical linear relationship exists between eyestalk length and carapace width (Purves *et al.*, 1994). In all four of the studies (Table 13.5) the remains of a wider range of crabs were found in the scats of the Cape clawless otter, which also included larger crabs in its diet than the other two carnivores did. These findings support the observation made by Rowe-Rowe (1977*a,b*) that the Cape clawless otter fed on crabs of all sizes, whereas the spotted-necked otter ate smaller crabs, and avoided very large ones. Available data (Table 13.5) indicate that the crabs eaten by *L. maculicollis* and *At. paludinosus* are of similar size.

In only two studies has the relative abundance of sympatric South African otters and the water mongoose been estimated. In part of the Natal Drakensberg Park in KwaZulu-Natal, Rowe-Rowe (1992*b*) used sign (spraint sites, holts, tracks) as well as animals seen, to estimate numbers. Estimated abundances were one *A. capensis* per 3–4 km of stream, one *L. maculicollis* per 6–11 km, and one *At. paludinosus* per 2 km. Combined abundance of all three carnivores was one per 1.0–1.5 km of stream. The water mongoose would not have been confined entirely to the aquatic habitat and immediate riparian area, but would have ranged further afield (Maddock & Perrin, 1993), having an overall home range of *ca* 2 km² (Maddock, 1988).

Working also in the 240 000 ha montane Natal Drakensberg Park, Carugati (1995) studied sympatric *A. capensis* and *L. maculicollis* along four separate 5 km stretches on three rivers. There appear to be differences in the total numbers of otters and in the proportions of the two species: one 5 km stretch was used by *ca* eight otters (in a ratio of three *A. capensis*: one *L. maculicollis*), whereas another 5 km stretch (where there were dams artificially stocked with trout) was used by nine otters (in a ratio of one *A. capensis*: two *L.*

maculicollis). The lowest number of otters using a 5 km stretch was four (three
A. capensis: one *L. maculicollis*). It should be pointed out that these are not
density estimates, but estimates of the numbers using the 5 km stretches of
river: the extent of home ranges beyond the study areas was not determined.

At Lake Victoria the density of the dominant spotted-necked otter has been
estimated at one per kilometre of shoreline (Proctor, 1963; Kruuk & Goudswaard, 1990). It occurs there together with *A. capensis*, which appears to be
present in much lower numbers. The ratio of *L. maculicollis*: *A. capensis* scats
found was 12 : 1 (Kruuk & Goudswaard, 1990). At Lake Muhazi in Rwanda,
Lejeune & Frank (1990) estimated that *L. maculicollis* occurred at one per
0.5 km of shoreline. Although Cape clawless otters occurred in the vicinity of
the lake, they were rare and foraged in marshes or small streams.

Conclusions

The Cape clawless otter and the spotted-necked otter are dependent on aquatic
habitats, using rivers, streams, lakes, or ponds almost exclusively, seldom
venturing far from water. The Cape clawless otter makes more use of small,
shallow streams (good crab habitat) than does the spotted-necked otter, which
in turn shows a preference for deeper, open water. The Cape clawless otter is
particularly well-adapted for the capture and consumption of crabs, whereas
the spotted-necked otter has evolved more as a piscivorous carnivore. The
water mongoose is more terrestrial than the otters are, and is not entirely
dependent on aquatic habitats. Its use of streams, lakes, or ponds is limited to
the shores, and it seldom ventures into water more than a few centimetres
deep. The water mongoose exploits a wider range of habitats than do the otters,
including marshes, very small water bodies, temporary stagnant water and
terrestrial habitats.

Although very little is known about the Congo clawless otter, its adaptations
suggest that both crabs and fish probably feature in its diet. The unwebbed,
clawless, hairless forefeet (Kingdon, 1977) suggest manual dexterity. The
molars and premolars are not as broad and flat as those of *A. capensis*, while the
carnassial shear is not as well-developed as that in *L. maculicollis*. The skull of
A. congica appears to be slightly smaller than that of *A. capensis* (Harris, 1968),
although the overall body size is similar. If the four freshwater amphibious
carnivores were to be arranged in descending order of adaptation for feeding
on, and dependence on, crabs, the order would probably be *Aonyx capensis* >
Aonyx congica > *Lutra maculicollis* > or < *Atilax paludinosus*. Similarly, the
order of adaptation for feeding on, and dependence on, fish, would probably

be *Lutra maxulicollis* > *Aonyx congica* > *Aonyx capensis* > *Atilax paludinosus*. The studies reviewed suggest that none of our basic hypotheses can be rejected: there were differences between the species in food, foraging behaviour and distribution, but on all counts there were overlaps between them.

ACKNOWLEDGEMENTS
We thank Michelle Hamilton for typing the manuscript, Ant Maddock for commenting on it and Heidi Snyman for preparing the map. The research by one of us (M.J.S.) was supported by Mazda Wildlife Fund and World Wildlife Fund (South Africa).

References

Arden-Clarke, C. H. G. (1983). *Population Density and Social Organisation of the Cape Clawless Otter*, Aonyx capensis Schinz, *in Tsitsikama Coastal National Park.* MSc thesis: University of Pretoria, Pretoria.

Baker, C. M. (1989). Feeding habits of the water mongoose (*Atilax paludinosus*). *Z. Säugetierk.* **54**: 31–39.

Baranga, J. (1995). The distribution and conservation status of otters in Uganda. *Habitat* **11**: 29–32.

Butler, J. R. A. & du Toit, J. T. (1994). Diet and conservation status of Cape clawless otters in eastern Zimbabwe. *S. Afr. J. Wildl. Res.* **24**: 41–47.

Carpaneto, G. M. & Germi, F. P. (1989). The mammals in the zoological culture of the Mbuti pygmies in north-eastern Zaire. *Hystrix* **1**: 1–83.

Carugati, C. (1995). *Habitat, Prey, and Area Requirements of Otters* (Aonyx capensis *and* Lutra maculicollis) *in the Natal Drakensberg.* MSc thesis: University of Natal, Pietermaritzburg.

D'Inzillo Carranza, I. (1997). *Activity Rhythms and Space Use by Spotted-necked Otters in the Drakensberg.* MSc thesis: University of Natal, Pietermaritzburg.

Fryer, G. & Iles, T. D. (1972). *The cichlid fishes of the Great Lakes of Africa: their biology and evolution.* Oliver & Boyd, Edinburgh.

Harris, C. J. (1968). *Otters. A study of Recent Lutrinae.* Weidenfeld & Nicolson, London.

Kingdon, J. (1977). *East African mammals: an atlas of evolution in Africa* 3A (Carnivores). Academic Press, London.

Kruuk, H. & Goudswaard, P. C. (1990). Effects of changes in fish populations in Lake Victoria on the food of otters (*Lutra maculicollis* Schinz and *Aonyx capensis* Lichtenstein). (*sic*) *Afr. J. Ecol.* **28**: 322–329.

Lejeune, A. (1989). Ethologie de loutres (*Hydrictis maculicollis*) au lac Muhazi, Rwanda. *Mammalia* **53**: 191–202.

Lejeune, A. (1990). Ecologie alimentaire de la loutre (*Hydrictis maculicollis*) au lac Muhazi, Rwanda. *Mammalia* **54**: 33–45.

Lejeune, A. & Frank, V. (1990). Distribution of *Lutra maculicollis* in Rwanda: ecological constraints. *IUCN Otter Spec. Group Bull.* No. 5: 8–16.

Ligthart, M. F., Nel, J. A. J. & Avenant, N. L. (1994). Diet of Cape clawless otters in part of the Breede River system. *S. Afr. J. Wildl. Res.* **24**: 38–39.

Louw, C. J. & Nel, J. A. J. (1986). Diets of coastal and inland-dwelling water mongoose. *S. Afr. J. Wildl. Res.* **16**: 153–156.

MacDonald, J. T. & Nel, J. A. J. (1986). Comparative diets of sympatric small carnivores. *S. Afr. J. Wildl. Res.* **16**: 115–121.

Maddock, A. H. (1988). *Resource Partitioning in a Viverrid Assemblage.* PhD thesis: University of Natal, Pietermaritzburg.

Maddock A. H. & Perrin, M. R. (1993). Spatial and temporal ecology of an assemblage of viverrids in Natal, South Africa. *J. Zool., London.* **229**: 277–287.

Procter, J. (1963). A contribution to the natural history of the spotted-necked otter (*Lutra maculicollis* Lichtenstein) in Tanganyika. *E. Afr. Wildl. J.* **1**: 93–102.

Purves, M. G., Kruuk, H. & Nel, J. A. J. (1994). Crabs *Potamonautes perlatus* in the diet of otter *Aonyx capensis* and water mongoose *Atilax paludinosus* in a freshwater habitat in South Africa. *Z. Säugetierk.* **59**: 332–341.

Purves, M. G. & Sachse, B. (1998). *The utilization of freshwater crabs by co-existing otter (Aonyx capensis Schinz) and water mongoose (Atilax paludinosus Cuvier) in the Drakensberg, South Africa.* Unpublished report.

Roberts, A. (1951). *The mammals of South Africa.* Central News Agency, Johannesburg.

Rowe-Rowe, D. T. (1977a). Food ecology of otters in Natal, South Africa. *Oikos* **28**: 210–219.

Rowe-Rowe, D. T. (1977b). Prey capture and feeding behaviour of South African otters. *Lammergeyer* **23**: 13–21.

Rowe-Rowe, D.T. (1977c). Variations in the predatory behaviour of the clawless otter. *Lammergeyer* **23**: 22–27.

Rowe-Rowe, D.T. (1978). The small carnivores of Natal. *Lammergeyer* **25**: 1–48.

Rowe-Rowe, D.T. (1990). Action plan for African otters. In *Otters: an action plan for their conservation:* 41–51. (Eds Foster-Turley, P., Macdonald, S.M. & Mason, C.F.). IUCN, Gland.

Rowe-Rowe, D.T. (1991). Status of otters in Africa. *Habitat* **6**: 15–20.

Rowe-Rowe, D.T. (1992a). *The carnivores of Natal.* Natal Parks Board, Pietermaritzburg.

Rowe-Rowe, D.T. (1992b). Survey of South African otters in a freshwater habitat, using sign. *S. Afr. J. Wildl. Res.* **22**: 49–55.

Sheppey, K. & Bernard, R.T.F. (1984). Relative brain size in the mammalian carnivores of the Cape Province of South Africa. *S. Afr. J. Zool.* **19**: 305–308.

Skelton, P.H. (1994). Diversity and distribution of freshwater fishes in east and southern Africa. *Annls Mus. r. Afr. centr. (Sci. Zool.)* **275**: 95–131.

Skinner, J.D. & Smithers, R.H.N. (1990). *The mammals of the Southern African Subregion.* (2nd edn). University of Pretoria, Pretoria.

Smithers, R.H.N. & Wilson, V.J. (1979). Checklist and atlas of the mammals of Zimbabwe Rhodesia. *Mus. Mem. natn. Mus. Monum. Rhodesia* No. 9: 1–147.

Somers, M.J. & Purves, M.G. (1996). Trophic overlap between three syntopic semi-aquatic carnivores: Cape clawless otters, spotted-necked otters, and water mongooses. *Afr. J. Ecol.* **34**: 158–166.

Van der Zee, D. (1981). Prey of the Cape clawless otter (*Aonyx capensis*) in the Tsitsikama Coastal National Park, South Africa. *J. Zool., London.* **194**: 467–483.

Verwoerd, D.J. (1987). Observations on the food and status of the Cape clawless otter *Aonyx capensis* at Betty's Bay, South Africa. *S. Afr. J. Zool.* **22**: 33–39.

Welcomme, R.L. (1979). *Fisheries ecology of floodplain rivers.* Longmans, London.

Whitfield, A.K. & Blaber, S.J.M. (1980). The diet of *Atilax paludinosus* (water mongoose) at St Lucia, South Africa. *Mammalia* **44**: 315–318.

14

Feeding ecology of the smooth-coated otter *Lutra* *perspicillata* in the National Chambal Sanctuary, India

S.A. Hussain and B.C. Choudhury

Introduction

Three species of otter occur in the Indian sub-continent: the Eurasian otter, *Lutra lutra* Linnaeus, the smooth-coated otter, *L. perspicillata* Geoffroy, and the original small-clawed otter, *Aonyx cinerea* Illegar. The smooth-coated otter is distributed throughout the Indian sub-continent, from the Himalayas southward. Outside the Indian sub-continent, its range extends to Mynamar, Indonesia, Kampuchea, Laos People's Republic, Malaysia, Vietnam, south-western China and Brunei, with an isolated race, *L. perspicillata maxwelli*, in the marshes of southern Iraq (Pocock, 1941: 265–312; Prater, 1971; Mason & Macdonald, 1986). Despite its wide distribution, no detailed ecological studies on this species have been undertaken so far. It is listed as insufficiently known in the IUCN red data book (Groombridge, 1994) and is protected under Schedule II of the Indian Wildlife (Protection) Act, 1972. Previous work on Indian otters involved observations on captive animals (Desai, 1974; Acharjyo & Mishra, 1984) with occasional notes on their occurrence from different parts of the country (e.g. Hinton & Fry, 1923; Pocock, 1939; Chitampalli, 1979) and a few studies on their feeding habits (e.g. Wayre, 1978; Foster-Turley, 1992; Kruuk *et al.*, 1994).

It is believed that the existing populations of all the three Indian species are rapidly declining due to loss of habitat and intensive trapping (Hussain & Choudhury, 1997). Because of limited knowledge on the ecology of Oriental otter species, practical conservation measures are difficult to develop. In 1988, the Wildlife Institute of India undertook a project to study the ecology of smooth-coated otters in the National Chambal Sanctuary (NCS) with a view to gathering basic scientific data that could be helpful in developing a sound management strategy for otter conservation. This chapter gives an account of the feeding ecology of otters in the National Chambal Sanctuary.

Study area

The Chambal is a clear and fast-flowing river that originates from the Vin-
dhyan mountain range, central India. Lying between 24° 55' and 26° 50'N, 75°
34' and 79° 18'E, it flows north-east and joins the Yamuna river to form the
greater Gangetic drainage system. A 600 km stretch of the Chambal river has
been protected as the National Chambal Sanctuary for the conservation of the
Gangetic gharial, *Gavialis gangeticus* Gmelin (Reptilia, Crocodylia). The area
lies within the semi-arid zone of north-western India at the border of Madhya
Pradesh and Rajasthan states. In the Sanctuary, the Chambal averages 400 m in
width, 1–26 m in depth, with discharge ranging between 27 000 m^3/s to
500 m^3/s (Hussain, 1993). The temperature in the area ranges from 2–46° C
and the annual precipitation ranges between 500 and 600 mm. The natural
vegetation of much of the Sanctuary area is ravine thorn forest (Champion &
Seth, 1968). Evergreen riparian vegetation is completely absent. The severely
eroded river banks and adjacent ravine lands have sparse ground cover (Hussain, 1993). Apart from the gharial and smooth-coated otter, the other aquatic
fauna of the Sanctuary include the marsh crocodile, *Crocodylus palustris*, seven
species of freshwater turtles, the Ganges river dolphin, *Platanista gangetica*,
and 78 species of wetland birds (Hussain, 1993).

The intensive study site used is the stretch of river between 160–170 km.

Methods

Methods of spraint collection
During 1989 and 1990 spraint samples were collected from a 195 km stretch of
the Sanctuary. Collections were made from the entrance of dens, communal
sprainting sites adjacent to dens, and feeding areas. Attempts were made to
collect fresh spraints, but during the monsoon all spraints were collected. From
each site only a sample was collected. Date and location were recorded for each
spraint at the time of collection. Samples were sun-dried at the camp and
stored in paper bags. The samples were later soaked in water mixed with a
detergent for 5–6 h. After washing in running tap water over a 1 mm sieve, the
cleaned samples were dried in shade and kept in plastic bags for sorting of prey
remains. Prey remains from the cleaned spraint were sorted using a hand lens
and, when required, a compound microscope. All remains were identified by
comparing them with a reference collection of prey parts made during the
study. Efforts were made to identify the prey up to species level but this was not

always possible, hence some prey remains were identified only up to genus level.

Determination of sample size

To examine the representativeness of the samples, initially 140 samples were collected during the period December 1987 to March 1988. After processing, the samples were examined randomly. The cumulative number of prey species found in the spraints and the number of samples examined were plotted following the method described by Mason & Macdonald (1980). An asymptotic was reached at about 20 samples. Thus for each month 25 (25% more) samples were selected for analysis. During monsoons (July–October), when it was difficult to find a sufficient sample size, all the spraints found were examined. A total of 553 spraint samples were examined.

Estimation of prey availability

We express prey (fish and other items) densities as catch/unit effort (Snodgrass et al., 1994). To get an estimate of fish densities, sampling was carried out 10 times in each season, primarily by netting (cast and gill nets of 1–5 cm mesh size), within the home range of a known otter family. All samples were taken at the feeding sites during crepuscular time between 4:00 and 6:00 and between 17:00 and 19:00. It is likely that there is some bias in the catch due to the variation in 'trappability' of different species in a fast-flowing river. Thus it is difficult to relate the data to actual fish densities, as encountered by the otters. However, it is believed that these data will give some insight into prey selectivity by otters. For each catch effort the number of species and the frequency were recorded. The result was later converted for analysis into catch/unit effort. The densities of invertebrate prey species such as shrimp, water beetle and damselfly larvae were estimated from samples obtained with a mesh net scoop of 1 mm size and 30 cm diameter. Densities of crabs and frogs were estimated by monitoring and counting their number in a 1 km stretch of the river in the intensive study site (160–170 km). The avian population was assessed by direct counting in the intensive study site. For each non-fish prey category, two counts were made in every month.

Estimation of prey size and quality

The approximate size of the seven fish species was estimated by comparing the known size of the prey parts such as scales, vertebral columns and pectoral spines, with parts recovered from scats. For this, 28 fish species of different size were boiled and, after washing through a sieve, scales and bones were collected as reference material for comparison with prey parts recovered from the scats.

Calorific values of 12 fish species that were eaten throughout the year were estimated by using a Gallenkemp Ballistic Bomb Calorimeter. Freshly caught fish were oven dried at 60 °C and ground in an ordinary grinder. For each sample three replicates were made, from which mean ash-free calorific values were calculated.

Preference for major prey species

The χ^2 goodness-of-fit test was used to determine whether there was any significant difference between the expected utilization of different species and the observed frequency of their usage (Neu *et al.*, 1974). If a statistically significant difference was found, then the data were further examined by using the Bonferroni confidence interval as described by Byers *et al.* (1984) to determine which species were preferred.

Dietary overlap with marsh crocodile

Twenty-five scat samples of the marsh crocodile or mugger, *Crocodylus palustris*, were collected from the intensive study site between November 1989 and April 1990. The samples were soaked in water for 5–6 h. After washing in running tap water through a 1 mm sieve, the cleaned samples were dried in shade and prey remains from them were sorted and identified by using the same methods adopted for otter spraint analysis.

Analysis of data

The data on the diet composition were tabulated by frequency of occurrence (the number of occurrences of an item is expressed as a percentage of the total number of occurrences of all items in all the samples, the sum of the frequencies being 100%). Evidence of more than one individual of a species in a spraint was treated as a single occurrence. Such frequency calculations give a reasonably accurate analysis of the relative importance of the different prey items (Erlinge, 1969). To avoid discrepancy between the frequency at which a prey is consumed and the bulk of it eaten (i.e. over-representation of minor items and under-estimation of major items), the method proposed by Wise *et al.* (1981) and Mason & Macdonald (1986) was adopted. Based on the proportion of prey parts of each species found in the cleaned sample, the importance of each prey part or item was rated on a score of 1–10. The score for each item was then multiplied by the dry weight of spraint. The resulting figure was summed for each item in each sample and expressed as a bulk percentage. No attempt was made to determine the total biomass of prey species occurring in the spraint, since other studies (e.g. Lockie, 1959; Floyd *et al.*, 1978) have reported that estimating biomass from the faecal sample may result in misleading estimates of food consumption.

For statistical analysis, data were examined for between-year variation (1989 and 1990) and for within-year variation (month and season; winter = November to February, summer = March to June, monsoon = July to October).

Results

Diet composition

A total of 2040 key prey remains were recovered from 553 spraint samples collected during 1989 and 1990. The relative frequency of prey remains and the bulk percentage of prey consumed over the study period is given in Table 14.1. The relative frequency of occurrence of prey remains in the spraints was 93.8% fish, 3.8% invertebrates and 2.4% amphibians and birds (Fig. 14.1). Approximately 3% of fish remains (63) and 0.9% of birds could not be identified from the spraints. In terms of bulk, 97.7% of the diet was fish, and 2.3% invertebrates, amphibians and birds. Of this latter 2.3%, crabs, shrimps and insects formed 1.5%, whereas birds and amphibians constituted only 0.8%. Apart from these, 46 occurrences of molluscs, 12 of vegetative matter (grassroots or grass) and 7 occurrences of mammalian hair were also recorded. All the mammalian hair was identified as otter hair, which might have been ingested by the otters while grooming. Mollusc shells and vegetative matter were ingested as secondary ingestion – this was later confirmed by analysing the gut contents of a major prey, Rita rita. It was not clear whether the consumption of insects was direct or due to secondary ingestion. Of the 21 types of undigested items identified from spraints, 12 were fish, 3 were invertebrates, 3 were birds, 1 was amphibian and the remainder were unidentified birds and fish.

Number of prey species used by otter

The number of prey species in a spraint varied from 1–7 (mean = 3.78 (±0.043): Fig. 14.2). Almost equal numbers of species were recorded in 1989 (19) and in 1990 (18). Of the 12 fish species, Channa spp. constituted a minor part of the diet in the monsoon and was recorded in one sample in 1989 only. Xenetodon cancila was found only during the monsoon in both years. The number of prey species used in a month varied from 11–15 in 1989 (mean = 13.33 (±0.41)) and from 10–15 in 1990 (mean = 11.83 (±0.5): Fig. 14.3). The number of species used in different months was similar between the

Table 14.1. *The diet (relative frequency) of smooth-coated otters in the National Chambal Sanctuary*

Species	Prey remains (n)	Relative frequency (%)	Bulk percentage
Fish			
Rhinomugil corsula	468	22.90	38.5
Rita rita	458	22.5	33.3
Puntius spp.	283	13.9	8.50
Labeo calbasu	183	9.0	5.9
Oxygaster bacaila	150	7.4	3.0
Mystus seenghala	110	5.4	2.9
Mystus tengara	96	4.7	2.2
Ompok bimaculatus	36	1.8	0.7
Mastacembelus spp.	33	1.6	0.6
Notopterus spp.	28	1.4	0.6
Xenetodon cancila	5	0.3	0.1
Channa spp.	1	0.05	0.02
Unidentified fish	63	3.1	1.4
Invertebrates			
Crabs	46	2.25	1.9
Shrimps	27	1.32	0.4
Insects	5	0.25	0.04
Birds			
Tadorna tadorna	12	0.59	0.22
Anas poicilorhynca	5	0.25	0.1
Columba livia	10	0.49	0.2
Unidentified birds	18	0.88	0.2
Amphibians			
Frog	3	0.15	0.03

Total number of prey remains = 2040; number of spraints analysed = 553.

years (Spearman rank correlation coefficient = 0.934, $p < 0.05$). The number of prey species used varied considerably among seasons (Kruskal–Wallis one-way ANOVA, $\chi^2 = 10.339$, $p = 0.0057$), being higher in winter (mean = 14.12 (± 0.5)) than in summer (mean = 12.12 (± 0.47)) or monsoon (mean = 11.50 (± 0.59): Fig. 14.4).

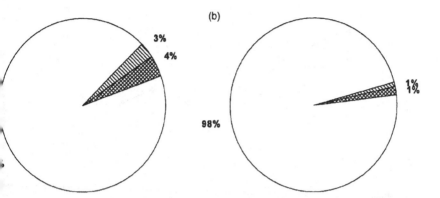

Figure 14.1. Diet composition of smooth-coated otter in the National Chambal Sanctuary. (a) Relative frequency (%); (b) bulk percentage; $n = 553$ spraints; □, fish; ▨, invertebrates; ▧, amphibians and birds.

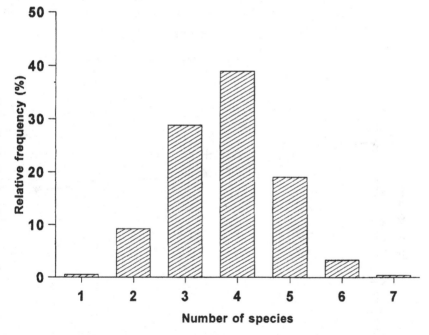

Figure 14.2. Relative frequency of the number of prey species recorded from spraints of the smooth-coated otter. $n = 553$ spraints.

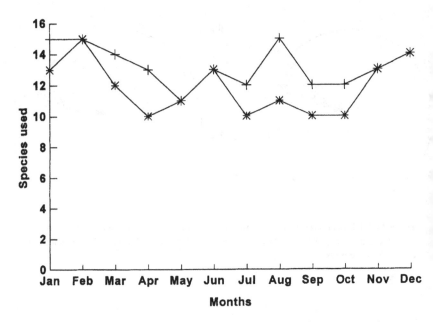

Figure 14.3. Monthly variation in the number of prey species used by the smooth-coated otter. +, 1989; *, 1990.

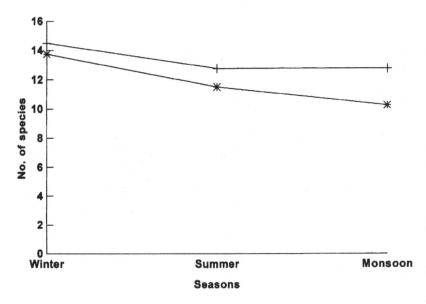

Figure 14.4. Seasonal variation in the number of prey species used by the smooth-coated otter. +, 1989; * 1990.

Table 14.2. *Number and relative frequency of different families of fish consumed by smooth-coated otters*

Family	Species	n	Relative frequency (%)	Bulk percentage
Bagaridae	*R. rita, M. seenghala, M. tengara*	664	35.9	38.4
Cyprinidae	*L. calbasu, O. bacaila, Puntius* spp.	616	33.3	17.4
Mugilidae	*R. corsula*	468	25.3	38.5
Siluridae	*O. bimaculatus*	36	1.9	0.7
Mastacembelidae	*Mastacembelus* spp.	33	1.9	0.6
Notopteridae	*Notopterus* spp.	28	1.5	0.6
Belonidae	*X. cancila.*	5	0.3	0.1
Ophiocephalidae	*Channa* spp.	1	0.05	0.02
Unidentified	—	63	3.1	1.4

n = number of prey remains

Dietary importance of fish

Of the 12 identified fish species, seven species (*Rhinomugil corsula, Rita rita, Puntius* spp., *Labeo calbasu, Mystus seenghala, M. tengara* and *Oxygaster bacaila*) were eaten throughout the year and accounted for 94% of the diet. *Rh. corsula* and *R. rita* constituted the bulk of the diet (72%), followed by *Puntius* spp., *L. calbasu, O. bacaila, M. seenghala* and *M. tengara*. Of the remaining five species, two (*Xenetodon cancila* and *Channa* spp.) were used only during the monsoon whereas the other three (*Mastacembelus* spp., *Notopterus* spp. and *Ompok bimaculatus*) were used occasionally.

Cyprinidae provided the most species (i.e. > 3) followed by Bagaridae (3 spp.). Other families provided one species each. Although Mugilidae provided only one species (*Rh. corsula*) this accounted for 38.5% of the bulk of the diet. The contribution of Bagaridae through *R. rita* alone was greater than that of all the Cyprinidae species combined. Thus *R. rita* formed the next most important dietary component (33.3% of the bulk). The contributions of families such as Siluridae, Mastacembelidae, Belonidae, Ophiocephalidae and Notopteridae were insignificant (Table 14.2).

Difference between frequency and bulk consumed

The members of the family Bagaridae were used as prey more frequently

Table 14.3. Seasonal importance of seven major fish species to otters during 1989 and 1990

Species	Monsoon			Winter			Summer		
	n	Relative frequency (%)	Bulk percentage	n	Relative frequency (%)	Bulk percentage	n	Relative frequency (%)	Bulk percentage
1989									
Rh. corsula	59	18.8	18.8	99	24.5	56.9	86	22.3	32.3
R. rita	70	22.3	40.2	79	19.6	17.9	91	23.6	38.6
Puntius spp.	51	16.2	16.0	45	11.5	5.7	64	16.6	10.7
L. calbasu	31	9.8	9.6	34	8.4	4.8	24	6.2	3.5
O. bacaila	33	10.5	4.8	7	1.7	0.7	41	10.6	4.7
M. seenghala	17	5.4	3.1	28	6.9	3.5	19	4.9	3.3
M. tengara	14	4.5	2.1	23	5.7	2.5	13	3.4	1.8
1990									
Rh. corsula	46	19.3	16.3	93	24.9	55.8	85	26.2	31.9
R. rita	56	23.4	46.4	78	20.9	20.8	84	25.9	46.7
Puntius spp.	33	13.8	9.8	45	12.6	5.81	45	13.9	6.8
L. calbasu	30	12.6	11.8	45	12.6	5.73	19	5.9	3.5
O. bacaila	15	15.0	1.9	11	2.9	1.7	43	13.3	5.0
M. seenghala	11	11.0	2.9	26	6.9	2.9	9	2.8	1.8
M. tengara	15	15.0	3.3	22	5.9	2.5	9	2.8	0.9

n = number of prey remains

Table 14.4. *Densities (catch/unit effort) of seven major fish species in different seasons during 1989 and 1990*

Species	Winter		Summer		Monsoon	
	Total catch	Density	Total catch	Density	Total catch	Density
Rh. corsula	16	0.80	8	0.40	4	0.20
R. rita	11	0.55	12	0.56	6	0.30
Puntius spp.	26	1.30	28	1.65	20	1.00
L. calbasu	27	1.35	20	0.90	19	0.95
O. bacaila	50	2.50	48	4.00	32	1.60
M. seenghala	16	0.80	10	0.45	6	0.30
M. tengara	9	0.45	7	0.30	5	0.25

(relative frequency 35.9%) than any other family. *Rh. corsula*, the only member of the family Mugilidae taken as prey, was used less frequently (25.3%) but the quantity of it consumed (38.5%) was similar to that of the Bagaridae (38.4%). Cyprinidae were also eaten more frequently (33.3%) but the quantity eaten was less (17.4% of the bulk) as compared to Bagaridae or Mugilidae. The remainder of the families were represented almost equally in frequency and bulk analysis, both being very low (Table 14.2). During both years, the numbers and the quantity of *Rh. corsula* consumed were highest in winter, lower in summer and lowest in the monsoon. In contrast, the numbers of *R. rita* eaten were highest during summer but the quantity eaten was highest during the monsoon. Similarly, the numbers and quantity consumed of *Puntius* spp. and *O. bacaila* were highest in summer, lower in the monsoon and winter (Table 14.3). The monthly variation in the frequency at which a species was consumed (relative frequency) and its quantity (bulk percentage) was similar between the two study periods (Spearman rank correlation coefficient = 0.983, $p < 0.05$; $n = 24$ months). *Rh. corsula* and *R. rita* showed seasonal shift in the diet between different seasons (Fig. 14.5), i.e. *Rh. corsula* was eaten more during winter, less in summer and least in the monsoon, whereas *R. rita* was eaten more during summer and the monsoon than in winter.

Prey distribution and densities
Twenty-eight fish species were recorded during the sampling. Nine species that were reported elsewhere from the Chambal river were not recorded during the sampling. Table 14.4 summarizes the results of 60 catch efforts, 30 each during

Table 14.5. *Relative abundance of prey items other than fish in different seasons dur,*
1989 and 1990

	Winter		Summer		Monsoon	
Prey	1989	1900	1989	1990	1989	1990
Shrimp (catch/unit effort)	6.62	6.30	2.20	2.35	0.78	0.8
Crab (encounter rate/km)	4.88	4.77	3.18	3.40	1.34	1.1
Birds (density/river km)	90.0	78.8	31.6	36.0	5.6	7.6

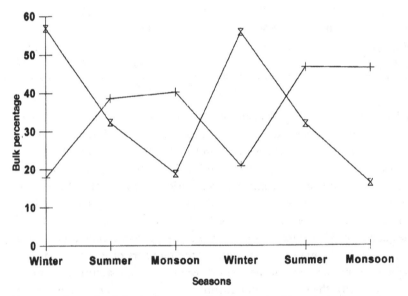

Figure 14.5. Seasonal variation in the consumption of two major fish species by the
smooth-coated otter in the National Chambal Sanctuary. Ⴈ , *Rhinomugil corsula*; +, *Rita rita.*

1989 and 1990. The densities of different fish species were similar between 1989
and 1990 (Spearman rank correlation coefficient = 0.89, $p < 0.05$) but there
was a significant variation in the abundance of different species in different
seasons (Friedman's test, $\chi^2 = 30.452$, d.f. = 2, $p < 0.05$). *O. bacaila* was the
most abundant fish, followed by *Puntius* and *Labeo* spp. *Rh. corsula* was more
abundant during winter than summer and the monsoon, whereas *R. rita* was
more abundant in summer. The abundance of two catfish, *M. seenghala* and *M.
tengara*, also appeared to be higher during winter. Table 14.5 gives abundance
of non-fish prey categories such as invertebrates and birds in the study area.

Table 14.6. *Calorific value of major prey species and their quantity consumed by smooth-coated otters*

Species	Cal. value (kcal/g)	*n*	Relative frequency (%)	Bulk percentage
Rh. corsula	4.3175	468	26.8	38.5
R. rita	4.3752	458	26.2	33.3
Puntius spp.	3.6900	283	16.2	8.5
L. calbasu	4.2751	283	10.5	5.9
O. bacaila	4.5505	150	8.6	3.0
M. seenghala	4.5907	110	6.3	2.9
M. tengara	4.5000	96	5.5	2.1

n = number of prey remains.

Table 14.7. *Pearson's correlation coefficient between availability and use of fish and non-fish prey*

Species	Correlation	*p* value	Category	Correlation	*p* value
Rh. corsula	0.88685	0.305	Birds	0.8370	0.037
R. rita	0.99590	0.057	Crab	0.89939	0.014
Puntius spp.	0.96958	0.157	Shrimp	0.93756	0.005
L. calbasu	0.91225	0.268	—	—	
O. bacaila	0.91225	0.542	—	—	
M. seenghala	0.95632	0.188	—	—	
M. tengara	0.85566	0.346	—	—	

Prey size, quality and quantity eaten

Otters consumed prey ranging from 4–46 cm in length (mean = 16 (±1.09), $n = 97$). The mean length of *O. bacaila* was the lowest (9.2 cm, range = 6–12 cm) and *L. calbasu* was the highest (26 cm, range = 12–45 cm). The largest prey in terms of length was *R. rita*, followed by *L. calbasu* and *M. seenghala*. The rest of the prey were 5–20 cm long. The calorific value of major fish species did not vary significantly among species (mean = 4.328 (±0.11) kcal/g, $n = 7$). It was lowest in *Puntius* spp. and highest in *M. tengara*. There was no relationship between the calorific value of a fish species and the quantity of it that was eaten (Spearman rank correlation coefficient = −0.196, $p > 0.05$: Table 14.6).

Table 14.8. *Seasonal importance of prey species other than fish to otters in 1989 and 1990*

Species	Monsoon			Winter			Summer		
	n	Relative frequency (%)	Bulk percentage	n	Relative frequency (%)	Bulk percentage	n	Relative frequency (%)	Bulk percentage
Birds	0	0	0	26	3.4	1.0	19	2.7	0.8
Crab	5	0.5	0.3	23	2.9	1.4	18	2.5	1.3
Shrimp	0	0	0	23	2.9	0.8	4	0.6	0.2

n = number of prey remains.

Table 14.9. *Conclusion on relative abundance of seven major prey species and their use by otters during 1989 and 1990*

Species	Density	Expected usage	Observed usage (n)	Expected prop. of usage	Bonferroni confidence interval	Conclusion
Rh. corsula	0.47	119.761	468	0.069	$0.239 \leq pi \leq 0.296$	P
R. rita	0.50	127.405	458	0.073	$0234 \leq pi \leq 0.290$	P
Puntius spp.	1.27	323.609	283	0.185	$0.138 \leq pi \leq 0.186$	UIA
L. calbasu	1.07	272.647	183	0.156	$0.085 \leq pi \leq 0.124$	NP
O. bacaila	2.70	687.988	150	0.394	$0.068 \leq pi \leq 0.104$	NP
M. seenghala	0.52	132.501	110	0.076	$0.047 \leq pi \leq 0.079$	UIA
M. tengara	0.33	84.87	96	0.048	$0.040 \leq pi \leq 0.070$	UIA

n, number of prey remains; P, used more than available, Prefered; NP, used less than its availability, Not preferred; UIA, used in proportion to its availability, neither preferred nor avoided.

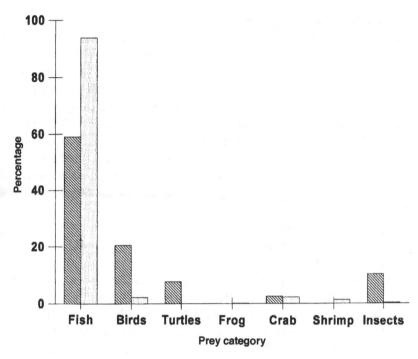

Figure 14.6. Comparison of the diets of the smooth-coated otter and the Indian marsh crocodile or mugger. ▧, mugger, $n = 25$ scats analysed; □, otter, $n = 553$ sraints analysed.

Relationship between availability and use of prey species

All seven major prey species showed an increase in percentage consumed with an increase in relative abundance (catch/unit effort). But it was not significant in most cases, probably due to the low sample size (Table 14.7). Consumption of all the non-fish species showed high correlation with their availability. All these prey items were more available to otters during winter, less so in summer and the monsoon. Shrimps showed the highest correlation between availability and use, followed by crabs and birds (Table 14.7). Frogs and insects were eaten only during the monsoon and the quantity eaten was very low. Crabs were eaten almost throughout the year, whereas birds and shrimps were eaten only during winter and summer (Table 14.8). Thus, other than fish, crabs were the subsidiary food item for otters, followed by birds and shrimps. Out of 27 occurrences of shrimps, 85.2% were recorded in winter and 14.8% in summer. There were 45 records of birds, of which only 27 could be identified. Among these, 12 were common shelduck, *Tadorna tadorna*, 5 were spotbilled duck, *Anas poicilorhynca*, and 10 were blue rock pigeon, *Columba livia*.

Preference for major prey species

Seven major fish species that constituted 94% of the total diet were considered for analysis. Analysis, using the Bonferroni confidence interval, of the abundance of major fish species and their use by otters shows that *Rh. corsula* and *R. rita* were used more frequently than expected from their abundance, indicating higher preference for them. *Puntius* spp., *M. seenghala* and *M. tengara* were consumed in proportion to their abundance, whereas *L. calbasu* and *O. bacaila* were used less than expected from their abundance, i.e. they were not preferred (Table 14.9). Subsequent analysis for different seasons showed that this pattern of consumption was maintained throughout the year, and thus that the dietary preferences of the otters did not change from one season to another.

Dietary overlap with marsh crocodile

Six marsh crocodile or mugger, *Crocodylus palustris*, ranging in size from 1.5–3 m were sighted within a 5 km stretch of the intensive study area. Analysis of 25 scats of mugger collected from this stretch between November 1989 and April 1990 revealed that 59% of the muggers' diet consisted of fish, 20.5% birds, 7.7% turtles, 2.6% crabs and 10.3% insects (Fig. 14.6). Of the fish diet, 74% was composed of unidentified Cyprinidae, 22% unidentified Bagaridae and 4% unidentified fish.

Discussion

The overall results of this study indicate that the smooth-coated otter is predominantly a fish eater and is similar in food habits to the American river otter, *Lutra canadensis*, and the Eurasian otter, *Lutra lutra*. Otters have evolved two distinct foraging modes: piscivory and invertebrate feeding. Piscivorous otters are represented by *Lutra* species and the giant river otter, *Pteroneura brasiliensis*, whereas the invertebrate feeders are the clawless and small-clawed otters (*Aonyx* spp.) and the sea otter, *Enhydra lutris* (Estes *et al.*, 1981; Chanin, 1985; Estes, 1989).

An understanding of the temporal relationship between predators and their prey is important in identifying potential limiting factors. If otters are to remain as permanent residents in an area, fish must be available all year round (Melquist & Hornocker, 1983). In the present study, the percentages of various prey items found in the spraints were considered as proportions of the prey species ingested. Changes in the diet that were recorded from spraints were found to be related to changes in the abundance, indicating that the composition of the diet of otters depended on the availability (catchability and abundance) of prey species. The items that occur in the otter's diet vary according to the species of otter involved, time of the year and place. They depend on what species are available, and it is also possible that otters have a preference for certain types of prey (Chanin, 1985). The number of species used by smooth-coated otters in the study area is similar to those used by the American river otter and the Eurasian otter. Melquist & Hornocker (1983) have identified seven fish species as potential prey of the American river otter in west central Idaho. Tiler *et al.* (1989) have identified 19 fish species as prey of the smooth-coated otter in the Narayani river in the Royal Chitwan National Park, Nepal. The seven fish species that were consumed throughout the year have been considered as the 'principal diet' (Petrides, 1975). Any changes in the abundance of these seven species may alter the food habits of otters in the study area.

Rhinomugil corsula and *Rita rita* were the preferred prey eaten throughout the year. This may be due to greater availability of these species. *Rh. corsula* is mostly found in shoals of 4–200 along the water's edge and mostly congregates near waterfalls and rapids. Otters in groups were often observed to hunt this fish in semicircles moving against the river current. *Rita rita* is found under and between stones and in cracks and crevices and is a sluggish bottom-dwelling catfish. Thus both these species are more easily available to otters than other fish species. Nevertheless, food selection and preferences cannot always be discerned from the otter's diet. It is therefore quite difficult to detect such preferences. Even if considerable care is taken to analyse the diet, it is rarely possible to obtain more than rough indications of the relative importance of the prey consumed (Chanin, 1985).

On the basis of our sampling, *Oxygaster bacaila* and *Puntius* spp. appeared to be the most abundant prey, but were poorly represented in monsoon and winter spraints. The apparent reason for this may be that these fish are small in size (5–10 cm) and have very thin scales, and thus are often digested easily. Even when the scales are present in spraints they are often difficult to identify. But during summer the fish become large enough to be detected in spraints. Larsen (1983) in his study on the American river otter has commented that soft parts of prey are difficult to identify from carnivore diet, a shortcoming that has not been addressed adequately. Better techniques are needed to quantify consumption rate of soft tissues or to improve estimates of soft tissues based on identifiable remains.

Labeo calbasu, a large carp, constituted 6% of the total diet. It is mostly found in stagnant water and is available throughout the year. It was eaten less in proportion to its availability but more in winter when the prey species available were smaller. This may be due to its higher abundance and high calorific value. However, from our comparative data of prey quality and the quantity consumed it appears that the quality of prey is not a determinant in prey selection. As compared to *Rh. corsula* and *R. rita* which form 72% of the diet, *L. calbasu* is insignificant. *Mystus seenghala* and *M. tengara* were used in proportion to their availability; these two species and *Ompok bimaculatus* have little economic value. *Mastacembelus* spp., which are eel-like slow-moving fish, constitute a minor component of the diet (0.6%). *Notopterus* spp., *Xenetodon cancila* and *Channa* spp. were occasionally eaten.

According to optimal foraging theory the predator ought to choose the most 'profitable' prey (MacArthur & Pianka, 1966; Schoener, 1971; Pulliam, 1974; Werner & Hall, 1974; Charnov *et al.*, 1976). For otters, the most profitable prey type would appear to be the most available prey that can be easily found, pursued and captured. In such a case, if a large prey is not easily available, the predator may increase its search effort. It may move larger distances in search of potential prey or may go for smaller, easily available and more abundant prey that might be more profitable. When food abundance decreases (absolutely or relatively), animals adopt one or both of the following strategies: increase the searching time with a concomitant increase in energy expenditure, or, decrease the selectivity of food items within the constraint of digestive anatomy and physiology (Schoener, 1971; Emlen, 1973; Pyke *et al.*, 1977; Krebs, 1978). Smaller prey such as crab, shrimp and frog constitute a smaller proportion of the diet of other otter species (e.g. Wise *et al.*, 1981 and Weber, 1990 for *Lutra lutra*; Melquist & Hornocker, 1983 for *Lutra canadensis*). This is also true for the smooth-coated otter. Exploitation of subsidiary prey such as bird, crab and shrimps in winter may be due to two reasons: first, such prey are more abundant, and second, the energetic need of otters is increased by postnatal investment while their movement over larger areas is restricted in winter by dependent offspring (Hussain, 1993; Hussain & Choudhury, 1995). Therefore, they have to satisfy themselves in the shortest possible time. Consequently the diversity of prey species consumed is greater in winter than in other seasons. McNab (1989) has stated that body mass, food habit, activity level and (possibly) climate influence the level of basal metabolic rate in carnivores. Thus, a sharp increase in the consumption of non-fish diet in winter appears to be a strategy for meeting additional energy requirements during winter for thermoregulation and for rearing pups.

The National Chambal Sanctuary is a protected area where fishing has been banned since 1979. During this study it was observed that the diet of the otter overlapped with that of the mugger at least with respect to fish. In the study area, large-scale interspecific competition was obvious, as other aquatic fauna such as gharial, mugger, Ganges river dolphin and several species of fish-eating birds also live within and around the otter habitat. However, in terms of density, otters are rarer than other aquatic species (Hussain, 1991; Hussain et al., 1993). We estimated that there were about 5–6 otter families with a mean group size of 4.62 (±0.11) (Hussain, 1993, 1996; Hussain & Choudhury, 1988, 1998) living in the Sanctuary. Further study on resource partitioning will be helpful in understanding the level of competition among these species. Although there was evidence of occasional illegal fishing in the areas where otters live, it is unlikely to be of a magnitude that could deplete the fish stock.

ACKNOWLEDGEMENTS
This study was sponsored by the Wildlife Institute of India through the project 'Ecology of aquatic mammals in National Chambal Sanctuary'. We are extremely grateful to Drs Sheila Macdonald and Chris Mason, University of Essex, England, and Ajith Kumar, Salim Ali Centre for Ornithology and Natural History, for their guidance throughout the study. Our colleagues Ruchi Badola, Qamar Qureshi, Raghu Chundawat and Charu Dutt Mishra gave valuable comments on the earlier version of the manuscript. Mr Rajesh Thapa and his team helped on computer. We are grateful to Dr Martyn Gorman, University of Aberdeen, and Dr Nigel Dunstone, University of Durham, for inviting us to present this paper in the Symposium. Funds for participating in the Symposium were provided to S.A.H. by the Zoological Society of London. Thanks are also due to the two anonymous referees who took much trouble to improve the quality of this paper.

References

Acharjyo, L. N. & Mishra, C. G. (1984). A note on the longevity of two species of Indian otters in captivity. *J. Bombay nat. Hist. Soc.* **80**: 636.

Byers, C. R., Steinhorst, R. K. & Krausman, P. R. (1984). Clarification of a technique for analysis of utilization-availability data. *J. Widl. Mgmt* **48**: 1050–1053.

Champion, H. G. & Seth, S. K. (1968). *A revised survey of the forest types of India.* Manager of Publication, Delhi.

Chanin, P. R. F. (1985). *The natural history of otters.* Croom Helm, London & Sydney.

Charnov, E. L., Orians, G. H. & Hyatt, K. (1976). Ecological implications of resource depression. *Am. Nat.* **110**: 247–259.

Chitampalli, M. B. (1979). Miscellaneous notes. 1. On the occurrence of the common otter in Maharashtra (Itiadoh Lake–Bhandara District) with some notes on its habits. *J. Bombay nat. Hist. Soc.* **76**: 151–152.

Desai, J. H. (1974). Observation on the breeding habits of the Indian smooth otter *Lutrogale perspicillata* in captivity. *Int. Zoo Yb.* **14**: 123–124.

Emlen, J. M. (1973). *Ecology: an evolutionary approach.* Addison-Wesley, Reading, MA.

Erlinge, S. (1969). Food habits of the otter (*Lutra lutra* L.) and the mink (*Mustela vison* Schreber) in a trout water in southern Sweden. *Oikos* **20**: 1–7.

Estes, J. A. (1989). Adaptations for aquatic living by carnivores. In *Carnivore behavior, ecology, and evolution*. 242–282. (Ed. Gittleman, J. L.). Chapman and Hall, London.

Estes, J. A., Jameson, R. J. & Johnson, A. M. (1981). Food selection and some foraging tactics of sea otters. In *Proceedings of the worldwide furbearers conference*. 606–641. (Eds Chapman, J. A. & Pursley, D.). Worldwide Furbearers Conference, Inc., Frostburg, MA.

Floyd, T. J., Mech, L. D. & Jordan, P. A. (1978). Relating wolf scat content to prey consumed. *J. Wildl. Mgmt* 42: 528–532.

Foster-Turley, P. (1992). *Conservation Ecology of Sympatric Asian Otters* Aonyx cinerea *and* Lutra perspicillata. PhD thesis: University of Florida.

Groombridge, B. (Ed.). (1994). *1994 IUCN red list of threatened animals*. IUCN, Gland, Switzerland, and Cambridge, UK.

Hinton, M. A. C. & Fry, T. B. (1923). Bombay Natural History Society's mammal survey of India, Burma and Ceylon. *Report No. 37, Nepal. J. Bombay nat. Hist. Soc.* 29: 399–428.

Hussain, S. A. (1991). *Ecology of gharial in National Chambal Sanctuary*. MPhil. thesis: Centre for Wildlife and Ornithology, Aligarh Muslim University, Aligarh.

Hussain, S. A. (1993). *Aspects of the Ecology of Smooth Coated Otters* Lutra perspicillata *in National Chambal Sanctuary*. PhD thesis: Centre for Wildlife and Ornithology, Aligarh Muslim University, Aligarh.

Hussain, S. A. (1996). Group size, group structure and breeding in smooth-coated otter *Lutra perspicillata* Geoffroy (Carnivora, Mustelidae) in National Chambal Sanctuary. *Mammalia* 60(2): 289–297.

Hussain, S. A. & Choudhury, B. C. (1995). Seasonal movement, home range and habitat use by smooth-coated otter *Lutra perspicillata* in National Chambal Sanctuary. *Habitat* 11: 45–55.

Hussain, S. A. & Choudhury, B. C. (1997). Distribution and status of the smooth-coated otter *Lutra perspicillata* in National Chambal Sanctuary, India. *Biol. Conserv.* 80: 199–206.

Hussain, S. A. & Choudhury, B. C. (1998). A preliminary survey of the status and distribution of smooth-coated otter (*Lutra perspicillata*) in the National Chambal Sanctuary. In *Proceedings of the First international seminar on Asian otters, October, 1988, Bangalore, India*. (Ed. Estes, J. A.). (In press).

Hussain, S. A., Sharma, R. K. & Choudhury, B. C. (1993). A note on the morphometry of Ganges river dolphin with comments on its mortality in fishing nets. *J. Bombay nat. Hist. Soc.* 90: 501–505.

Krebs, J. R. (1978). Optimal foraging: decision rules for predators. In *Behavioural ecology: an evolutionary approach*: 23–63. (Eds Krebs, J. R. & Davies, N. B.). Blackwell Scientific Publications, Oxford.

Kruuk, H., Kanchanasaka, B., O'Sullivan, S. & Wanghongsa, S. (1994). Niche separation in three sympatric otters *Lutra perspicillata, Lutra lutra* and *Aonyx cineria* in Huai Kha Khaeng, Thailand. *Biol. Conserv.* 69: 115–120.

Larsen, D. N. (1983). *Habitats, Movements and Foods of River Otters in Coastal Southern Alaska*. MSc thesis: University of Alaska.

Lockie, J. D. (1959). The estimation of the food of foxes. *J. Wildl. Mgmt* 23: 224–227.

MacArthur, R. H. & Pianka, E. R. (1966). On optimal use of a patchy environment. *Am. Nat.* 100: 603–609.

McNab, B. K. (1989). Basal rate of metabolism, body size, and food habits in the order Carnivora. In *Carnivore behavior, ecology, and evolution*: 335–354. (Ed. Gittleman, J. L.). Chapman and Hall, London.

Mason, C. F. & Macdonald, S. M. (1980). The winter diet of otters (*Lutra lutra*) on a Scottish sea loch. *J. Zool., Lond.* 192: 558–561.

Mason, C. F. & Macdonald, S. M. (1986). *Otters: ecology and conservation*. Cambridge University Press, Cambridge.

Melquist, W. E. & Hornocker, M. G. (1983). Ecology of river otters in west central Idaho. *Wildl. Monogr.* No. 83: 1–60.

Neu, C. W., Byers, C. R. & Peek, J. M. (1974). A technique for analysis of utilization-availability data. *J. Wildl. Mgmt* 38: 541–545.

Petrides, G. A. (1975). Principal foods versus preferred foods and their relations to stocking rate and range condition. *Biol. Conserv.* 7: 161–169.

Pocock, R. I. (1939). Notes on some British Indian otters, with description of two new subspecies. *J. Bombay nat. Hist. Soc.* **41**: 514–517.

Pocock, R. I. (1941). *The fauna of British India, including Ceylon and Burma. Mammalia* **2**. Taylor & Francis, London.

Prater, S. H. (1971). *The book of Indian animals* (3rd edn.) Bombay Natural History Society Pub., Bombay, India.

Pulliam, H. R. (1974). On the theory of optimal diets. *Am. Nat.* **108**: 59–74.

Pyke, G. H., Pulliam, H. R. & Charnov, E. L. (1977). Optimal foraging: a selective review of theory and tests. *Q. Rev. Biol.* **52**: 137–157.

Schoener, T. W. (1971). Theory of feeding strategies. *Annu. Rev. Ecol Syst.* **2**: 369–404.

Snodgrass, W. A., Bryan, A. L., Lide, R. F. & Smith, G. M. (1994). *Factors affecting the occurrence and assemblage structure of fishes in isolated wetlands of the upper coastal plain, U. S. A.* Unpublished revised second draft. Savannah River Ecology Laboratory, University of Georgia, USA.

Tiler, C., Evans, M., Hardman, C. & Houghton, S. (1989). Diet of the smooth Indian otter (*Lutra perspicillata*) and of fish eating birds: a field survey. *J. Bombay nat. Hist. Soc.* **86**: 65–70.

Wayre, P. (1978). Status of otters in Malaysia, Sri Lanka and Italy. In *Otters: proceedings of the first working meeting of the Otter Specialist Group, Paramaribo, Surinam, March 1977*: 152–155. International Union for the Conservation of Nature and Natural Resources, Morges, Switzerland.

Weber, J. M. (1990). Seasonal exploitation of amphibians by otters (*Lutra lutra*) in north-east Scotland. *J. Zool., Lond.* **220**: 641–651.

Werner, E. E. & Hall, D. J. (1974). Optimal foraging and the size selection of prey by the bluegill sunfish (*Lepomis macrochirus*). *Ecology* **55**: 1042–1052.

Wise, M. H., Linn, I. J. & Kennedy, C. R. (1981). A comparison of the feeding biology of mink *Mustela vison* and otter *Lutra lutra*. *J. Zool., Lond.* **195**: 181–213.

15

Population trends of hippopotami in the rivers of the Kruger National Park, South Africa

P. C. Viljoen and H. C. Biggs

Introduction

Hippopotamus (*Hippopotamus amphibious*) populations in South Africa, estimated to total about 5000, are considered to be stable (Eltrinham, 1993). In the 19 485 km^2 Kruger National Park (KNP) hippopotami constitute about 5.6% of the KNP's estimated total large herbivore biomass of 2136 kg/km^2 (Viljoen *et al.*, 1994). An estimated 93.8% of the KNP hippopotami population of about 2600 occur in six major rivers, while most of the hippopotami ($\bar{x} = 81.3\%$) were in three rivers; the Letaba, Olifants and Sabie. Hippopotami near the park boundary are able to move in and out of the park, as boundary fences do not cross the major rivers.

Several surveys have reported on hippopotami densities in various parts of their distribution range (Mackie, 1976; Viljoen, 1980; Karstad & Hudson, 1984; Tembo, 1987; Ngog Nje, 1988; Jacobson & Kleynhans, 1993) while a recent study emphasized important aspects of the species' social behaviour (Klingel, 1991). However, population trends in both space and time have received little attention. Scotcher (1978) described seasonal movements in response to changing water levels and Viljoen (1995) reported on the effect of a drought on the distribution of hippopotami in the KNP's Sabie River. This drought, which occurred during the study period (1991–92), was the severest recorded in the region (Zambatis & Biggs, 1995).

Hippopotami were first culled in the KNP during 1962 when 104 animals were destroyed during an experimental culling operation in the Letaba River (Pienaar *et al.*, 1966). The original KNP environmental management policy allowed for the culling of hippopotami in the Letaba, Olifants, Sabie and Crocodile Rivers (Joubert, 1986a). Hippopotamus populations in these rivers were managed according to an 'optimum population number' for each river (Letaba/Olifants = 1400; Sabie = 500, altered to 700 in 1991). No culling operations were undertaken in either the Olifants or Letaba rivers during the study period, although culling took place in the Sabie River in 1987 and 1988. A total

of 102 hippopotami were culled in the Sabie River and 201 in the Olifants/
Letaba River system in 1982 (unpublished KNP records).

This chapter aims to describe the population trends of hippopotami, as
determined during annual aerial censuses, in the Sabie, Letaba and Olifants
Rivers for an 11 year period. While the influence of culling and mortalities,
resulting from anthrax (*Bacillus anthracis*) and drought, were investigated, the
possible effect of fluctuations in river flow and grazing potential on distribution were also examined.

Study area

The climate (Gertenbach, 1980) and vegetation (Gertenbach, 1983) of the KNP
have been thoroughly described. Most of the KNP has an undulating, gently
undulating or flat and positive relief (Venter & Bristow, 1986). Two distinct
seasons, rainy and dry, are present. The rainy season (October–March) is
characterized by hot and humid conditions with temperatures reaching 44 °C
or more, while the dry season (April–September) is predominantly a mild and
dry period (Gertenbach, 1980; Venter & Gertenbach, 1986). Average rainfall
decreases from south to north and west to east, ranging from 440 mm to
700 mm per annum. The vegetation of the region consists mainly of deciduous
savannas and a total of 35 landscape types have been identified on the basis of
floristic composition and abiotic environment (Gertenbach, 1983).

Two primary river systems, the Limpopo System in the north and the
nKomati System in the south, drain the KNP from west to east. Four major
rivers, the Sabie, Olifants, Letaba and Luvuvhu, flow through the KNP while
the southern and northern boundaries are formed by the Crocodile and
Limpopo rivers, respectively. Although all these rivers were previously considered to be perennial (Joubert, 1986b), the Letaba has been largely seasonal in
recent years. The Shingwedzi River is an important seasonal river between the
Letaba and Luvuvhu Rivers. During the early 1970s, four dams were construc-
ted in the Letaba River, a tributary of the Olifants River. Pools in seasonal rivers
and pans are also potential water resources to be utilized by hippopotami.

Both the Letaba and Olifants Rivers are situated entirely within the 450–
500 mm rainfall zone in the KNP, while most of the Sabie River is within the
550–650 mm zone. A small section of the Sabie River, in the western part, is in
the region receiving more than 650 mm rain. Alternating *Combretum
apiculatum* woodland/*Colospermum mopane* tree savanna, which characterizes
most of the Letaba and Olifants Rivers' area between the western KNP boundary and the central part of the Park, is replaced by *Colospermum mopane*

scrubveld further east. The Sabie River flows through mostly *Combretum* spp./*Terminalia serivea* woodland and *Sclerocarya birrea*/*Acacia nigrescens* open tree savanna.

Venter (1991) described river geomorphology with particular reference to different river reaches, while Van Niekerk & Heritage (1993) gave a detailed review of the Sabie River's geomorphology. Vogt (1922) studied recent geomorphological changes in the Sabie and Letaba Rivers. The Sabie River has a greater diversity of geomorphological features and is also characterized by denser vegetation in the river channel than the Letaba or the Olifants Rivers. The Olifants River has the highest estimated annual runoff (2284 million m^3), the Letaba River the lowest (819 million m^3), while the Sabie River (849 million m^3) is marginally higher than the Letaba River (SA Department of Water Affairs and Forestry, unpublished data).

Methods

Hippopotami in the KNP's five major rivers, the Crocodile, Sabie, Olifants, Letaba and Luvuvhu/Limpopo Complex, have been counted annually since 1984. Standardized survey methodology, as in all KNP monitoring projects (Joubert 1983), has been used. Although aerial counts of hippopotami are usually considered to be underestimates (Marshall & Sayer, 1976), it is the only method that can be effectively applied in the KNP's river systems. A Bell 206 Jetranger helicopter was used for attempted total counts of hippopotami. The survey crew consisted of the pilot (front right seat) and two observers (front left seat and rear right seat) and a data recorder (left rear seat) who also assisted as an observer. Sightings, which included group size and number of calves (individuals estimated to be 1 year or younger), were recorded directly onto 1 : 10 000 maps noting exact localities while differentiating between hippopotami seen in or out of the water. Although hippopotamus groups can be classified into two main groups, groups consisting of females and their offspring and bull groups (Klingel, 1991), it was impossible to sex individuals reliably from the air.

The surveys were conducted at a height of 20–55 m above the surface at an airspeed of 30–65 km/h, except when the helicopter had to hover briefly to allow for the counting of large herds. Flight paths generally followed the river's centre line but a criss-cross pattern was flown in areas where the river is wide or braided.

Counts were restricted to warm, sunny days between mid-morning and early afternoon (from about 10:00 to 14:00). This ensured that hippopotami were not grazing away from the rivers and that the majority were in shallows, outside the water or on the river/sandbanks and therefore easier to count. The surveys were conducted during the mid-dry season (June–August) when pool depths are shallow as a result of reduced water flow and the relatively clearer water allows easier counting of submerged hippopotami.

A digitizer tablet was used to transfer recorded observations to a computer after each survey flight. The latitude and longitude of each observation was later calculated. A Global Positioning System (GPS) and palmtop computer system (Viljoen & Retief, 1994) was implemented in 1994 to enable real-time data collection.

Hippopotamus carcasses were located by KNP field staff during two drought-related die-offs. KNP ranger diaries were also consulted for additional information on mortalities.

Venter's (1991) physical classification of the KNP's rivers was adapted to identify five sections for each river. Densities, expressed as the number of hippopotami per kilometre of river front, were calculated according to these river sections. To allow for comparisons with hippopotamus densities else-where, grazing densities (Laws, 1981), based on the general assumption that hippopotami forage within a 3.2 km zone from the river (Pienaar *et al.*, 1966; Lock, 1972; Mackie, 1976), were also calculated. The surface area thus calculated was 669.3 km^2 for the Letaba River, 519.2 km^2 for the Olifants and 546.0 km^2 for the Sabie. Where rivers also form park boundaries (section A of the Sabie River and a part of sections A and B of the Olifants River), those parts of the zones lying outside the boundary were excluded prior to 1994 since a boundary fence prevented almost all hippopotamus movements to neighbour-ing areas. In addition, grazing densities for 10 km zones were also calculated. A Geographical Information System (Spans, Interna Tydac Technologies, Inc.) was used to calculate surface areas associated with rivers.

The grazing potential of areas adjoining the KNP rivers was estimated from annual routine monitoring data on veld condition and herbaceous biomass (> 500 sample points in the KNP) by applying a technique described by Trollope *et al.* (1989) and Trollope (1990). Since 1989, veld condition, in terms of its potential to produce grass forage for consumption by larger herbivores, has been assessed by using multiple regression models.

Areal extent of each forage potential class within a 10 km buffer zone was determined for interpolated forage potential maps. These areas were multi-plied by the class value and a new area-weighted composite forage potential was calculated for each section for each year (1988–1994). This can be con-sidered to be the overall forage value of the river section per year.

The possible effect of river flow rates on hippopotamus densities was investigated for the dry season only, since this period is considered to be of particular importance to hippopotami as KNP rivers then reach their lowest flow.

Results

Flight times during the six most recent surveys (1989–1994) had a mean duration of 114 min for the Sabie River (range 97–131 min), 93 min (range 81–108 min) and 85 min (range 67–97 min) for the Letaba and Olifants Rivers, respectively. Mean search times were 65 min/km for the Sabie, 52 min/km for the Olifants and 65 min/km for the Letaba Rivers. A mean total of 2000 (range 1675–2244) hippopotami were counted annually during the 11 year period in the three KNP rivers. The majority of hippopotami were in the Olifants River ($\bar{x}$= 691.0; range 601–850; SD = 65.7) followed by the Letaba ($\bar{x}$= 675.2; range 559–793; SD = 77.2) and Sabie rivers ($\bar{x}$= 633.7; range 499–769; SD = 96.7). Most hippopotami (64.7%; n = 4354) were counted outside the water; significantly more hippopotami (76.1%) were seen outside the water in the Sabie River (χ^2 = 8.8; p < 0.05; d.f. = 2).

Hippopotamus densities were highest in the Letaba River ($\bar{x}$= 6.9 animals/km; range 5.7–8.1; SD = 0.8; Figs. 15.1 and 15.2). The mean hippopotamus density in the Olifants River was 6.4 animals/km (range 5.6–7.9; SD = 0.6; Fig. 15.3) and 5.7 animals/km in the Sabie River (range 4.5–6.9; SD = 0.9; Figs 15.4 and 15.5). Mean hippopotami grazing densities (Table 15.1) were the highest for the Olifants River (1.3 animals/km), followed by the Sabie and Letaba Rivers (1.2 and 1.0 animals/km, respectively).

Letaba River
Both the highest (11.9 animals/km; SD = 1.5) and lowest (2.7 animals/km; SD = 2.2) mean hippopotamus densities per river section occurred in the Letaba River sections D and A, respectively (Figs. 15.2 and 15.5). On average, most of the Letaba River's hippopotami (76.5%) were in sections D (42.3%) and B (34.2%). The maximum density never exceeded 5.0 animals/km in section C.

The four dams in the Letaba River (Table 15.2) had a mean total of 212.9 hippopotami (31.3% of the river's population). Engelhard Dam in section C (Fig. 15.2) had most of the hippopotami counted in dams ($\bar{x}$= 81.9; range 62–97). Shimuwini and Mingerhout dams had as many as 130 (in 1988) of section B's hippopotami ($\bar{x}$= 48.5%; range 28.1–72.3%). All of section A's hippopotami were in Black Heron Dam during the 1993 and 1994 surveys (six and two animals, respectively).

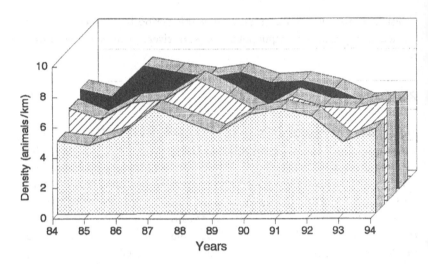

Figure 15.1. Hippopotamus density trends in three Kruger National Parks rivers (1984–1994). ▥, Sabie River; ▨, Olifants River; ■, Letaba River.

There was no significant difference in hippopotamus density between survey years (two-way analysis of variance; F-ratio = 0.94; $p > 0.05$; d.f. = 10). However, densities varied very significantly between river sections (two-way analysis of variance; F-ratio = 110.47; $p < 0.01$; d.f. = 4). Hippopotamus densities did not vary significantly between sections A and E, but the variance was significant between all other sections (Table 15.3).

The Letaba River stopped flowing at the KNP's western boundary for periods of 1 month or longer during 7 of the 11 years (54.5%). The number of months when no flow was recorded exceeded 5 in 3 years (1986, 1993 and 1994). Although the Letaba River's flow varied greatly (total annual runoff range = 30.2–300.7 million m³), no correlation existed between hippopotamus counts and annual dry-season (May–October) flow rates ($r = 0.03$). However, a marginal relationship existed between the numbers of months per year with no or minimal total flow (≤ 1.0 million m³; $\bar{x} = 5.0$ days; sd = 2.0) and hippopotamus counts in the Letaba River's sections A ($r = -5.1$; $p = 0.11$) and B ($r = -0.6$; $p = 0.06$). An increase in the number of months with reduced flow thus resulted in a decrease of hippopotamus counts. Hippopotamus counts increased in section D ($r = 0.8$; $p = 0.004$) during years with reduced flow.

Olifants River

An average of 46.2% of the Olifants River's hippopotami were in sections C (29.0%) and D (17.2%), which also had the highest mean densities, 11.1 and

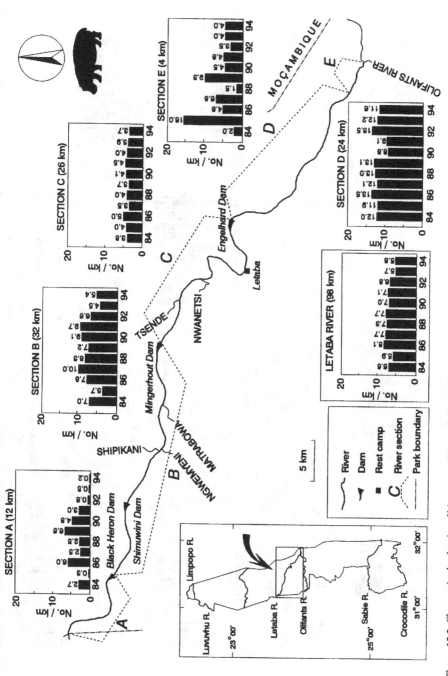

Figure 15.2. Changes in the density of hippopotami (animals/km) in five sections of the Letaba River, Kruger National Park (1984–1994).

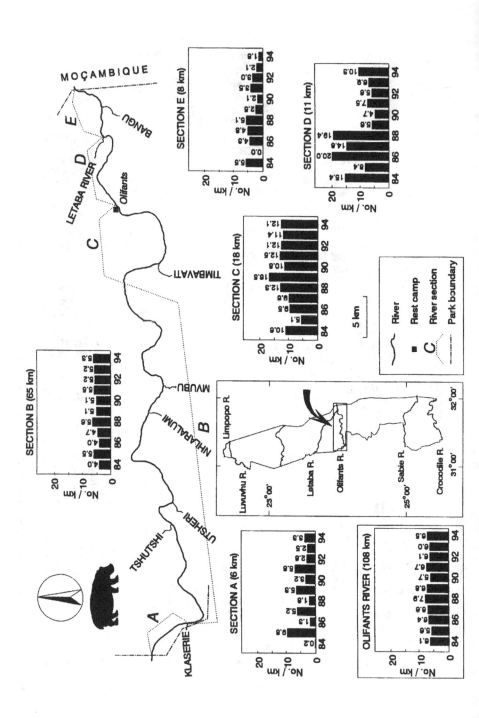

MOÇAMBIQUE

BANGU

LETABA RIVER

Olifants

TIMBAVATI

MVUBU

NHLARALUMI

TSHUTSHI

UTSHERI

KLASERIE

SECTION E (8 km)

No. / km

84	86	88	90	92	94
5.5	4.8	5.1	2.5	2.1	3.5	3.0	2.1	1.8

SECTION D (11 km)

No. / km

84	86	88	90	92	94
8.4	15.4	20.0	14.8	19.4	5.6	4.7	7.5	5.6	6.8	10.9

SECTION C (18 km)

No. / km

84	86	88	90	92	94
10.6	5.1	9.5	9.5	12.5	18.5	10.6	12.5	12.1	11.4	12.1

SECTION B (65 km)

No. / km

84	86	88	90	92	94
4.0	6.5	4.0	4.7	5.6	6.1	6.5	5.2	5.2	5.9

SECTION A (6 km)

No. / km

84	86	88	90	92	94
0.2	9.8	1.3	5.2	1.3	1.8	5.5	5.2	5.5	2.5	5.9

OLIFANTS RIVER (108 km)

No. / km

84	86	88	90	92	94
5.1	5.6	5.4	7.6	7.9	5.6	5.7	6.7	6.1	6.0	6.5

Limpopo R.

Luvuvhu R.

23°00'

Letaba R.

Olifants R.

25°00'

Sable R.

Crocodile R.

31°00'

32°00'

5 km

River
Rest camp
C River section
Park boundary

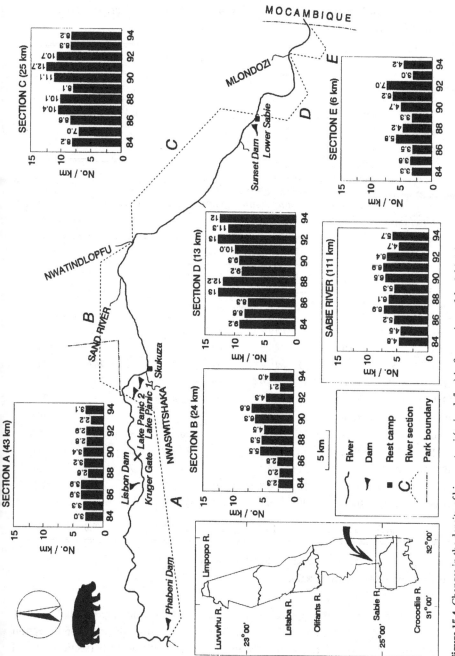

Figure 15.4. Changes in the density of hippopotami (animals/km) in five sections of the Sabie River, Kruger National Park (1984–1994).

Table 15.1. *Mean hippopotamus grazing densities in a 3.2 km zone for three Kruger National Park rivers (1984–1994)*

River	Area[†] (km²)	Density (animals/km²)		
		$\bar{x}$	Range	SD
Letaba	669.3	1.01	0.84–1.18	0.12
Olifants	519.2	1.32	1.15–1.62	0.13
Sabie	546.0	1.16	0.91–1.41	0.17

[†] Grazing areas calculated as a 3.2 km zone on both sides of each river

Table 15.2. *Numbers of hippopotami counted in dams situated in the Letaba River, Kruger National Park (1984–1994)*

	Total in dams[†]								
	Black Heron		Shimuwini		Mingerhout		Engelhard		
Year	n	%[‡]	n	%[‡]	n	%[‡]	n	%[‡]	All
1984	27	84.4	40	17.9	71	31.8	77	77.8	215
1985	31	100.0	19	40.3	67	56.3	85	81.7	203
1986	48	66.7	49	19.7	62	24.9	97	74.6	256
1987	25	92.6	51	15.9	68	21.2	75	83.3	219
1988	26	86.7	48	18.0	82	30.8	82	78.1	238
1989	51	64.6	30	13.0	62	26.8	62	64.6	205
1990	39	67.2	22	7.5	60	20.5	81	76.4	202
1991	46	97.2	40	12.9	60	19.4	82	70.1	217
1992	3	33.3	11	5.2	95	44.8	79	76.0	188
1993	6	100.0	10	8.3	86	59.7	94	94.0	198
1994	2	100.0	29	16.7	83	47.7	87	89.7	201
$\bar{x}$	32.2	81.2	34.5	16.0	72.4	34.9	81.9	78.8	212.9

[†] Hippopotami counted in dams are included in river totals.
[‡] Percentage of total for river section.

10.8 animals/km respectively (Figs 15.3 and 15.5). The highest single density recorded during the study, 20.0 animals/km, occurred during 1986 in section D. The two relatively short sections at the KNP's western (section A) and eastern (section E) boundaries had the lowest mean densities (3.8 and 3.2 animals/km, respectively).

Table 15.3. *Multiple range analyses summarizing differences in hippopotamus densities between river sections in three Kruger National Park rivers (1984–1994)*

River	River section	LS mean[†]	Differences between river sections[‡]
Letaba	A	32.182	B, C, D
	B	230.909	A, C, D, E
	C	104.364	A, B, D, E
	D	285.545	A, B, C, E
	E	22.182	B, C, D
Olifants	A	22.545	B, C, D
	B	324.636	A, C, D, E
	C	200.182	A, B, D, E
	D	118.455	A, B, C, E
	E	25.182	B, C, D
Sabie	A	133.909	B, C, E
	B	100.182	A, C, D, E
	C	235.182	A, B, D, E
	D	137.727	B, C, E
	E	26.727	A, B, C, D

[†]Least-squares mean.
[‡]Significant difference between river sections.

Hippopotamus densities did not vary significantly between survey years (two-way analysis of variance; F-ratio $= 0.52$; $p > 0.05$; d.f. $= 10$) but differences between river sections were significant (two-way analysis of variance; F-ratio $= 99.34$; $p < 0.001$; d.f. $= 4$). Hippopotamus densities varied significantly between sections B, C and D (Table 15.3).

Sabie River
The majority of hippopotami in the Sabie River, 58.9%, occurred in sections D ($\bar{x}$ density $= 10.6$ animals/km; sd $= 1.7$) and C ($\bar{x}$ density $= 9.4$ animals/km; sd $= 1.6$). The only densities exceeding 10 animals/km in the Sabie River were in these two sections; the maximum was 13.0 animals/km in 1992 (Fig. 15.4). Section A had the lowest mean density (3.1 animals/km; sd $= 0.5$) while densities never exceeded 7 animals/km in any of the other sections.

Hippopotami were also encountered in dams near the Sabie River (≤ 5 km

Table 15.4. *Numbers of hippopotami counted in dams near the Sabie River, Kruger National Park (1984–1994)*

	Total in dams[†]							
	Phabeni (1.1 km)[‡]		Lisbon (1.0 km)[‡]		Lake Panic 1 & 2 (0.7–1.2 km)[‡]		Sunset (0.3 km)[‡]	
Year	n	%[§]	n	%[§]	n	%[§]	n	%[§]	All
1984	27	20.9	36	27.9	1	1.8	n.c.		64
1985	26	18.2	29	20.3	5	10.6	n.c.		60
1986	30	18.1	26	15.7	4	6.1	n.c.		60
1987	28	16.9	27	16.3	21	15.8	5	1.9	81
1988	25	22.5	12	10.8	17	13.3	3	1.2	57
1989	29	21.3	1	0.7	18	16.7	4	2.0	52
1990	18	12.2	2	1.4	17	11.3	3	1.1	40
1991	27	22.1	15	12.3	17	10.5	15	4.7	74
1992	26	20.6	12	9.5	11	10.6	1	0.4	50
1993	15	16.0	4	4.3	9	18.0	4	1.9	32
1994	25	18.8	1	0.8	12	12.4	3	1.5	41
$\bar{x}$	25.1	18.9	15.0	10.9	12.0	11.5	4.8	1.8	55.5

[†]Totals for dams included in river totals.
[‡]Distance from the Sabie River.
[§]Percentage of total for river section.
n.c., no separate count available for Sunset Dam.

from the river) during all surveys (Table 15.4, Fig. 15.4). The total number of hippopotami in these five dams ranged between 32 and 81 ($\bar{x} = 55.5$; 5.5–12.1% of river total). Up to 48.8% ($\bar{x} = 29.8\%$) of hippopotami in section A alone (1984) have been recorded in Lisbon and Phabeni dams.

Differences in hippopotamus density were significant both between years (two-way test of variance; F-ratio = 2.98; $p < 0.05$; d.f. = 10) and sections (two-way test of variance; F-ratio = 94.098; $p < 0.01$; d.f. = 4). Sabie River hippopotamus density varied significantly between all river sections except between sections A and D (see Table 15.3).

Group sizes and calves
Single individuals constituted 30.4%, 32.6% and 32.7% of hippopotami counted in the Letaba, Olifants and Sabie Rivers, respectively (Fig. 15.5). The

majority of hippopotami occurred in groups of 2–5 individuals (Letaba River = 36.0%; Olifants River = 36.5%; Sabie River = 40.9%). This ratio did not differ significantly ($\chi^2 = 0.17$; $p > 0.05$; d.f. = 2). The ratio of hippopotami that occurred in the six different group-size categories (Fig. 15.5) differed significantly during the study period for the Letaba ($\chi^2 = 97.85$; $p < 0.01$; d.f. = 50) and Sabie rivers ($\chi^2 = 63.10$; $p < 0.05$; d.f. = 50) but not for the Olifants River ($\chi^2 = 53.18$; $p > 0.05$; d.f. = 50). Groups exceeding 40 individuals occurred in the Letaba and Olifants rivers ($n = 7$; $n = 2$, respectively).

The mean percentages of calves were 6.1% (range 0.8–12.4%; SD = 4.0), 4.9% (range 0.9–11.9%; SD = 3.6) and 7.8% (range 1.4–13.1%; SD = 2.8) for the Letaba, Olifants and Sabie Rivers, respectively (Fig. 15.6). There were no significant differences in the percentage of calves between years and individual river sections for each river (Letaba River: $\chi^2 = 201.3$; $p > 0.05$; d.f. = 10; Olifants River: $\chi^2 = 199.6$; $p > 0.05$; d.f. = 10; Sabie River: $\chi^2 = 74.4$; $p > 0.05$; d.f. = 10). The lowest mean percentage of calves for all three rivers combined (1.4%; range 1.4–10.8%) was recorded during 1993, the year following the 1991/92 drought (Fig. 15.6). Marshall & Sayer's (1976) population model indicated a calving rate of 66% of the mature female hippopotamus population in eastern Zambia.

Mortalities

A localized drought in 1988 near the lower reaches of the Letaba and Olifants Rivers resulted in at least 50 hippopotamus mortalities in sections D and E of the latter river. The effect of this die-off was visible in the hippopotamus densities of these sections during the census of the following year (1989) (Fig. 15.3). Thirteen Letaba River hippopotami died at the end of the 1990 dry season in sections A and B, most probably as a result of anthrax (V. de Vos, personal communication) while a further five were found further downstream. The relative vulnerability (percentage deaths/percentage availability) of the KNP's larger herbivores including hippopotami to anthrax has been described by de Vos (1990) as 2.1%. Seven hippopotamus carcasses recorded along the Letaba River in 1991 were also suspected anthrax-related mortalities (V. de Vos, personal communication).

A total of 70 hippopotamus mortalities were recorded in the Letaba River (sections A and B) between March and October 1992 (1991/92 drought) of which 49 (70.0%) occurred prior to the 1992 census. The majority of these carcasses (58.6%) were located during June–July 1992. Predominantly adult animals died (77.1%) while calves ($\leq$ 1 year old) constituted 8.3% of the aged carcasses ($n = 48$). An additional six hippopotamus mortalities were also recorded in Engelhard Dam during the 1991/92 drought.

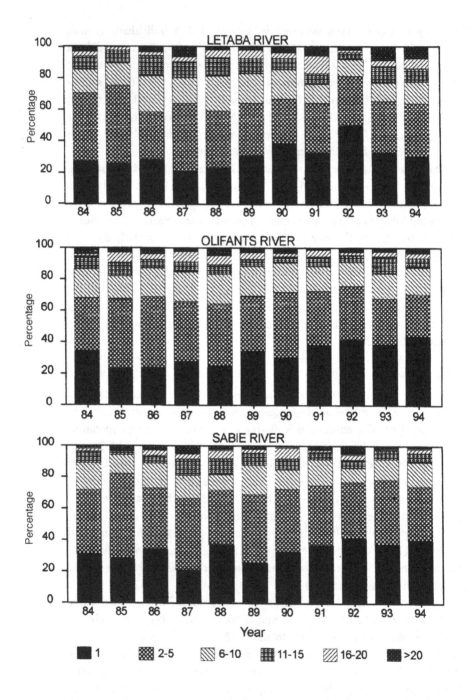

Figure 15.5. Hippopotamus group sizes in three Kruger National Park rivers (1984–1994).

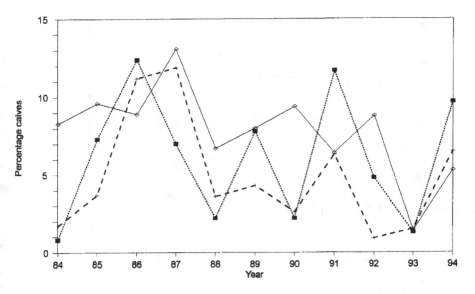

Figure 15.6. Percentages of hippopotamus calves recorded in three Kruger National Park rivers (1984–1994). Calves classified as individuals estimated to be ≤1 year old. --■--, Letaba River; --+--, Olifants River, ◇, Sabie River.

Although fewer than three hippopotamus mortalities were reported for the Olifants River during the 1991/92 drought, 14 and 11 carcasses were located in sections B and C, respectively during 1992. Ten carcasses were found at the Letaba/Olifants confluence at the end of the 1993 dry season. Twelve hippopotamus mortalities were recorded for the Sabie River (Viljoen, 1995).

Discussion

Densities

Hippopotami are known to have a preference for areas of slow and relatively shallow water combined with gently sloping banks where animals can lie half-immersed (Clarke, 1953; Laws & Clough, 1966; Field, 1970; Olivier & Laurie, 1974). The highest mean densities of hippopotami in the Letaba River (7.2 animals/km, section B; 11.9 animals/km, section D) were in river sections characterized by an irregular, highly divided riverbed with banks of rocks and stands of dense trees and reed beds. Three dams are also present in these sections (see Table 15.2; Fig. 15.2). High mean densities in the Olifants River (11.1 animals/km, section C; 10.3 animals/km, section D) were similarly

Table 15.5. Population trends of hippopotami in different river sections of the Letaba River, Kruger National Park (1984–1994)

	River section[†]											
	Section A		Section B		Section C		Section D		Section E		All	
Year	n	% diff.	n	% diff.	n	% diff.	n	% diff.	n	% diff.	n	% diff.
1984	32	—	223	—	99	—	289	—	8	—	651	—
1985	3	−90.6	119	−46.6	104	5.1	285	−1.4	64	700.0	575	−11.7
1986	72	2300.0	249	109.2	130	2.50	323	13.3	19	−70.3	793	37.9
1987	27	−62.5	321	28.9	90	−30.8	290	−10.2	27	42.1	755	−4.8
1988	30	11.1	266	−17.1	105	16.7	312	7.6	6	−77.8	719	−4.8
1989	79	163.3	231	−13.2	96	−8.6	315	1.0	37	516.7	758	5.4
1990	58	−26.6	292	26.4	106	10.4	212	−32.7	18	−51.4	686	−9.5
1991	36	−37.9	309	5.8	117	10.4	219	3.3	19	5.6	700	2.0
1992	9	−75.0	212	−31.4	104	−11.1	324	47.9	14	−26.3	663	−5.3
1993	6	−33.3	144	−32.1	100	−3.8	293	−9.6	16	14.3	559	−15.7
1994	2	−66.7	174	20.8	97	−3.0	279	−4.8	16	0.0	568	1.6
$\bar{x}$	32.2	208.2	230.9	5.1	104.4	1.0	285.5	1.4	22.2	105.3	675.2	−0.5
%[‡]	−14.6		−0.9		−0.4		−0.9		−2.4		−1.4	

[†] See Fig 15.2.

[‡] Average percentage population change ($\bar{r} \times 100$).

associated with irregular, divided channels and stands of trees and reeds (section C). However, deep pools in mostly a single channel with few reeds distinguished section D. Sabie River sections with single or moderately irregular, divided channels and dense reedbanks, sandbanks and islands had the highest mean densities (11.1 animals/km, section C; 10.8 animals/km, section D).

Relative densities of hippopotami remained more stable in the Sabie River than in the Letaba and Olifants rivers (Tables 15.5–15.7). The Letaba River's hippopotamus population decreased by an average percentage of −1.4 ($\bar{r}= -0.014$). Minor increases, indicated by an average percentage growth of 0.2 ($\bar{r}= 0.002$) and 1.5 ($\bar{r}= 0.016$) respectively, occurred in the Olifants and Sabie Rivers.

The most important changes in the Letaba River's sections occurred in sections E and A (Table 15.5); the latter had a marked decrease in hippopotamus numbers ($\bar{r}= -0.146$). Olifants River hippopotamus numbers reflected marked changes in sections A and E (Table 15.6) while increases also occurred in these sections ($\bar{r}= 0.114$ and $\bar{r}= 0.028$, respectively). The only decrease was in section D ($\bar{r}= -0.085$). Most of the Sabie River's hippopotamus changes were in sections B and E (Table 15.7). The only population decrease in this river was in section A ($\bar{r}= -0.024$).

Mean hippopotamus densities in the three KNP rivers are significantly lower than estimated during total counts in many other areas. Tembo (1987) found a density of 38 animals/km along Zambia's Luangwa River in the South Luangwa National Park. During an earlier survey Ansell (1965) estimated a hippopotamus density of about 18.8 animals/km in the Luangwa River. Ngog Nje (1988) calculated densities of up to 114 and 135 animals/km in two of Bénoué National Park's (NP) rivers. A density of 16.1 animals/km was estimated for a section of the Mara River (Olivier & Laurie, 1974) while Karstad & Hudson (1984) reported a density of 19.6 animals/km in the Mara River within the Masai Mara National Reserve and 8.4 animals/km outside the reserve. Norton (1988) reported a density as high as 73.8 animals/km along a 51.5 km stretch of the Luangwa River. Considering a 3.2 km grazing zone both sides of rivers, Mackie's (1976) data from the Lundi River in Zimbabwe's Gonarezhou Game Reserve indicated a density of 4.5 animals/km^2. This is considerably more than the densities calculated for the three KNP rivers (Table 15.1) where the highest density was 1.6 animals/km^2 (Olifants River). Laws (1981) considered the hippopotamus populations in lakes and rivers on the western Rift Valley, formerly exceeding 31 animals/km^2, to be the highest known. Hippopotami accounted for over 50% of the large herbivore biomass in parts of the Queen Elizabeth National Park, Uganda (Lock, 1972).

Table 15.6. *Population trends of hippopotami in different river sections of the Olifants River, Kruger National Park (1984–1994)*

| | River section[†] | | | | | | | | | | | |
| | Section A | | Section B | | Section C | | Section D | | Section E | | All | |
Year	n	% diff.	n	% diff.	n	% diff.	n	% diff.	n	% diff.	n	% diff.
1984	1	—	257	—	190	—	169	—	44	—	661	—
1985	59	5800.0	359	39.7	91	-52.1	92	-45.6	1	-97.7	601	-9.1
1986	8	-86.4	257	-28.4	171	87.9	220	139.1	34	3300.0	690	14.8
1987	31	287.5	303	17.9	173	1.2	163	-25.9	38	11.8	708	2.6
1988	11	-64.5	363	19.8	222	28.3	213	30.7	41	7.9	850	20.1
1989	32	190.9	328	-9.6	297	33.8	61	-71.4	20	-51.2	738	-13.2
1990	19	-40.6	329	0.3	194	-34.7	52	-14.8	17	-15.0	611	-17.2
1991	35	84.2	357	8.5	225	16.0	82	57.7	28	64.7	727	19.0
1992	17	-51.4	340	-4.8	217	-3.6	62	-24.4	24	-14.3	660	-9.2
1993	15	-11.8	336	-1.2	205	-5.5	76	22.6	17	-29.2	649	-1.7
1994	20	33.3	342	1.8	2.17	5.9	113	48.7	14	-17.6	706	8.8
$\bar{x}$	22.6	614.1	324.6	4.4	200.2	7.7	118.5	11.7	25.3	315.9	691.0	1.5
%[‡]	11.4		2.0		4.6		-8.5		2.8		0.2	

[†] See Fig 15.3.
[‡] Average percentage population change ($\bar{r} \times 100$).

Table 15.7. *Population trends of hippopotami in different river sections of the Sabie River, Kruger National Park (1984–1994)*

| | River section[†] | | | | | | | | | | | |
| | Section A | | Section B | | Section C | | Section D | | Section E | | All | |
Year	n	% diff.	n	% diff.	n	% diff.	n	% diff.	n	% diff.	n	% diff.
1984	129	—	56	—	205	—	119	—	20	—	529	—
1985	143	10.9	47	-16.1	174	-15.1	112	-5.9	23	15.0	499	-5.7
1986	166	16.1	66	40.4	214	23.0	108	-3.6	21	-8.7	575	15.2
1987	166	0.0	133	101.5	261	22.0	166	53.7	35	66.7	761	32.3
1988	111	-33.1	128	-3.8	253	-3.1	158	-4.8	25	-28.6	675	-11.3
1989	136	22.5	108	-15.6	203	-19.8	119	-24.7	20	-20.0	586	-13.2
1990	147	8.1	151	39.8	278	36.9	121	1.7	28	40.0	725	23.7
1991	122	-17.0	162	7.3	318	14.4	130	7.4	37	32.1	769	6.1
1992	126	3.3	104	-35.8	267	-16.0	169	30.0	42	13.5	708	-7.9
1993	94	-25.4	50	-51.9	208	-22.1	147	-13.0	18	-57.1	517	-27.0
1994	133	41.5	97	94.0	206	-1.0	166	12.9	25	38.9	627	21.3
$\bar{x}$	133.9	2.7	100.2	16.0	235.2	1.9	137.7	5.4	26.7	9.2	633.7	3.4
%[‡]	-2.4		4.5		1.7		3.0		2.2		1.6	

[†] See Fig. 15.4.

[‡] Average percentage population change ($\bar{r} \times 100$).

Scotcher (1978) reported seasonal fluctuations in hippopotamus numbers as a result of movement between the Ndumu River (KwaZulu-Natal) and seasonal pans that depended on the degree of flooding.

Day living space, rather than productivity of the habitat, was a limiting factor for Mara River's hippopotami (Olivier & Laurie, 1974), as appeared also to be the case in Ndumu Game Reserve (Scotcher, 1978).

Several population increases in many areas have been reported. Tembo (1987) calculated an annual population increase of 7% in the Luangwa River. Karstad & Hudson (1984) found that the rates of increases in the Masai Mara varied from 0.9% to 10.3% but an earlier annual rate of increase as high as 16.5% was reported by Darling (1961). Norton's (1988) subsequent survey indicated a marked population increase since the 1950s. An average hippopotamus population increase of 2.8% during an 11 year period occurred in Bénoué NP (Ngog Nje, 1988).

Effect of river flow

Dry-season (April–September) river flow varied in all three KNP rivers during the study (Figs 15.7–15.9). Variations in river flow were particularly noticeable in the Sabie River, where peaks in flow were typically recorded early in the dry season (May) in both the Olifants and Sabie Rivers. Higher rates of flow in the Letaba River occurred mostly towards the end of the dry season (September or later).

Changes in dry-season river flow rates had no correlation with hippopotamus numbers in the various sections and years in either the Olifants ($r = 0.12$) or Sabie Rivers ($r = 0.06$). Two additional Sabie River hippopotamus counts during the 1991/92 drought did not detect significant changes in density between river sections (Viljoen, 1995).

Jacobson & Kleynhans (1993) described the role of storage weirs in providing permanent waters that proved to be important for the survival of hippopotami. The four Letaba River dams, which had an average of 31.3% of this river's hippopotami during the study period, clearly contributed to providing additional aquatic habitat to hippopotami during periods of reduced flow.

Effect of grazing potential

River sections D and E of the Letaba River (Fig. 15.2; Table 15.8) and C and D of the Olifants River (Fig. 15.3) were combined to investigate the effect of grazing potential (Trollope et al., 1989; Trollope, 1990) on hippopotamus density at the confluence of these two rivers. The maximum distance between the two rivers is 18 km in this area, which was therefore considered to be utilized by hippopotami from both rivers. The remaining sections of both

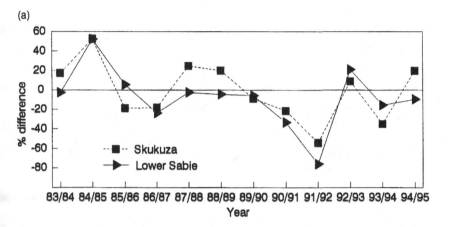

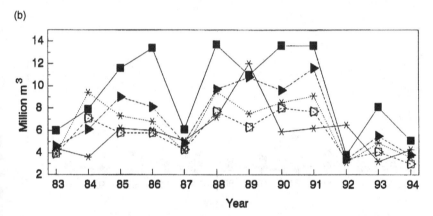

Figure 15.7. (a) Rainfall trends recorded at two rainfall stations (Skukuza and Lower Sabie) in the vicinity of the Sabie River (1983/84–1994/95), expressed as a percentage of the long-term mean annual rainfall. (b) Total monthly dry-season (May–Oct) flow measured for the Sabie River at Perry's Farm (1984–1994) upstream of the study area. Flow data supplied by the Department of Water Affairs and Forestry. ■, May; -- ▶--, June; --✳--, July; □, August; --▷--, September; —✳—, October.

rivers were investigated separately. No evidence of a significant relationship between the grazing potential and hippopotamus density (animals/km^2) for the confluence could be found ($R^2 = 35.9$). The combined confluence area appeared to have had a higher hippopotamus density than might be expected from the grazing potential.

Hippopotamus density (animals/km^2) in the non-confluence region of the Letaba River (sections A–C) was strongly related to the grazing potential of the surrounding areas ($R^2 = 70.4$). River sections A–C of the Letaba and Olifants

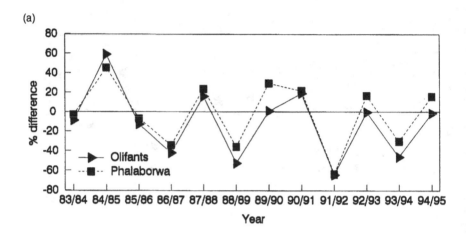

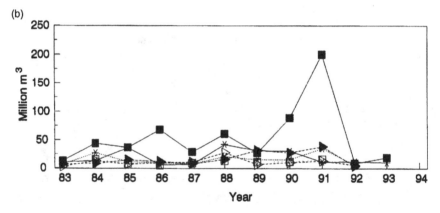

Figure 15.8. (a) Rainfall trends recorded at two rainfall stations (Phalaborwa and Olifants) in the vicinity of the Olifants River (1983/84–1994/95), expressed as a percentage of the long-term mean annual rainfall. (b) Total monthly dry-season (May–Oct.) flow measured for the Olifants River at Phalaborwa Barrage (1984–1994) upstream of the study area. Flow data supplied by the Department of Water Affairs and Forestry. ■-, May; --▶--, June; --✻--, July; □, August; --▷--, September; —✻—, October.

Rivers had a significant relationship ($R^2 = 77.5$ and $R^2 = 66.4$, respectively) when regressed on grazing potential in appropriate years. Higher hippopotamus counts were therefore associated with higher estimates of grazing potential in any year and any section away from the confluence of these two rivers. However, in the case of the Olifants River, surface area alone (sections A–C) gave an even better relationship ($R^2 = 86.0$).

Grazing potential influenced Sabie River hippopotamus counts. The overall grazing potential of the Sabie River sections in the various years could be used

(a)

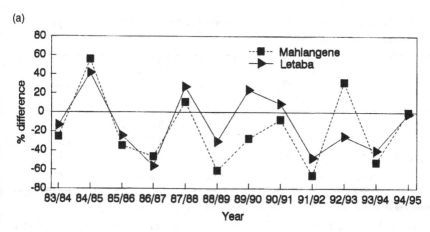

(b)

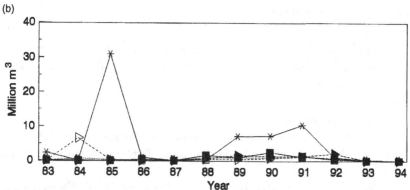

Figure 15.9. (a) Rainfall trends recorded at two rainfall stations (Mahlangene and Letaba) in the vicinity of the Letaba River (1983/84–1994/95), expressed as a percentage of the long-term mean annual rainfall. (b) Total monthly dry-season (May–Oct.) flow measured for the Letaba River at Letaba Ranch (1984–1994) upstream of the study area. Flow data supplied by the Department of Water Affairs and Forestry. ■, May; --▶--, June; --✳--, July; □, August; --▷--, September; —✳—, October.

with some success to roughly predict the hippopotamus totals ($R^2 = 52.7$). Hippopotamus density in river section C, the second highest mean hippopotamus density in this river, had, however, a peripherally higher density than predicted by the grazing potential, particularly during the low rainfall period of 1991/92 (Figs. 15.4 and 15.7; Table 15.9).

The lumping of two sections each of the Letaba and Olifants Rivers (D–E) to represent the confluence area formed by these two rivers was possibly not fully representative of the area utilized by hippopotami of both rivers. It is also not clear to what extent hippopotami move between the rivers in the vicinity of

Table 15.8. *Mean hippopotamus grazing densities in a 10 km zone for three Kruger National Park rivers (1984–1994)*

| River | River sections | Area[†] (km^2) | Density (animals/km^2) | | |
			$\bar{x}$	Range	SD
Letaba	A–C	1422	0.24	0.15–0.30	0.05
Olifants	A–e	1103	0.36	0.28–0.43	0.04
Confluence[‡]	D–E	1034	0.44	0.29–0.58	0.08
Sabie	A–E	1535	0.41	0.33–0.50	0.06

[†] Grazing areas calculated as a 10 km zone on both sides of each river.
[‡] Sections D and E of both the Letaba and Olifants Rivers were combined to calculate densities for the confluence region of these two rivers.

the confluence. Decreases in hippopotamus density in one river's sections were not clearly followed by corresponding increases in the other river's sections (Figs 15.1 and 15.2).

Population composition

The mean percentages of hippopotamus calves in the KNP rivers (4.9–7.8%) were mostly similar to the calf ratio reported by Tembo (1987) of 4.6% for the Luangwa River.

Olivier & Laurie (1974) concluded that the Masai Mara's hippopotamus population is unstable as a result of climatic influences: the population became more dispersed and the mean group size decreased after a rise in the water level. Berry (1973) encountered herd sizes of up to 74 in the Luangwa River, while groups exceeding 100 individuals have occurred in Bénoué NP (Ngog Nje, 1988). Such large groups should be considered temporary congregations rather than herds.

Group size depends on density and environmental conditions (Klingel, 1991). Ngog Nje (1988) found a negative correlation between group size and water levels in Bénoué NP. Nursery groups of females and offspring were unstable in composition and moved in response to changing water levels (Karstad & Hudson, 1986). These authors also found no evidence that behavioural mechanisms played a role in limiting population growth; dominant males followed groups of females to adjacent pools while assuming a sub-dominant position. Once water levels were again at a suitable depth, the males returned. Such flexibility does not preclude territoriality (Karstad & Hudson, 1986). Although hippopotamus group sizes were influenced by drought condi-

Table 15.9. *Calculated grazing potential for areas surrounding three Kruger National Park rivers (1989–1994)*

		River section				
River	Year	A	B	C	D	E
Letaba	1989	966	1280	1028	1034[†]	
	1990	1032	1644	985	1190[†]	
	1991	926	1563	1169	1454[†]	
	1992	852	1790	1326	1043[†]	
	1993	708	1341	1222	1138[†]	
	1994	737	1169	1029	1175[†]	
Olifants	1989	220	2307	657	[†]	
	1990	274	2361	683	[†]	
	1991	312	2537	756	[†]	
	1992	198	2319	544	[†]	
	1993	198	1179	1029	[†]	
	1994	737	2164	513	[†]	
Sabie	1989	1779	1444	2118	749	289
	1990	1565	1547	2148	833	218
	1991	1493	1493	1920	722	311
	1992	1501	1169	1577	513	290
	1993	1520	1355	1182	1057	225
	1994	1414	1456	1664	852	212

[†] Sections D and E of both the Letaba and Olifants Rivers were combined to calculate densities for the confluence region of these two rivers.
The grazing potential (Trollope *et al.*, 1989; Trollope, 1990) was calculated for a 10 km zone on both sides of each river.

tions in the Letaba and Olifants Rivers (1991/92), KNP hippopotamus groups, unlike those in other areas, were smaller during the drought (Fig. 15.5). A large proportion (49.1%) of the Letaba River hippopotami were encountered as single animals during this period.

Removal of hippopotami
Hippopotami were twice culled during the study in the Sabie River, 100 animals (13.1%) being shot in 1987 and 68 (10%) in 1988. (Table 15.10). Most

Table 15.10. *Numbers of hippopotami removed in different sections of the Sabie River, Kruger National Park, during two culling operations (1987 and 1988)*

	River section[†]								
	A		B		C		D		
Year	n	%[‡]	n	%[‡]	n	%[‡]	n	%[‡]	Total
1987	48[§]	28.9"	19	14.3	27	10.3	6	3.6	100
1988			21	16.4	28	11.1	19	12.0	68
Total	48		40		55		25		168

[†]See Figure 15.4.
[‡]Percentage of hippopotami removed from section.
[§]Total includes hippopotami removed from dams (Phabeni Dam = 11; Lisbon Dam = 8).
"Total includes hippopotami removed from dam (Lake Panic = 8).

of the hippopotami were removed from section A ($n = 48$) during the first cull, the highest percentage of a river section's total culled (28.9%).

In addition to the hippopotami removed during the 1988 culling operation, 21 animals were also removed from the Nhlanganzwani Dam 7.5 km south of the Sabie River's section D. Hippopotami in this dam could have been moving between the dam and the Sabie River or the Crocodile River 13.8 km further south.

The effect of hippopotamus removal during culling operations is reflected in the population trends of hippopotami in most of the river sections concerned. Decreases in population density were noted during the 1988 survey in sections A (−33.1%) and D (−4.8%) and particularly in sections C and D in 1989 (−19.8% and −24.7%, respectively) following the 1988 cull (Fig. 15.4; Table 15.10).

A total of 10 hippopotami were captured at the Lake Panic 1 and 2 dams in 1992 for relocation to a conservation area outside the KNP. An unknown number of hippopotami have also been captured at Lisbon Dam during the study period. A lower hippopotamus density (25.4% decrease) was recorded in the river section associated with Lake Panic 1 and 2 (section A) during the following year (Table 15.7; Fig. 15.2).

Conclusions

Hippopotami maintained relatively stable numbers in all three KNP rivers. However, changes in hippopotamus density in these rivers appeared to have been largely the result of a redistribution between river sections and possibly associated dams. Drought conditions, during which the KNP had only 44.1% of the long-term mean annual rainfall (Zambatis & Biggs, 1995), clearly had an effect on population trends. The dams in or near rivers (Letaba and Sabie Rivers, respectively) played an important role in maintaining relatively high hippopotamus densities in most parts of the Letaba River during the drought or periods of reduced river flow. In addition to die-offs related to drought, the two anthrax outbreaks were also responsible for mortalities.

The calculated grazing potential appeared to explain hippopotamus densities, particularly in the Letaba River. There was no evidence that dry-season flow rates determined hippopotamus density, although the availability of suitable pools in rivers could have played a role. Smithers (1984) concluded that hippopotami prefer structurally suitable pools to pools that are closer to grazing areas.

Results of this study support Smuts & Whyte's (1981) remarks that KNP hippopotami appear to maintain relatively stable populations and a close adjustment to the long-term carrying capacity of the environment.

ACKNOWLEDGEMENTS

The National Parks Board of Trustees is thanked for the opportunity to conduct the study. We are most grateful to many colleagues in Environmental Management and Scientific Services who participated as observers during censuses. Helicopter pilots Piet Otto, Hugo van Niekerk and Danie Terblanche are also thanked. Mr. Ian Whyte kindly made earlier census data available. Drs Freek Venter and Andrew Deacon were responsible for valuable comments on several aspects of river ecology. Dr Douw Grobler collected data on hippopotamus mortalities, Gerhard Strydom and Mr Philip Odendaal of the Department of Water Affairs and Forestry made river flow data available. Naledi Maré assisted with GIS analyses. Zanne Viljoen very kindly helped with the preparation of the manuscript.

References

Ansell, W. F. H. (1965). Hippo census on the Luangwa river 1963–1964. *Puku* 3: 15–27.

Berry, P. S. M. (1973). A hippo count on the upper Luangwa River. *Puku* 7: 193–195.

Clarke, J. R. (1953). The hippopotamus in Gambia, West Africa. *J. Mammal.* 34: 299–315.

Darling, F. F. (1961). An ecological reconnaissance of the Mara plains in Kenya colony. *Wildl. Monogr.* 5: 1–41.

de Vos, V. (1990). The ecology of anthrax in the Kruger National Park, South Africa. *Salisbury med. Bull.* No. 68 (Spec. Suppl.): 19–24.

Eltringham, S. K. (1993). The common hippopotamus. In *Pigs, peccaries, and hippos. status survey and conservation:* 43–55. (Ed. Oliver, W. L. R.). IUCN/SSC, Gland.

Field C. R. (1970). A study of the feeding habits of the hippopotamus (*Hippopotamus amphibius* Linn.) in the Queen Elizabeth National Park, Uganda, with some management implications. *Zool. Afr.* 5: 71–82.

Gertenbach, W. P. D. (1980). Rainfall patterns in the Kruger National Park. *Koedoe* 23: 35–43.

Gertenbach, W. P. D. (1983). Landscapes of the Kruger National Park. *Koedoe* 26: 9–121.

Jacobson, N. H. G. & Kleynhans, C. J. (1993). The importance of weirs as refugia for hippopotami and crocodiles in the Limpopo River, South Africa. *Water SA* 19: 301–306.

Joubert, S. C. J. (1983). A monitoring programme for an extensive national park. In *Management of large mammals in African conservation areas:* 202–212. (Ed. Owen-Smith, N). Haum, Pretoria.

Joubert, S. C. J. (1986a). *Masterplan for the Kruger National Park.* Unpublished typescript, National Parks Board, Skukuza.

Joubert, S. C. J. (1986b). The Kruger National Park – an introduction. *Koedoe* 29: 1–11.

Karstad, E. L. & Hudson, R. J. (1984). Census of the Mara River hippopotamus (*Hippopotamus amphibius*), southwest Kenya, 1980–1982. *Afr. J. Ecol.* 22: 143–147.

Karstad, E. L. & Hudson, R. J. (1986). Social organization and communication of riverine hippopotami in southwestern Kenya. *Mammalia* 50: 153–164.

Klingel, H. (1991). The social organization and behaviour of *Hippopotamus amphibius.* In *African wildlife: research and management:* 73–75. (Eds Kayanja, F. I. B. & Edroma, E.). ICSU, Kampala.

Laws, R. M. (1981). Experience in the study of large mammals. In *Dynamics of large mammal populations:* 19–45. (Eds Fowler, C. W. & Smith, T. D.). John Wiley, New York.

Laws, R. M. & Clough, G. (1966). Observations on reproduction in the hippopotamus *Hippopotamus amphibius* Linn. *Symp. zool. Soc. Lond.* No. 15: 117–140.

Lock, J. M. (1972). The effects of hippopotamus grazing on grasslands. *J. Ecol.* 60: 445–467.

Mackie, C. (1976). Feeding habits of the hippopotamus on the Lundi River, Rhodesia. *Arnoldia, Rhodesia* 7: 1–16.

Marshall, P. J. & Sayer, J. A. (1976). Population ecology and response to cropping of a hippopotamus population in eastern Zambia. *J. appl. Ecol.* 13: 391–403.

Ngog Nje, J. (1988). Contribution à l'étude de la structure de la population des hippopotames (*Hippopotamus amphibius* L.) au Parc National de la Bénoué (Cameroun). *Mammalia* 52: 149–158.

Norton, P. M. (1988). Hippopotamus numbers in the Luangwa Valley, Zambia, in 1981. *Afr. J. Ecol.* 26: 337–339.

Olivier, R. C. D. & Laurie, W. A. (1974). Habitat utilization by hippopotamus in the Mara river. *E. Afr. Wildl. J.* 12: 249–271.

Pienaar, U. de V., van Wyk, P. & Fairall, N. (1966). An experimental cropping scheme of hippopotami in the Letaba River of the Kruger National Park. *Koedoe* 9: 1–33.

Scotcher, J. S. B. (1978). Hippopotamus numbers and movements in Ndumu Game Reserve. *Lammergeyer* 24: 5–12.

Smithers, R. H. N. (1984). *The mammals of the southern African subregion.* University of Pretoria, Pretoria.

Smuts, G. L. & Whyte, I. J. (1981). Relationships between reproduction and environment in the hippopotamus *Hippopotamus amphibius* in the Kruger National Park. *Koedoe* 24: 169–185.

Tembo, A. (1987). Population status of the hippopotamus on the Luangwa River, Zambia. *Afr. J. Ecol.* 25: 71–77.

Trollope, W. S. W. (1990). Development of a technique for assessing veld condition on the Kruger National Park using key grass species. *J. Grassl. Soc. South. Afr.* 7: 46–51.

Trollope, W. S. W., Potgieter, A. L. F. & Zambatis, N. (1989). Assessing veld condition in the Kruger National Park using key grass species. *Koedoe* 32: 67–93.

Van Niekerk, A. W. & Heritage, G. L. (1993). *Geomorphology of the Sabie River: Overview and Classification.* Unpublished report No. 2/93, Centre for Water in the Environment, University of the Witwatersrand, Johannesburg.

Venter, F. J. (1991) *Physical Characteristics of the Reaches of Perennial Rivers in the Kruger National Park.* Unpublished typescript, National Parks Board, Skukuza.

Venter, F. J. & Bristow, J. W. (1986). An account of the geomorphology and drainage of the Kruger National Park. *Koedoe* 29: 117–124.

Venter, F. J. & Gertenbach, W. P. D. (1986). A cursory review of the climate and vegetation of the Kruger National Park. *Koedoe* 29: 139–148.

Viljoen, P. C. (1980). Distribution and numbers of the hippopotamus in the Olifants and Blyde Rivers. *S. Afr. J. Wildl. Res.* 10: 129–132.

Viljoen, P. C. (1995). Changes in number and distribution of hippopotamus (*Hippopotamus amphibius*) in the Sabie River, Kruger National Park, during the 1992 drought. *Koedoe* 38: 115–121.

Viljoen P. C. & Retief, P. F. (1994). The use of the Global Positioning System for real-time data collecting during ecological aerial surveys in the Kruger National Park, South Africa. *Koedoe* 37: 149–157.

Viljoen, P. C., Rochat, M. A. & Wood, C. A. (1994). *Ecological Aerial Survey in the Kruger National Park: 1993.* Unpublished typescript, National Parks Board, Skukuza.

Vogt, I. (1992). *Short-term Geomorphological Changes in the Sabie and Letaba Rivers in the Kruger National Park.* MSc dissertation: University of the Witwatersrand, Johannesburg.

Zambatis, N. & Biggs, H. C. (1995). Rainfall and temperatures during the 1991/92 drought in the Kruger National Park. *Koedoe* 38: 1–16.

16

Reproductive strategies of female capybaras: dry-season gestation

E. A. Herrera

Introduction

The life histories and reproductive strategies of vertebrates are closely linked to the pattern of seasonality of the environment in which they live (May & Rubenstein, 1984). In mammals, this allows females to synchronize the energetically costly gestation and lactation with the period in the year when resources are abundant. Depending on the particular strategy and especially on the development of the young at birth, the timing may maximize resources for the mother during gestation or for the young after birth (May & Rubenstein, 1984; Bronson, 1989; Clutton-Brock, 1991). In general, however, lactation requires a greater energy investment than gestation (Oftedal, 1984*a*).

In some tropical regions, seasons may be virtually non-existent or at least much less marked than in temperate ecosystems. In others, seasonality may be more apparent in rainfall than in temperature and photoperiod. Such rainfall has an important effect on primary production, causing herbivores to adjust their annual cycle of reproduction to the cycle of rains. Thus, there are examples of tropical mammalian herbivores that reproduce only once a year (e.g. the Greater kudu, *Tragelaphus strepsiceros*, at the end of the rains: Perrin & Allen Rowlandson, 1995), although the reproductive season may be longer than in temperate regions (e.g. *Mastomys* rats: Leirs *et al.*, 1994).

The reproductive strategy includes other aspects such as litter size, weight at birth and precocity of the young. More subtle aspects such as the resorption of foetuses and 'manipulation' of the sex ratio of offspring in relation to the mother's condition (Trivers & Willard, 1973) are also part of the reproductive strategy of females. Kozlowski & Stearns (1989) suggest that bet-hedging and selective abortion (or resorption) are alternative, non-exclusive, litter-size manipulation tactics that females may use to optimally allocate their reproductive resources. When bet-hedging, the female starts with the greatest possible litter size and then 'decides' whether to reduce it or not by resorption of embryos according to her assessment of resource availability during gestation. With selective resorption, the female assesses the foetuses themselves and 'decides' how many to carry to term in relation to their individual conditions. Trivers & Willard (1973) predict that in a species where (i) males have a greater variance in reproductive success than females, (ii) reproductive success in males is correlated with their body weight, and (iii) adult body weight is affected by maternal investment, females in good condition should have a greater proportion of males in their litters than females in poor condition. Tests of the Trivers–Willard hypothesis have met with mixed success (e.g. Austad & Sunquist, 1986 and Gosling 1986 in favour; Armitage, 1987 and Hoefs & Nowlan, 1994 against).

In this chapter, I analyse the reproductive strategy of female capybaras, *Hydrochoerus hydrochaeris*, especially reproductive seasonality and litter-size manipulation, in relation to environmental seasonality and characteristics of the female (age, condition). Since the first two premises of the Trivers–Willard hypothesis are met in capybaras (Herrera & Macdonald, 1993) and the third one is not implausible, a second objective of this chapter is to test this hypothesis in capybaras.

Capybara biology

Capybaras are very large (*ca.* 50 kg), grazing rodents common in the Neotropical seasonally flooded savannas or Llanos of Venezuela. Their daily routine consists of resting in the morning, wallowing around midday and grazing in the evening and night. Capybaras are invariably found near permanent water holes, which they need for thermoregulation, for escape from predators, for mating, and for feeding since many of the grasses on which they feed are semi-aquatic.

In common with other caviomorphs (South American Hystricomorpha), capybaras are highly precocious (Weir, 1974), weighing 1.5–2 kg at birth (Ojasti, 1973) and able to graze within days or perhaps hours of birth. Gestation length is 5 months (Zara, 1973) and average litter size is 4 (range 1–8). Young from several females join in crèches shortly after birth and females appear to nurse indiscriminately any soliciting young (Macdonald, 1981). Nursing seems to be relatively unimportant since it is seldom observed and nursing bouts are short (Ojasti, 1973). In captivity, weaning occurs at 6 weeks or 3 months (Zara, 1973; Parra *et al.*, 1978). Weir (1974) says that other caviomorphs assume the nursing position long after lactation has ceased, indicating that actual weaning may occur earlier than apparent weaning. Sexual maturity is reached at 1–1.5 years.

Capybaras form stable social groups with an average composition of 3–4 adult males, 6–7 females and a variable number of young (Herrera & Macdonald, 1987). Males are permanent group members (as opposed to 'floaters') and a rigid dominance hierarchy among them is the salient feature of the social system. The dominant male obtains more matings than any one subordinate, but subordinates as a group may mate more than all dominants (Herrera & Macdonald, 1993). There is a significant correlation between body weight and rank in the hierarchy, with dominant males being heavier than subordinates. However, on average, there is little or no sexual dimorphism in body weight in this species (Ojasti, 1973; Herrera, 1992).

Because of the capybaras' grazing habits and need for water, their reproductive strategy would be expected to be strongly affected by seasons. Indeed, in capybara groups, there is a clear surge in social and mating activity at the beginning of the rains (Herrera, 1986), with a corresponding peak in density of newborn animals 5 months later (September–October). Ojasti (1973) reports that capybaras reproduce seasonally, with 61% of females captured in the wet season being pregnant but only between 20 and 30% being so in the dry months. This is not surprising since the climate in the region is markedly seasonal with around 6 months of rain and 6 months of drought (see 'Study area', below). It is however, interesting that a significant proportion of females are pregnant in the dry months, despite the important lack of resources of this period, which may cause up to 30% density-independent mortality (Ojasti, 1978).

Study area

This study was carried out on Hato El Frío (7° 46' N, 68° 57' W), an 80 000 ha cattle ranch located within the seasonally flooded savannas (Llanos) of Venezuela. In the Llanos, the dry season lasts from November into early April, and the wet season from May until the end of October, the period when 90% of the annual 1600 mm of rain falls (Ramia, 1972). Although the area is apparently flat, there are three physiographical units that differ in their relative heights, soil types and vegetation. At El Frío, a system of dykes, which double as roads, 2–3 m high by several kilometres long, has produced larger reservoirs that limit the effect of the wet-season floods while retaining water for the dry season.

El Frío sustains a large capybara population, numbering between 15 000 and 25 000 animals. An annual cull, based on a quota obtained from Venezuela's Ministry of the Environment, extracts about 20% of the population every year in a commercial operation for meat production (Ojasti, 1991). The slaughter is carried out in February–March each year. This practice started more than 150 years ago (Humboldt & Bonpland, 1805–37), although it was not until 1968 that it was regulated by the government. Recently, however, because of an apparent reduction in population size due to heavy poaching, smaller or no quotas have been allowed.

Methods

In order to collect data for this project, I took advantage of the annual slaughter, when a large number of capybara carcasses, both males and females, are available for study. During the slaughter, cowboys kill animals with a single selection criterion: size, the lower limit being 35 kg, estimated by eye. Although they attempt to avoid pregnant females, they have little success at this, except perhaps for heavily pregnant ones. This bias in the sample will be taken into account when interpreting results. Fieldwork was carried out in 1990, after 3 years without a slaughter at El Frío. A previous study (Herrera, 1992) has shown that after such a period of time the population can recover to some extent from the effects of the slaughter, at least with respect to mean body size and age distribution.

Two samples were taken during the 1990 slaughter. One consisted of any females killed, from which the proportion of pregnant females was calculated. To increase the number of pregnant females studied, a second sample, selected after pregnancy was detected (by eye) was also collected. From each female, the following data were recorded: body weight to the nearest kilogram, length of one humerus to the nearest millimetre (to assess skeletal size) and age class from the degree of ossification of the humerus epiphyses, following Ojasti (1973). Weights were recorded after cowboys had taken out the digestive and reproductive tracts. Although this was done mainly for practical reasons, the eviscerated weight was considered a better estimate of size and physical condition of the animal, since it is not influenced by the amount of food in the gut or by the weight of foetuses in the case of pregnant females. An index of physical condition was obtained from the residuals of the regression between humerus length and body weight (Berger & Peacock, 1988; Berger, 1992). The best regression was linear because of the small range of weights used.

From pregnant females, the following data were recorded: number of foetuses; weight (nearest 0.1 g or less, depending on size) and length of each foetus (nearest millimetre); sex for foetuses of at least 50 mm (López-Barbella, 1987); and number of resorptions. These were detected either because the foetus was small (relative to others in the uterus), white and without structures, or because the placenta was small and hard. The study of gestation in

Table 16.1. *Comparison of eviscerated weights, humerus lengths and condition indices (see the text) of pregnant and non-pregnant females at El Frío Ranch (Venezuela) in 1990*

	Pregnant	Non-pregnant	ANCOVA	
Humerus length (mm)	174.6 (±6.7)	176.3 (±5.9)	Age	$F = 5.93^{**}$
	$n = 192$	$n = 157$	Pregnancy	$F = 4.88^{*}$
				d.f. = 3, 1, 344
Weight (kg)	42.1 (±4.3)	39.1 (±4.3)	Age	$F = 11.04^{***}$
	$n = 196$	$n = 153$	Pregnancy	$F = 53.54^{***}$
				d.f. = 3, 1, 343
Condition index	1.59 (±3.80)	−1.98 (±3.43)	Age	$F = 3.38^{*}$
	$n = 190$	$n = 152$	Pregnancy	$F = 86.11^{***}$
				d.f. = 3, 1, 337

An ANCOVA to test for significance controlling for female age-class is shown.
$^{*}p < 0.05$; $^{**}p < 0.01$; $^{***}p < 0.001$.

capybaras by López-Barbella (1987) permits an estimation of conception date from the size and degree of development of embryos.

Results

From a total of 361 females collected, 290 were from the general sample of females and 71 were those selected because they appeared pregnant. From the first sample it was found that 126 or 43.4% were pregnant. In the following analyses, the total ($n = 197$) sample of pregnant females is used and compared to the sample of non-pregnant ($n = 164$). Sample sizes vary because field conditions did not allow all data to be collected for all females.

Although the skeletal size of pregnant females was on average lower than that of non-pregnant ones, their eviscerated weight was significantly higher than that of the latter (Table 16.1). As a consequence, the condition index was also higher for pregnant females than for the rest (Table 16.1). There were no differences, however, in age class distribution between the two groups of females (Fig. 16.1; $\chi^2 = 5.15$, d.f. = 3, N.S.).

The average number of apparently healthy embryos (hereafter called embryo count) was 4.4 (± 1.5) ($n = 191$, Fig. 16.2) so that the female with zero embryos in Fig. 16.2 corresponds to a female with four resorbed foetuses and no live ones. There was no difference in embryo count among females of

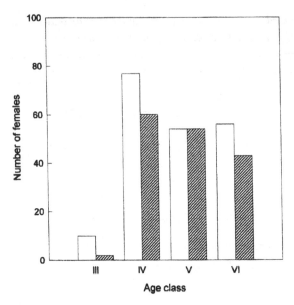

Figure 16.1. Distribution of age-classes (see the text) of pregnant (open bars) and non-pregnant (hatched bars) females collected in February 1990 at El Frío Ranch, Venezuela.

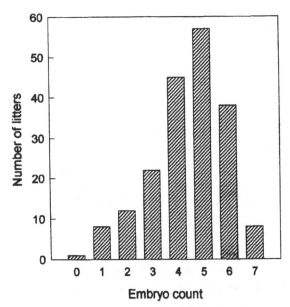

Figure 16.2. Histogram of numbers of apparently healthy embryos (embryo count) in uteri obtained from female capybaras collected in February 1990 at El Frío ranch, Venezuela.

different age classes (Kruskal–Wallis test, $H = 4.25$, d.f. $= 3$, N.S.) or body condition (partial Kendall's rank correlation coefficient between condition index and embryo count, controlling for mean foetus length, $r_{Kp} = -0.084$, $n = 151$, N.S.).

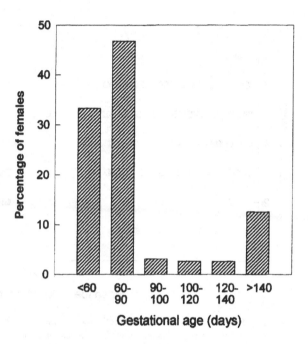

Figure 16.3. Distribution of gestational ages of litters of embryos obtained from female capybaras collected in February 1990 at El Frío ranch, Venezuela. $n = 191$.

Seasonality

Mean foetus length in litters was used as an indicator of gestation time. No attempt was made to convert these values to estimates of time since conception because of imprecision in the measurement of very small embryos and because the 'calibration curve' of foetus length and gestation time of López-Barbella (1987) gives a very low resolution since it was made with only five points. These values will nevertheless be used to classify gestations into time brackets.

A third of all the embryos were conceived within the 60 days previous to the sample being taken, i.e. between December and January (dry season), and would have been born between May and June, i.e. during the wet season (Fig. 16.3). Another 48%, with between 60 and 90 days gestation, would have had their litters by the end of April, when the rains start. The rest, less than 20%, would have given birth before the end of the dry season, having mated at the end of the previous wet season or at the beginning of the dry season.

Embryo count was not affected by time of conception (Fig. 16.4; Spearman rank correlation coefficient between mean foetus length and embryo count: $r_s = -0.019$, $n = 158$, N.S.). Neither was there any apparent effect of the progress of gestation on condition or weight of the females (correlation between mean foetus length and female eviscerated weight: $r = 0.013$, d.f. = 149, N.S.; and between mean foetus length and female condition index: $r = 0.019$, d.f. = 153, N.S.).

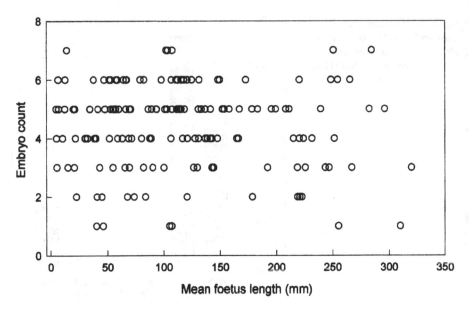

Figure 16.4. Numbers of apparently healthy embryos (embryo count) plotted against the mean length of foetuses (an indicator of gestational age); from pregnant female capybaras collected in February 1990 at El Frío ranch, Venezuela.

Embryo resorption

As many as 74, or 38.7%, of 191 pregnant females had at least one resorbed embryo in their uterus. The number of resorptions varied between 0 and 4 with a mean of 0.54 ($\pm$ 0.82) when those females with no resorption are included, and 1.38 ($\pm$ 0.74) excluding such females. The mean total embryo count (live embryos plus resorptions) was 4.97 ($\pm$ 1.21), with no difference between females with and without resorptions (mean total embryo count with resorptions: $\bar{x} = 4.99$ ($\pm$ 1.23), $n = 74$; without: $\bar{x} = 4.97$ ($\pm$ 1.21), $n = 117$; Student's $t = 0.11$, d.f. = 189, N.S.). However, the mean live-embryo count was smaller for females with resorptions ($\bar{x} = 3.59$ ($\pm$ 1.43), $n = 74$) than for those without ($\bar{x} = 4.97$ ($\pm$ 1.20), $n = 117$; $t = 7.12$, d.f. = 189, $p < 0.0001$). There was a negative correlation between the number of live embryos and the number of resorptions (Fig. 16.5: $r_s = -0.489$, $n = 191$, $p < 0.0001$).

There was no relation between maternal age and number of resorptions (Kruskal–Wallis test, $H = 2.51$, d.f. = 3, N.S.) or between the mother's condition index and the number of resorptions (ANOVA among females with 1 to 4 resorptions, $F = 1.91$, d.f. = 4, 177; N.S.). Females in later stages of pregnancy did not show more resorptions than those in early stages (Spearman rank

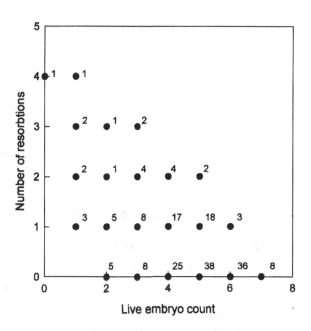

Figure 16.5. Numbers of resorbed embryos plotted against numbers of live embryos obtained from pregnant female capybaras collected in February 1990 at El Frío ranch, Venezuela. Numbers next to dots indicate the number of times each combination appeared in the sample.

correlation coefficient between mean foetus length and number of resorptions: $r_s = -0.076$, $n = 158$, N.S.).

Sex ratio

Overall sex ratio (number of males/embryo count) in uteri did not differ significantly from 0.5 (245 males and 218 females, $\chi^2 = 0.789$, d.f. = 1, N.S.; $n = 105$ litters). Sex ratio of embryos in each uterus showed a small but significant negative correlation with mean foetus length ($r_s = -0.192$, $n = 102$, $p < 0.03$; Fig. 16.6), indicating that older litters had fewer males.

To test the Trivers–Willard hypothesis, I calculated a partial Kendall's correlation coefficient between the mother's condition index and the sex ratio of the embryos in her uterus (keeping mean foetus length constant). The prediction is that the coefficient should be positive (females in better condition should have more males), but in fact it was found to be non-significant ($r_{Kp} = -0.094$, $n=98$, N.S.; Fig. 16.7). In case the effect was only noticeable at the later stages of pregnancy (after resorptions or abortions), this was calculated for litters at least 100 days old, and found to be non-significant again. To test further for a possible effect of the mother's condition on the sex of the young, the condition index (CI) of females, with all-male embryos (mean CI = 0.206, $n = 10$) was compared to that of females with all-female embryos (mean CI = 3.020, $n = 12$) and although lower, the difference only approached significance (Fig. 16.8; $t = 1.97$, d.f. = 16, $p = 0.066$). To test whether male

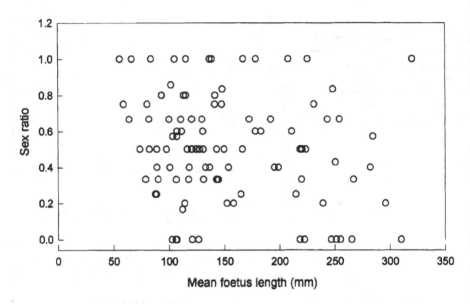

Figure 16.6. Sex ratio of embryos (number of males/embryo count) plotted against their mean length; from female capybaras collected in February 1990 at El Frío ranch, Venezuela.

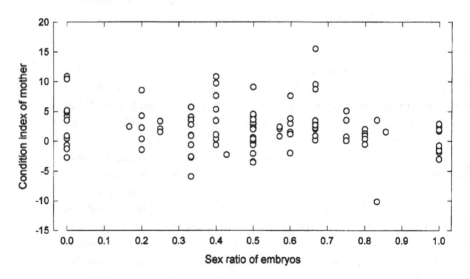

Figure 16.7. Condition index of female capybaras in relation to the sex ratio of the live embryos they carried. Samples collected in February 1990 at El Frío ranch, Venezuela.

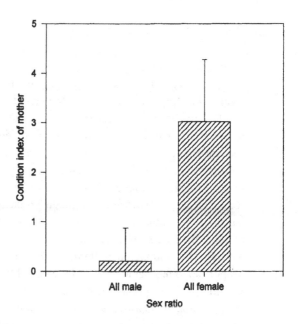

Figure 16.8. Condition index of female capybaras carrying all male or all female embryos. n (all male) = 10; n (all female) = 12. Bars are standard errors. Samples collected in February 1990 at El Frío ranch, Venezuela.

Table 16.2. *ANOVA using a randomized block design to compare weights of male and female embryos from capybara females collected at El Frío ranch (Venezuela) in 1990.*

Source	d.f	Sum of squares	Mean squares	F-ratio	p
Litter	78	2.2964×10^7	294413.69	202.32	0.000
Sex	1	2093.02	2093.02	1.438	0.231
Error	289	420541.16	1455.16		

'Blocks' are litters in order to control for different degrees of development.

offspring were more costly during gestation than females, I compared the weights of male and female embryos and found them to be not significantly different, using litters as blocks in a random block design ANOVA (to control for the different ages of litters; Table 16.2).

Discussion

Seasonality

The percentage of females found pregnant in the dry season (43.4%) was greater than expected, being exactly twice the value previously reported of 21.7% (calculated from data in table 21 of Ojasti, 1973). This is even more remarkable since this percentage is an underestimate of the actual value because early pregnancies can go undetected, and a certain number of females in late pregnancy are spared from the cull by the cowboys. This result raises two issues: first, the possibility that a number of these females may be having a second litter in the year; and second, the problem of seasonality itself. How can females have a litter at a time of year when water and grasses are dwindling rapidly? In a strongly seasonal environment where 90% of the rains fall between May and October (Ramia, 1972), it is remarkable that such a large proportion of females are able to invest in a litter in the dry season, let alone a second one. These points will be discussed in turn.

According to Ojasti (1973), a maximum of 61% of adult females are pregnant in the dry season. Ojasti's data were obtained from the same ranch in a similar situation, i.e. after a few years without a slaughter, and are therefore directly comparable. If, as seems likely, Ojasti's estimate of the proportion of pregnant females in the wet season is valid, then it is possible that a much smaller proportion are having a second litter in the year. This is hardly surprising given the scarcity of water and grasses at this time.

As to seasonality itself, a number of other factors can help to interpret the large number of pregnant females in the dry season. A relevant result in this sense is the finding that pregnant females were in better condition than non-pregnant ones and that gestational age appeared not to affect the physical condition of mothers. This suggests that those females that become pregnant in the dry season do so because they are in good condition, possibly as a result of storing fat during the wet season. Also, pregnant females were found to be of the same ages as non-pregnant ones, which indicates that females that are reproducing in the dry season are not first-time breeders (if there had been a bias towards younger females) or old females reproducing for the last time. If either of these had been the case, it could have been argued that females were breeding at the wrong time of year. The fact that they apparently are not means that having a litter in the dry season may be a strategy, not an 'error'.

There is one major factor allowing females to be in good condition in order to breed in the dry season. This is the water management system, which retains water into the dry season in large tracts of El Frío and maintains some green grass for most of the year. The effect of this should be more important in some parts of the ranch than in others, since only a proportion of the ranch is affected by the dykes, but unfortunately no exact locations with description of the habitat were recorded.

It can be inferred from the results (Fig. 16.3) that most pregnant females will give birth during the wet season, thus allowing the young access to good quality grass in the first few months of their life. For females in good condition, to have litters in the early wet season when grass is growing and vegetation offers cover against predators may in fact be an effective alternative to breeding in the wet season. Conversely, the smaller number of females who would give birth in the dry season would be exposed to the high predation pressure faced by newborn capybaras at this time of year (Ojasti, 1973; Herrera, 1986) while having to feed on low-quality grass at least until the rains start. It is not clear how this can be a successful strategy.

The litters

The mean embryo count found in this study ($\bar{x} = 4.4$) corroborates previous findings, although it is smaller than the average for the dry season found by Ojasti (1973: $\bar{x} = 5.3$). However, the wide range of months when litters would have been born invalidates a strict comparison between the two studies.

Embryo count did not affect the physical condition of mothers, as has been found by other authors (e.g. Samson & Huot, 1995). Apparently, females are able to maintain their condition regardless of the characteristics of the embryos they are carrying. There was no relation between embryo count and age, so older females are not producing larger litters.

A high proportion (38.7%) of females had at least one resorption, which is also higher than previously reported (16.8%) by Ojasti (1973). This is also greater than the values found for deer mice (*Peromyscus maniculatus*) and pocket gopher (*Thomomys bottae*) by Loeb & Schwab (1987). This does not appear to be the result of disease or weakness of females since females with resorbed embryos were in similar condition to females with complete litters. Roberts & Perry (1974) who studied viscachas (*Lagostomus maximus*) and chinchillas (*Chinchilla laniger*) found that resorption in these species and other caviomorphs was common: up to 49% ($n = 17$) in viscachas, although they do not give quantitative data for other species. They also conclude that resorbing embryos does not appear to be a result of disease or other problems such as overcrowding.

The facts that full embryo count (live embryos + resorptions) was similar for females with and without resorptions and that females with no resorptions had more embryos than the population average, seem to be more consistent with bet-hedging than with selective resorption but not conclusively so. The negative correlation between sex ratio and gestational age (as assessed by mean foetal length) suggests that male embryos are selectively resorbed.

Sex ratio of litters

The sex ratio in litters did not differ from 1 : 1, but it did tend to be biased towards females in more advanced gestations (Fig. 16.6). This may reflect a selective resorption of males. Since these are gestations from the period of the year when food may be scarce, this trend suggests that male embryos – possibly weaker than those gestated in the wet season – are being resorbed because they would benefit their mother less, in accordance with the Trivers–Willard hypothesis. This hypothesis, however, was not supported by the main result of the present study in that there was no correlation between sex ratio and female condition. The latter, if anything, showed a trend – albeit not quite significant – in the opposite direction: females with all-female litters may have been in better condition than females with all-male litters. There is one important caveat to be mentioned in this context. Although the capybara situation appears to comply with the premises of Trivers & Willard (1973), male embryos were no larger than female embryos (Table 16.2) and may not therefore have been more costly, clearly a factor affecting the females' 'decision' whether or not to alter the sex ratio of their litters. Furthermore, the difference in size between dominant and subordinate males is not large (Herrera & Macdonald, 1993), and it may be a product of post-weaning growth. Perhaps males from wet-season gestations are in better condition than those from the dry season. This would imply that females should give birth to more females in the dry season, and this is in fact what the data in Fig. 16.6 suggest: all-female litters are those that will be born within a month after the sample was taken, i.e. before the end of the dry season. Wright *et al.* (1995) propose an alternative to the Trivers–Willard hypothesis based on time of birth rather than the mother's condition. According to this hypothesis, males born early in the breeding season would be at a competitive advantage because they would be somewhat older and larger than those born later.

   The results of the present study point to a reproductive strategy of female capybaras based on timing within the dry–wet season alternation: (i) to become pregnant during the wet season, assuring food for the female during her costly gestation; (ii) to become pregnant in the dry season, ensuring that the newborn will have access to good quality forage when the new rains begin. The former is the most common strategy, possibly because gestation may be more costly than lactation. Although opposed to the general trend in mammals (Oftedal, 1984*a*), this pattern has been found in the capybara's closest relative, the guinea pig, *Cavia porcellus* (Oftedal, 1984b). The second alternative may be less common because it is more risky, since the beginning of the rains is not 100% predictable: any delay in the onset of rains would jeopardize the reproductive effort. As to sex-ratio manipulation within litters, there is a slight trend for females to have female-biased litters if they are to be born in the dry season. The small or perhaps non-existent difference in costs between male and female offspring does not favour a more definite strategy of *in utero* sex-ratio manipulation.

ACKNOWLEDGEMENTS
I express my gratitude to Iván Darío Maldonado and the late M. Fernández for their continuing support of my work at El Frío. For their help in the field, I must make special mention of C. Ramos, P. Borges and M. C. Yáber. I am also grateful to P. Alvizu, M. T. Badaracco, C. Bastidas, D. Dearden, C. Fonck, M. Guevara, A. Gols, Y. Higuerey, A. Ivernón, J.A. Isea, M. Morán, V. Pineda, A. Riera, V. Salas, C. Villalaba and S. Zipman. This work was supported by the Research Deanery of Universidad Simón Bolívar. I am indebted to David Macdonald for his hospitality and to the British Council and the Royal Society for their support of my sabbatical leave at Oxford during which this chapter was gestated. Paul Johnson and Eduardo Klein helped with the statistics and Guillermo Barreto and two anonymous referees made helpful suggestions that improved the manuscript.

References

Armitage, K. (1987). Do female yellow-bellied marmots adjust the sex ratios of their offspring? *Am. Nat.* 129: 501–519.

Austad, S. N. & Sunquist, M. E. (1986). Sex-ratio manipulation in the common opossum. *Nature, Lond.* 324: 58–60.

Berger, J. (1992). Facilitation of reproductive synchrony by gestation adjustment in gregarious mammals – a new hypothesis. *Ecology* 73: 323–329.

Berger, J. & Peacock, M. (1988). Variability in size-weight relationships of *Bison bison*. *J.*

Mammal. 69: 618–624.

Bronson, F. H. (1989). *Mammalian reproductive biology.* University of Chicago Press, Chicago.

Clutton-Brock, T. H. (1991). *The evolution of parental care.* Princeton University Press, Princeton, NJ.

Gosling, L. M. (1986). Selective abortion of entire litters in the coypu: adaptive control of offspring production in relation to quality and sex. *Am. Nat.* 127: 772–795.

Herrera, E. A. (1986). *The Behavioural Ecology of the Capybara.* Hydrochoerus hydrochaeris. DPhil thesis: University of Oxford.

Herrera, E. A. (1992). Effect of the slaughter on the age structure and body size of a capybara population. *Ecotropicos* 5: 20–25.

Herrera, E. A. & Macdonald, D. W. (1987). Group stability and the structure of a capybara population. *Symp. zool. Soc. Lond.* No. 58: 115–130.

Herrera, E. A. & Macdonald, D. W. (1993). Aggression, dominance, and mating success among capybara males (*Hydrochoerus hydrochaeris*). *Behav. Ecol* **4**: 114–119.

Hoefs, M. & Nowlan, U. (1994). Distorted sex ratios in young ungulates: the role of nutrition. *J. Mammal.* **75**: 631–636.

Humboldt, F. H. A. von & Bonpland, H. J. A. (1805–37). *Voyages aux régions équinoxiales du nouveau continent . . . etc.* Imprimerie J. Smith, Paris.

Kozlowski, J. & Stearns, S. C. (1989). Hypotheses for the production of excess zygotes: models of bet-hedging and selective abortion. *Evolution* **43**: 1369–1377.

Leirs, H., Verhagen, R. & Verheyen, W. (1994). The basis of reproductive seasonality in *Mastomys* rats (Rodentia: Muridae) in Tanzania. *J. trop. Ecol.* **10**: 55–66.

Loeb, S. C. & Schwab, R. G. (1987). Estimation of litter size in small mammals: bias due to chronology of embryo resorbtion. *J. Mammal.* **68**: 671–675.

López Barbella, S. (1987). Consideraciones generales sobre la gestación del chigüire, (*Hydrochoerus hydrochaeris*). *Acta cient. venez.* **38**: 84–89.

Macdonald, D. W. (1981). Dwindling resources and the social behaviour of capybaras (*Hydrochoerus hydrochaeris*)

(Mammalia). *J. Zool., Lond.* **194**: 371–391.

May, R. M. & Rubenstein, D. I. (1984). Reproductive strategies. In *Reproduction in mammals 4: reproductive fitness* (2nd edn): 1–23. (Eds Austin, C. R. & Short, R. V.). Cambridge University Press, Cambridge.

Oftedal, O. T. (1984a). Milk composition, milk yield and energy output at peak lactation: a comparative review. *Symp. zool. Soc. Lond.* No. 51: 33–85.

Oftedal, O. T. (1984b). Body size and reproductive strategy as correlates of milk energy output in lactating mammals. *Acta zool. fenn.* **171**: 183–186.

Ojasti, J. (1973). *Estudio biológico del chigüire o capibara.* FONAIAP, Caracas.

Ojasti, J. (1978). *The Relation Between Population and Production in Capybaras.* PhD thesis: University of Georgia, Athens, Georgia, USA.

Ojasti, J. (1991). Human exploitation of capybara. In: *Neotropical wildlife use and conservation*: 236–252. (Eds Robinson, J. G. & Redford, K. H.). University of Chicago Press, Chicago & London.

Parra, R., Escobar, A. & Gonzalez-Jiménez, E. (1978). El chigüire: su potencial biológico y su cría en confinamiento. In *Informe Anual IPA*: 83–94. Universidad Central de Venezuela, Maracay, Venezuela.

Perrin, R. & Allen-Rowlandson, T. S. (1995). The reproductive biology of the greater kudu, *Tragelaphus strepsiceros. Z. Säugetierk.* **60**: 65–72.

Ramia, M. (1972). Cambios en las sabanas del Hato El Frío producidos por diques. *Bol. Soc. venez. cienc. nat.* **30**: 57–90.

Roberts, C. M. & Perry, J. S. (1974). Hystricomorph embryology. *Symp. zool. Soc. Lond.* No. 34: 333–360.

Samson, C. & Huot, J. (1995). Reproductive biology of female black bears in relation to body mass in early winter. *J. Mammal.* **76**: 68–77.

Trivers, R. L. & Willard, D. E. (1973). Natural selection of parental ability to vary the sex ratio of offspring. *Science* **179**: 90–92.

Weir, B. J. (1974). Reproductive characteristics of hystricomorph rodents. *Symp. zool. Soc. Lond.* No. 34: 265–301.

Wright, D. D., Ryser, J. T. & Kiltie, R. A. (1995). First cohort advantage hypothesis: a new twist on facultative sex ratio adjustment. *Am. Nat.* **145**: 133–145.

Zara, J. L. (1973). Breeding and husbandry of the capybara *Hydrochoerus hydrochaeris* at Evansville Zoo. *Int. Zoo Yb.* **13**: 137–145.

17

The continuing decline of the European mink *Mustela lutreola*: evidence for the intraguild aggression hypothesis

T. Maran, D. W. Macdonald, H. Kruuk, V. Sidorovich and V. V. Rozhnov

Introduction

European mink, *Mustela lutreola* (Linnaeus 1761) look so similar to American mink, *Mustela vison* (Schreber, 1777) that the two were formerly distinguished only as subspecies (Ognev, 1931; Heptner *et al.*, 1967; for a review, see Novikov, 1939). This similarity offers not only a striking instance of convergence, but also a plausible explanation of the European mink's precipitous decline to the verge of extinction. In fact, phylogenetically the European mink is probably closest to the Siberian polecat, *Mustela sibirica* (Pallas, 1773), whilst the American mink is the most aberrant in the genus *Mustela* (Lushnikova *et al.*, 1989; Graphodatsky *et al.*, 1976; Youngman, 1982).

The European mink was formerly widespread in eastern Europe, but has declined over a long period and by 1990 was clearly endangered (Maran, 1994*b*). Here, after reviewing briefly the species' biology, we will present the results of an up-dated survey of the European mink's status. Against that background we will present a critical review of hypotheses that might explain its decline, dwelling in particular on two that we will subject to a preliminary test by presenting new data. These two implicate the American mink, first through the transmission of disease and second as a direct aggressor. Concerning the possibility that direct aggression between the endemic and alien mink disadvantages the native species, we have in mind that intraguild competition between the two must be evaluated in the context of the entire guild. Therefore, when, elsewhere, we present data on the possibility of competition for food between these congeners, we do so in the context also of the diets of European polecat (*Mustela putorius*) and otter (*Lutra lutra*) (see Chapter 11).

The European mink, of which six subspecies have been distinguished (Novikov, 1939; Heptner *et al.*, 1967), but not universally accepted (Ogney, 1931; Youngman, 1982), is an inconspicuous denizen of small, undisturbed water-courses with rapid currents and lush riparian vegetation, in forested areas (Novikov, 1939; Danilov & Tumanov, 1976*a,b*; Sidorovich *et al.*, 1995). Their prey includes fish, amphibians, small mammals and invertebrates (see

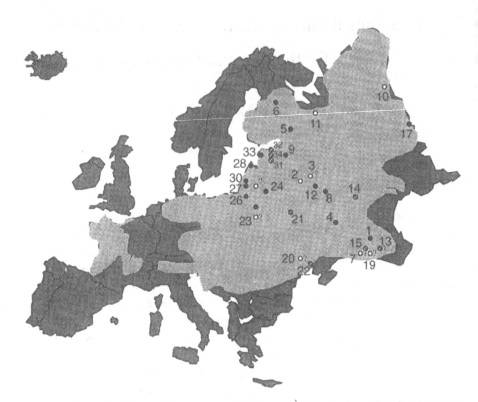

Figure 17.1. Historical and recent range of the European mink. Numbers refer to locations names in Table 17.1. ○, present 1990, present 1995; ◔, present 1990, extinct 1995; ●, extinct in 1990; ○?, present 1990, no data 1995; ◔, no data 1990, extinct 1995.

chapter 11, and see Danilov & Tumanov, 1976a,b; Sidorovich, 1992). European mink mate in the last 2 weeks of April, and a 43 day gestation leads, without delayed implantation, to the birth of 1–7 (mean 3.5) kits. The longest recorded lifespan for the European mink is 7 years (Maran, 1994a).

The only prehistoric records of European mink are from Vlaardingen, in the Netherlands, between 2300 and 2100 BC (van Bree, 1961a,b). From historical records the species' range once extended from the Ural Mountains to eastern Spain and from central Finland to the Black Sea (Novikov, 1939; Heptner *et al.*, 1967) (Fig. 17.1). However since the mid-nineteenth century its range has dwindled (Maran & Henttonen, 1995) and it was recently included as endangered in the IUCN Red Data Book (Groombridge, 1993). In the European Union the European mink is listed in Annex II (species whose conservation requires the designation of special areas) and in Annex IV (species of community interest in need of strict protection) of the Directive on the conservation of natural habitats and of wild fauna and flora. The IUCN *Action Plan for the Conservation of Mustelids and Viverrids* (Schreiber *et al.*, 1989) nominates the European mink as a priority.

Course of the decline

The European Mink disappeared from Germany in the mid-nineteenth century (Youngman, 1982), then from Switzerland (Gautschi, 1983) and, in the 1890s, from Austria (Novikov, 1939). Subsequently, in western Europe, an isolated enclave of European mink originally persisted between Brittany in France and Galicia in Spain (Blas Aritio, 1970; Chanudet & Saint-Girons, 1981; Braun, 1990; Camby, 1990; Palomares, 1991; Ruiz-Olmo & Palazón, 1991). However, recently they have disappeared from the northern part of this range and appear in widespread decline in what remains of their French range (C. Maizaret, personal communication; Moutou, 1994). In intriguing contrast, and although no data exist prior to 1951, European mink are reported to be spreading southwards in Spain (Ruiz-Olmo & Palazón, 1990).

In eastern Europe the European mink's situation is almost unremittingly bad. Between the 1930s and 1950s European mink disappeared from Poland, Hungary, the Czech and Slovak Republics and probably also Bulgaria (Bartá, 1956; Szunyoghy, 1974; Schreiber *et al.*, 1989; Romanowski, 1990). In the closing years of the twentieth century, the species teeters on, or over, the brink of extinction throughout most of the remainder of its range. In Finland, following a rapid decline between 1920 and 1950 only isolated specimens were seen until the early 1970s, when the species was judged extinct until one was trapped in 1992 (Henttonen *et al.*, 1991; Henttonen, 1992). In Latvia the only evidence for the species' survival is single specimens caught in 1984, 1991 and 1993 (Ozolins & Pilats, 1995). In Lithuania, where the last record is 1979 (Bluzma, 1990), an intensive search for European mink in 1989–1990 revealed none (Mickevicius & Baranauskas, 1992) If the European mink survives at all in Georgia, it is as an extreme rarity in the rivers flowing to the Black Sea in the north-west of Georgia, where it was common in the early twentieth century (A. N. Kudatkin, personal communication; Novikov, 1939). Following a decline in Moldova in the 1930s, by the early 1980s the species survived only on the lower reaches of the river Prut, along the Romanian border (Muntjanu cited in Maran, 1994*b*) where it has not been seen for 15 years and is now considered extinct (A. Mihhailenko, personal communication). In Romania itself, where European mink are legal game, as recently as 1970 2700 pelts were recorded by state hunting organizations. Subsequently, the species has plummeted to rarity in northern Romania, but is apparently more numerous but declining in the Danube delta (H. Almasan, D. Muriaru & O. Ionescu, personal communications). Although none have been seen in Estonia since 1992, unconfirmed sightings, and the capture of an apparent hybrid with a polecat in both 1994 and 1995, suggest that remnants may persist, as they may in the Ukrainian Carpathians (Turjanin, 1986) following a major decline in the late 1950s (Tumanov, 1992). By 1990 an estimated 100–150 European mink survived in the north-east of Belarus (Sidorovich, 1992). In the early decades of this century the European mink was a common and widespread carnivore almost everywhere in western Russia where it was a valuable furbearer (Novikov, 1939). Its decline there was first noticed in the 1950s and widely lamented by the 1970s (Ternovskij & Tumanov, 1973; Ternovskij 1975; Tumanov & Ternovskij, 1975; Danilov & Tumanov, 1976a). A survey by Tumanov & Zverjev (1986) revealed that by the mid-1980s European mink had declined dramatically throughout Russia, but in the Tver regions there were still reports of 4–6 European mink/10 km of river bank. Several recent reports indicate that the European mink decline is continuing in Russia and recent reports are gloomy (Sidorovich & Kozhulin, 1994; Sidorovich et al., 1995).

To conclude this litany of disaster, Maran (1992a) reported the answers to questionnaires distributed in 33 areas of the former Soviet Union, of which 16 reported extinction, 13 reported very critical status and only 4 considered the populations viable.

Summarizing the population trend, there has been a steady, long-term decline of the European Mink since the nineteenth century, which has led to the general extinction of the species in western Europe, except in Spain. This decline has accelerated very rapidly in the last three decades in eastern Europe.

Below, we present the results of a new survey.

Causes of the decline

The precipitate decline of the European mink caught conservation biologists unawares, and it was not until 1995 that Maran & Henttonen published the first synthesis of possible explanations. Here, before presenting preliminary data on two of them, we review five hypotheses, namely that the decline is caused by:

1   Habitat loss.
2   Pollution.
3   Overhunting.
4   Impact of the European polecat.
5   Impact of the American mink.

Habitat loss

The decline of the European mink has repeatedly been attributed to habitat loss. For example, Claudius (1866) stated that over one decade the European mink was exterminated from several districts of Germany by changes in land-use, and Löwis (1899) foresaw the rapid decline of the species in Lithuania in the wake of agricultural intensification. In the Ukraine and Moldova, Tumanov (1992) links the demise of the European mink with drainage, cultivation and land reclamation. Having evaluated competing hypotheses, Schubnikova, (1982) concluded that in at least several regions of Russia the European mink's decline was due to habitat degradation and loss, particularly in central Russia (Jaroslav, Vladimir, Ivanovo and Kostroma Regions) (see also Shashkov, 1977).

It is indisputable that urbanization and agricultural intensification have radically altered much of the post-war European landscape, generally to the detriment of wildlife (e.g. Macdonald & Smith, 1991). This is true for at least many areas in which European mink have declined and, in particular, natural riverbasins have been canalized. For example, in Estonia during the 1950s and 1960s almost half of all the wetlands were drained (Kask, 1970) and agricultural improvement affected at least 20% of the country (Mäemets, 1972). Furthermore, as European mink thrive in natural river systems (Novikov, 1939, 1970; Danilov & Tumanov, 1976b) and especially in undisturbed streams (Novikov, 1939, 1970; Sidorovich et al., 1995), it seems obvious that these twentieth century landscape changes will have been to their detriment. Nonetheless, beyond this loose historical correlative evidence, there have been no direct quantitative tests of the hypothesis that habitat loss caused their decline. Moreover, while there are sites, perhaps many, where radical degradation of habitat would seem to render rivers inhospitable for the mink, there are other places where the species has declined in apparently suitable habitat. Furthermore, in areas where the European mink is still present, its choice of habitat appears to include types of vegetation and banks that are still in abundance in areas where the species has gone. Therefore, to test this hypothesis rigorously, data are needed to elucidate which aspects of landscape change are detrimental to European mink.

Pollution

Pollution has been mooted as contributing to the European mink's decline by Schröpfer & Paliocha (1989). Riparian pollution is locally severe in Europe, including some areas where European mink do, or did, occur (Blas Aritio, 1970; Novikov, 1970). Furthermore, López-Martin et al. (1994) report organochlorine residues in European mink at levels that could perhaps impair

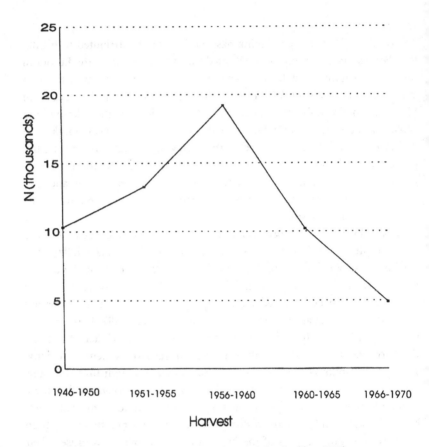

Figure 17.2. Harvest of mink in central regions of Russia (after Shashkov, 1977)

reproduction, although this report is from an area of Spain where the species is doing well. However, while this hypothesis remains largely untested, European mink have declined in some wilderness areas (such as north-eastern Belarus: see Chapter 11) which do not appear to be gravely polluted.

Overhunting

At least in the eastern part of its range the European mink has traditionally been trapped for its pelt. For instance, Schubnikova (1982) reports that as early as the seventeenth century 11 500 mink pelts were exported annually from the Arkhangelsk Region of Russia alone, while in the early decades of the twentieth century the annual Russian bag was 40 000–60 000 with a maximum of 75 000 in the winter of 1927–28. Overhunting was perceived as such a threat to the fur harvest that a moratorium on killing European mink was imposed around the former Leningrad and in the north Caucasus at the beginning of the present century (Novikov, 1939). Following World War II, intensive harvest of European mink continued apace in Russia and, after 15 years of increasing tallies, plummeted – a decline that Shashkov (1977) attributed to local extinctions due to overhunting (Fig. 17.2).

Scarcity appears to have affected the demand, and thus value, of European mink pelts. In the late nineteenth century in Russia the pelts of European mink and polecat were valued equally (Martenson, 1905) whereas by 1983 in the Soviet Union the value of an unspecified mixture of the two mink species had increased tenfold in real terms and fetched twice the price of polecats (Maran, 1991). Although it is difficult to separate the inflationary effects of fashion and scarcity, the end result is an increased incentive to trap European mink. The official harvest target in the 1960s and 1970s in the Jaroslav Region was 28–35% of the estimated total European mink population prior to the hunting season, and approximated the annual productivity of the species (Shashkov, 1977). However, Shashkov (1977) shows that, at least for the Kostroma Region, the actual harvest was double the official figure.

Clearly, prolonged, widespread hunting significantly in excess of a population's recruitment will inevitably lead to its decline. Hunting statistics provide compelling, if somewhat fragmentary, evidence that European mink have been seriously over-harvested in some regions. This is the first of several instances in which the spread of American mink in Europe both complicates the interpretation of data and may disadvantage the European mink. First, the hunting statistics for European mink may be confounded by confusion between the two species. Second, the expanding population of American mink may have sustained the incentive for trapping beyond the time at which numbers of European mink were so low that they alone were not economically harvestable. However, although over-hunting has clearly been rife, and continues to be so (Sidorovich *et al.*, 1995), and although it seems certain to have reduced their numbers in many places, there is no clear evidence that it has been the sole, or principal, cause of European mink disappearance anywhere. Indeed, on the grounds that other semi-aquatic mustelids, such as American mink, polecat and otter, have not been exterminated by comparably intense hunting, support for this hypothesis requires evidence that European mink are strongly selected by hunters or very much more susceptible to hunting pressure than are their congeners.

Impact of the European polecat, Mustela putorius
In a variant of the habitat-change hypothesis, Schröpfer & Paliocha (1989) argue that landscape changes associated with agricultural encroachment have

not, or not only, disadvantaged European mink directly, but also indirectly by favouring a competitor, the European polecat. While it may be true that polecats are better adapted to the new European landscape than European mink are, there are no data with which to evaluate the hypothesis that this has led to a competitive disadvantage for the mink. However, it is noteworthy that in relatively undisturbed habitat in Belarus, polecats and European mink co-existed both before and after the arrival of American mink and there is little overlap in diet (see Chapter 11).

A second potential hazard to European mink posed by polecats became apparent with the discovery of hybrids between them. Granqvist (1981) postulated that climatic warming in the early twentieth century may have facilitated a northward extension of the polecat's range, as it has that of the red fox, *Vulpes vulpes* (Hersteinsson & Macdonald, 1992), and led to greater contact between the two species and hence greater risk of genetic introgression. While the climatic change aspect of this hypothesis is weakened by the historical co-existence of these two species in southern Europe, it remains the case that genetic introgression of rare carnivores is a real threat (e.g. *Canis lupus, Canis rufus, Canis simensis, Felis silvestris*).

Hybridization between European and American mink appears out of the question; the chromosome number of the European mink is 38 and that of the American mink is 30 (Graphodatsky et al., 1976). Hybrids of polecat and European mink are relatively well-known (Ognev, 1931; Novikov, 1939; Heptner et al., 1967; Tumanov & Zverjev, 1986), but appear to have remained a rarity in the days prior to the mink's dramatic decline. For instance, of 500–600 pelts examined by Tumanov & Zverjev (1986) only 3–5 were suspected to be hybrids. Therefore, while there are no grounds to implicate this hybridization in triggering the mink's decline, the critical question is whether its incidence has increased where the mink are rare. Maran & Raudsepp (1994) report that during the last years of the European mink's existence in Estonia, a surprisingly high proportion of suspected mink–polecat hybrids were found: six between 1992 and 1994. The hypothesis that in fragmented populations such hybridization may pose an additional threat to European mink therefore merits further testing.

Impact of the American mink, Mustela vison

The success of the American mink throughout Europe provokes several potentially linked hypotheses for the demise of the European mink. These include (i) sustained trapping pressure of which the European mink has been a partly incidental victim (see above), (ii) scramble competition for shared (and possibly declining) food resources (see Chapter 11), (iii) transmission of disease against which the European mink has inadequate resistance, and (iv) intraspecific aggression. The first line of evidence for each of these is the spread in the American mink's range and contemporaneous shrinkage of the European mink's. American mink were brought to Europe in 1926 both for farming and for release, and are now widespread (Stubbe, 1993). Initially they came to France, in 1928 to Sweden, Norway and the UK in 1929, Denmark and Iceland in 1930. Between 1933 and 1963 they were deliberately released in many localities in the former Soviet Union (Heptner et al., 1967). Indeed, 20400 mink had been released by 1971 in nearly 250 sites (Pavlov & Korsakova, 1973).

The arrival of the American mink could have changed the epidemiological circumstances of European mink, most plausibly by introducing a pathogen to which the European mink had inadequate immunity. In view of the inconspicuousness of the species, the absence of evidence of ill mink is hardly a weakness of this hypothesis, which has been suggested by Henttonen & Tolonen (1983) and Henttonen (1992). We also present some further preliminary evidence, but there is no solid conclusion.

Male American mink weigh 0.9–2.0 kg, with a head–body length (HBL) of 37–47 cm. The corresponding figures for male European mink are 0.7–1.1 kg and 30–43 cm. Female American mink weigh ca. 0.7–1.0 kg with a HBL of 33–42 cm. The corresponding figures for female European mink are ca. 0.5–0.8 kg and 25–34 cm (Sidorovich, 1995). The male American mink is thus substantially heavier than its European counterpart, and even the female American mink is on a par with the male European mink (Danilov & Tumanov, 1976b; Sidorovich, 1992) and may be more robust in the face of harsh climates: the range of the European mink is confined below 66 °N in central Finland (Heptner et al., 1967), whereas the American mink is abundant in Iceland (Stubbe, 1993) and Norway (Bevanger & Henriksen, 1995). Furthermore, Maran (1989) reports that in captivity American mink were more versatile than European mink in using artificial environments, and this may reflect greater opportunism in adapting to diverse environments in the wild. While European mink are generally recorded from small, fast-flowing streams (Novikov, 1939, 1970) or, in snow-free periods, inland lakes and marshes (Danilov & Tumanov, 1976b; Sidorovich et al., 1995), American mink appear adapted to almost any body of water, including coastlines, offshore islands and large lakes (Gerell, 1967; Dunstone, 1993; Bevanger & Henriksen, 1995; Niemimaa, 1995).

The reproductive biology of the two species also differs in that the gestation of the American mink may vary, due to delayed implantation, from 30–92 days (Ternovskij, 1977), whereas that of the European mink always lasts 40–43 days

(Maran, 1994a). Furthermore, American mink litters are larger ($x = 5.4$ (± 0.35), $n = 38$, versus 4.3 (± 0.10), $n = 280$: Ternovskij & Ternovskaja, 1994).

There are also differences in behaviour, with captive European mink appearing more nervous of their keepers, and less socially interactive with conspecifics (Maran, 1989).

It is therefore possible that the European mink might be disadvantaged in contest competition against the American mink in three, non-exclusive, ways, various combinations of which have been espoused by Popov (1949), Danilov & Tumanov (1976b), Maran (1991, 1994b), Ryabov et al. (1991) and Sidorovich (1992). First, in contests for mates, and assuming interspecific sexual attraction, larger, more vigorous male American mink might exclude European males from females. Furthermore, if, as suspected, the American mink becomes reproductively active earlier in the year than European mink males do, and as Ternosvkij (1977) records that hybrid embryos are resorbed, early interspecific pregnancies would pre-empt the European males' reproduction and render the female European mink reproductively abortive for that year. The proposal that becoming reproductively active earlier enables American males to monopolize matings with European females begs the question of whether European mink females come into breeding condition earlier than their males. Second, the robust build and confident character of American mink may allow them to overwhelm European mink in direct contests over other resources, such as food or dens. Third, there is mounting evidence of interspecific aggression between carnivore species, which may constitute pre-emptive competition for resources or territory, and generally results in the larger of two species within a guild harassing the smaller (e.g. Hersteinsson & Macdonald, 1992). The more numerous the American mink became – and Sidorovich (1993) has emphasized their great reproductive capacity (up to 7.3 embryos/female) in expanding populations – the worse would be the impact of each of these variants upon European mink.

There are no published data with which to evaluate these variants of the interspecific competition hypothesis, so there we present the results of an experiment designed to elucidate the tenor of encounters between American and European mink in captivity.

Methods

1995 Questionnaire survey

In 1995 we conducted a questionnaire survey of 19 reserves in Russia that had reported European mink populations in 1990 (Maran, 1992a). With the aim of detecting changes in status, we asked the correspondents for evidence of the status of European mink in their reserve.

Disease transmission
A partial test of the hypothesis that American mink have transmitted a fatal disease to European mink would be to capture wild specimens of both species and expose them to each other in captivity. This test has been performed incidentally since 1983 at Tallinn Zoo, where both species of mink have been housed in adjoining pens, and sometimes successively in the same pen, in the course of developing captive breeding populations. We have explored the zoo records to enumerate the instances when European mink were exposed to American mink, and any subsequent illness.

Interspecific aggression
The mink were housed in an outdoor **L**-shaped enclosure comprised of three 5×5 m compartments, each separated by a metal wall. Each compartment contained four nest-boxes, four stumps and a pool. During the 'sympatric' phase two adult male and two adult female European mink were housed in one compartment, and a similar group of American mink was housed in another, with the intervening compartment empty. All the animals had been wild-caught in Estonia. During the 'allopatric' phase the interconnecting doors were open so that all eight mink had access to the entire 72 m^2 enclosure.

The animals were fed at midday on gruel and meat supplemented with rats and mice. Nocturnal observations were aided by electric light.

Observations were made from a small cabin in the middle of the enclosures. Interactions and activity were recorded every minute for 24 h during three, generally consecutive, days each month between September 1989 and July 1990.

Results

1995 survey
Of 19 reserves sent questionnaires, 13 replied (Table 17.1). Of these, five reported the mink population to be extinct since 1990. Of two which reported abundant European mink in 1990, one now judges it to be in decline, the other judges it extinct. Pinegeja Reserve, reporting a decline in 1990, judges that the population has now stabilized at a new low. In summary, our 1995 survey indicates that the European mink continues in a fast decline. Indeed, we have

Table 17.1. *The European mink,* Mustela lutreola, *in protected areas in the territory of the former Soviet Union*

Protected area	Area (ha)	Status 1990	Status 1995	Last evidence
1. Astrakhansky	62 500	Extinct	Extinct	?
2. Central forest	21 380	Good	Declining	1995
3. Darvinsky	112 630	Declining	No reply	1989
4. Hopersky	29 800	Extinct	Extinct	1940
5. Kivach	10 460	Extinct	Extinct	1975
6. Kostomuksky	47 567	Extinct	Extinct	?
7. Kavkavsky	263 300	Declining	No reply	1990
8. Mordovsky	32 140	Extinct	Extinct	1965
9. Nizhne svirsky	40 972	Extinct	Extinct	?
10. Pechoro Ilychsky	721 322	Declining	Declining	1990–92
11. Pinegeysky	41 224	Declining	Rare	1994
12. Prioksko Terrasny	4945	Extinct	Extinct	?
13. Severo Osetinsky	2990	?	Extinct	1950–60
14. Zhigulevsky	23 100	?	Extinct	1940
15. Teberda	85 840	Declining	Extinct	1981
16. Voronezhsky	31 053	Extinct	extinct	1950
17. Visimsky	3767	Extinct	Extinct	1958
18. Kunashir	?	?	No reply	?
19. Skrutsinsky	85	Good	No reply	?
20. Dunaiskiye Plavni	14 815	Declining	No reply	?
21. Kanevsky	1035	Declining	Extinct	1977
22. Tshernomorsky	87 348	Declining	Extinct	1989
23. Karpatsky	12 706	Declining	No reply	?
24. Berezinsky	35 000	Extinct	Extinct	1963
25. Pripjatsky	63 120	Extinct	Extinct	?
26. Bielovezhskayapushcha	87 600	Extinct	Extinct	?
27. Zhuvintas	5443	Extinct	Extinct	?
28. Slitere	15 440	Extinct	Extinct	?
29. Kruskalny	2902	Extinct	Extinct	?
30. Grini	1076	Extinct	Extinct	?
31. Endla	81 162	Declining	Extinct	1991
32. Lahemaa	64 911	Declining	Extinct	1992
33. Matsalu	48 640	Extinct	Extinct	1950
34. North Kôrvemaa	11 283	Good	Extinct	1992

failed to find a single reserve within the former USSR that reported a healthy population of the species.

Disease

Since 1983, 51 European mink and 40 American mink have been housed in close proximity at Tallinn Zoo. For eight European mink we could verify the number of different American mink that had occupied adjoining cages. Despite up to four such exposures, there was not one case of patent illness; all individuals survived and most subsequently bred. During 1987 eight European mink and nine American mink were involved in tests for interspecific aggression during which two individuals of either species were observed in the same cage for 15 min sessions. In each of 24 dyads, European mink were exposed to 15–75 min/day of direct contact with American mink over 34–62 days. During these observations (T. Maran, unpublished results) there was much physical contact and some fighting. However, no illness was observed subsequently in any of the experimental animals, and several became amongst the most successful breeders in the colony. In 1990, four European mink were housed in cages that had immediately beforehand housed American mink, and none developed illness.

The role of interactions within species

During 6 days of observation of the four American mink there were 45.3 ($\text{SE} = 13.35$) recorded behavioural events per day per animal, of which 10% were aggressive and 33% were approach. In contrast, during 5 days there were only 23.4 ($\text{SE} = 4.66$) recorded behavioural events per day per animal between the four European mink. However, the quality of interactions was similar, involving 11.7% aggressive and 46% approach. The American mink were markedly more active and more socially interactive than the European mink.

The flow of interactions amongst the quartet of each species is most easily compared visually (Fig. 17.3). There was no obvious difference between the species in the proportion of intraspecific interactions involving the initiation of aggression. In the quartet of American mink, 5.2% of 655 interactions initiated by females were aggressive, whereas amongst the European mink the comparable figure was 3.7% of 241 interactions. In the quartet of American mink, 17.1% of 432 interactions initiated by males were aggressive, whereas amongst the 230 interactions initiated by male European mink, 20.2% were aggressive.

Figure 17.3 also reveals no striking difference between the species in the flow of other categories of interaction. For example, in both species, most aggression initiated by females was directed towards females, whereas aggression

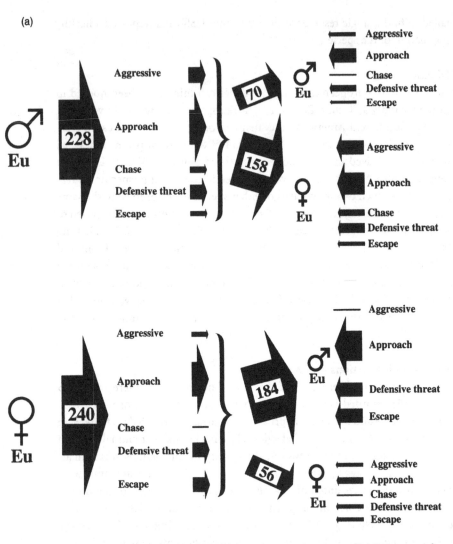

Figure 17.3. Sociograms showing the social dynamics amongst quartets of (a) European mink (Eu) and (b) American mink (Am) when housed separately. The thickness of the arrows is approximately proportional to the relative flow of interactions of a given class between each category of conspecific. Thus, in Fig. 17.3(a), of 228 events (of which the proportions comprising five behavioural classes are schematized) initiated by the two male European mink, 158 flowed from one male to the two females, whereas 70 flowed to the other male and of the latter the most common class of interaction was an approach.

(b)

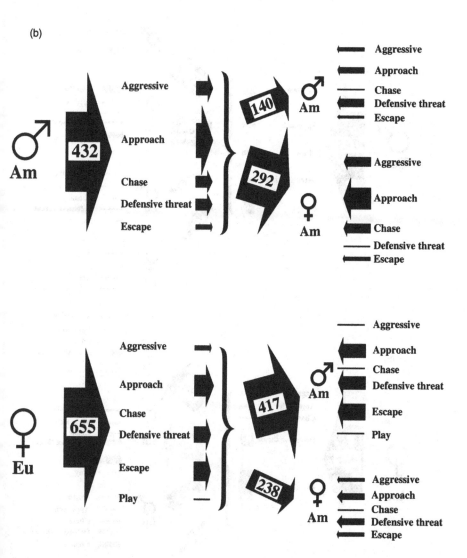

initiated by males was directed at both males and females. Furthermore, in both species, a greater proportion of chases involved males chasing females than chasing males.

The role of interactions between species

During the 10 months when both species were housed together, 5060 behavioural events were recorded during 22 days of observation, 4947 of which involved interactions between individuals. The remainder were mainly solitary play, which was more commonly observed in the males of both species

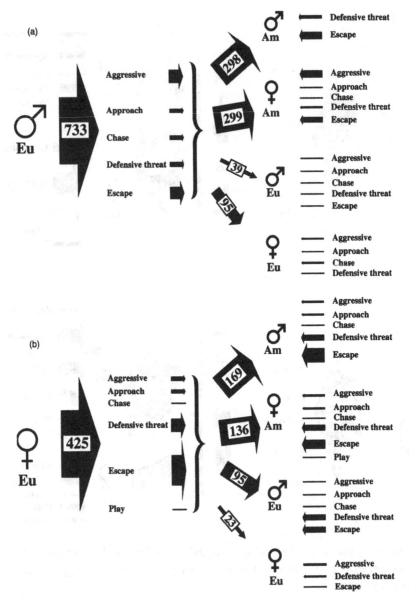

Figure 17.4. Sociograms showing the social dynamics amongst the eight mink, following the design of Fig. 17.3, during the period throughout which four individuals of each species were housed together. The flows of interactions are schematized separately for (a) male European mink (Eu), (b) female European mink, (c) male American mink (Am) and (d) female American mink. The thickness of the arrows indicates, for example, that of the 733 behavioural events initiated by male European mink, 298 involved male American mink; the proportional representation of these 298 amongst five behavioural classes is depicted by the five arrows indicating that these interactions rarely involved the European male chasing the American male, but frequently involved the European male escaping from the American male.

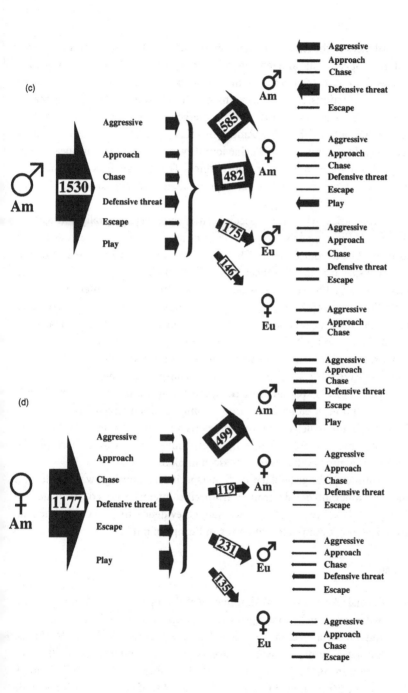

(41 instances in the American males compared with 11 in females, and 36 and 18 for European males and females, respectively. Overall, the behaviour of both species did not alter radically from that observed during the exclusive intra-specific interactions. In particular, American mink continued to be significant-ly more active and the majority of their interactions were intraspecific (of 2751 behaviours initiated by adult American mink, 67.5% were directed at con-specifics). In contrast, of 1158 behavioural events initiated by European mink only 21.8% were directed at conspecifics. The overall flow of behaviour patterns from adults of each sex of each species is schematized in Fig. 17.4(a)–(d).

First, we ask whether interspecific relationships were generally non-aggres-sive, neutral or aggressive. Male American mink were aggressive in 20.2% of their 1530 behavioural events. This aggression was largely directed at males, both conspecifics and European mink. Of their interactions with the male conspecifics 31.1% were aggressive, as were 20.0% of their interactions with male European mink. For male European mink, 19.1% of 733 behavioural events involved aggression. They were more interactive with both male and female American mink than with either sex of their own species, and a greater proportion of the interactions with both sexes of American mink were aggres-sive than were those involving either sex of conspecific. American mink males frequently (24.1%) played with conspecific females, but never played with European mink of either sex. European mink males interacted rather rarely. European mink males were more aggressive to conspecific females than were American mink males, whereas European mink females were more playful amongst themselves than were American mink females.

In summary, the general tenor of interspecific relationships was hostile, and did not differ obviously from intraspecific interactions. To elucidate the intraguild hostility hypothesis we sought evidence that male and female American mink dominated either sex of European mink.

Males

1   Do male American mink dominate male European mink? American mink males were much more aggressive to European mink males than *vice versa*, and often chased them (20.2% of their interactions). Of the 40.7% of male European mink behavioural events that were interactions with male American mink, 25.2% involved fleeing, whereas male American mink only fled from male European mink on 3.4% of occasions. We conclude that these male American mink did dominate the male European mink.

2   Do male American mink dominate female European mink? Of the

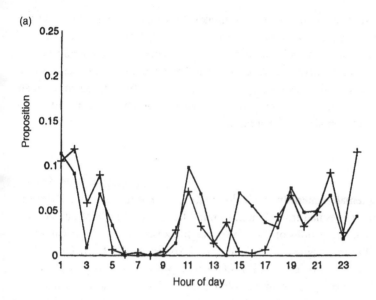

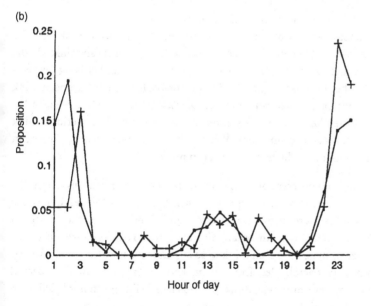

Figure 17.5. Specimen activity rhythms for (a) American mink and (b) European mink when the two species were housed together in June, showing the proportion of records, taken at 1 min intervals, during which males (+) and females (□) were, on average, active during each hour of the day.

interactions of male American mink with female European mink, 11.6% were aggressive and 39.7% involved chasing them. Indeed, 49.7% of female European mink's interactions with American mink involved fleeing from them. Only 5.8% of the male American mink's interactions with conspecific females were aggressive, as were 11.6% of those with female European mink. We conclude that these male American mink did dominate the female European mink.

Females

1   Do female American mink dominate male European mink? European mink males fled from American mink females on 22.4% of interactions. Of the interactions of male European mink with female American mink, 25.4% were aggressive, but only 3.3% involved chasing them. Only 13.0% of female American mink's interactions with male European mink involved fleeing from them. We conclude that while aggression flowed approximately symmetrically between these female American mink and male European mink, the male European mink were more inclined to flee from the female American mink than *vice versa.*

2   Do female American mink dominate female European mink? Of the interactions of female American mink and female European mink, 11.8%, 22.2% and 17.0% involved aggression, approaching and chasing, respectively. The equivalent figures for interactions of the female European mink with the female American mink were 9.5%, 7.4% and 2.2%. Indeed, 44.1% of female European mink's interactions with female American mink involved fleeing from them. We conclude that these female American mink dominated the female European mink.

The foregoing observations, together with other patterns displayed in Fig. 17.4, are compatible with the overall conclusion that within each sex, the American mink were aggressive towards, and dominant to, the European mink. Furthermore, male American mink appeared to harass female European mink, whereas male European mink were less assertive towards female American mink and, overall, fled from them. Female American mink appeared undaunted by male European mink, whereas female European mink fled from male American mink.

These overall conclusions combine data from different seasons, and we scrutinized the dynamics of interactions in each month of the study. These monthly observations on one unreplicated octet of mink are not the basis for generalizations, but serve to prompt questions for further study. The salient points include (i) soon after giving birth, one female American mink savagely attacked the other, (ii) there were indications that the males of each species showed sexual interest in the females of the other, (iii) there was hostility between breeding females of the two species, and one female European mink killed the kits of an American mink female.

The mink differed in their activity rhythms. Overall, European mink were largely nocturnal, whereas American mink, while tending to nocturnality, were more inclined to be active throughout the day (Fig. 17.5). In particular, American mink of both sexes had a supplementary peak of activity around 12:00 when they were fed. The result is that the American mink had generally eaten fully before the European mink emerged in the evening. The female European mink were even less inclined that the males to emerge by day. We compared, for each sex, the activity rhythms during the sympatric and allopatric phases, and there was no marked difference. The asynchrony in activity between the species meant that we made few observations of clashes over food, because the American mink had finished feeding before the European mink emerged.

Discussion

Our 1995 survey indicates that even since 1990 the distribution of the European mink has shrunk drastically. This trend suggests that the species faces imminent extinction.

The explanation for this precipitate decline remains unknown. Variations on at least five non-exclusive hypotheses have been proposed. All are plausible and the evidence allows few to be rejected categorically. Sidorovich *et al.* (see Chapter 11) reject the hypothesis, at least for our field study area in northeastern Belarus, that a declining prey base is responsible. Similarly, in our Belarussian study area European mink are declining in wilderness areas around natural floodplains with little evidence of pollution. Here, we report that there was no evidence of European mink in Tallinn Zoo succumbing to disease transmitted by American mink, but this evidence is scarcely grounds for rejecting the possibility. However, we also report here preliminary evidence for spontaneous hostility between American mink and European mink. If the evidence of our unreplicated experiment can be generalized, it suggests that American mink of both sexes go out of their way to harass European mink. There is even a possibility that shared social odours exacerbate this hostility when females are in oestrus. Furthermore, although domination was predominantly by the larger American mink over the smaller European species, we nonetheless recorded one instance of a female European mink killing the 3 week old kits of American mink.

There is increasing evidence of significant intraguild aggression amongst carnivores. Red foxes may deliberately kill pine martens, *Martes martes*, in Scandinavia (Storch *et al.*, 1995), coyotes, *Canis latrans*, kill kit foxes, *Vulpes velox*, in California, and lions kill cheetah cubs in Tanzania (Caro, 1994). In this context Hersteinsson & Macdonald (1922) argued that harassment by red foxes determines the southern limit to the range of Arctic foxes, and they suggested that the larger, more robust, red fox behaved towards Arctic foxes, *Alopex lagopus*, rather as if they were inferior conspecifics. There is evidence of red foxes killing Arctic fox cubs. It is plausible that a similar relationship exists between the American and European mink (and indeed, the interactions of both with polecats merit investigation). Our evidence suggests that American mink do not treat European mink exactly as if they were diminutive conspecifics – one female savagely attacked her female conspecific and merely harassed the female European mink. Nonetheless, the preliminary evidence presented here is compatible with the idea that the two species act to disadvantage each other when the opportunity arises, and the differential in body size, timorousness and activity patterns is such that the balance of disadvantage may generally be borne by the European mink. Further assessment of this hypothesis must consider the two species of mink in the context of the guild of semi-aquatic predators that also includes the European polecat and otter.

The question arises as to whether intra-guild hostility from the American mink is a sufficient explanation for the widespread extinction and invariable rarity of the European mink. Evaluating this possibility is difficult because much of the European mink's eastern range is remote and little populated with people (albeit many of them very skilled and active fur trappers). However, while the American mink population was largely seeded between the early 1930s and early 1960s (when at least 16 000 of them were deliberately released in the USSR), it seems European mink were already in decline (Tumanov & Zverjev, 1986). The simplest, if somewhat woolly, explanation would be that habitat loss, riparian engineering, pollution and ever more intensive hunting (backed by greater human mobility and encouraged by economic need) all set in train the European mink's decline. These same factors may in various combinations, continue to threaten the species, reducing their populations to a state of frailty in which intraguild hostility from the American mink becomes the last straw. This argument closely parallels the tightrope hypothesis proposed by Barreto *et al.* (1998) to explain the catastrophic impact of American mink on water voles, *Arvicola terrestris*, in Britain (see Chapter 19). They argue that the reason that American mink so effectively eradicate water voles, sometimes with a year of colonizing a section of river, is that agricultural intensification has reduced the vole's habitat to a 'tightrope' from which any detrimental factor will displace them. In the same way, there may be places where habitat loss and over-hunting weakened the European mink's population and American mink delivered the *coup de grâce*. On the other hand, there are places, and our study area is one of them, where the habitat remains wild and where despite intensive trapping, American mink flourish but European mink are declining fast. We therefore conclude that there is, as yet, no satisfactory explanation for the European mink's plight.

Whatever the cause of its decline, our data confirm that the European mink faces extinction. What has been done about it? There have been several attempts to release European mink into sites free of American mink. First, between 1981 and 1989 European mink bred at the Institute of Biology at the Siberian Branch of the Academy of Sciences of the then USSR were released well north of their geographical range on two islands (Kunashir and Iturup) in the Kuril Archipelago, amidst much debate (Shvartz & Vaisfeld, 1993). After 10 years significantly fewer mink survive than were released. Second, in 1988, 108 European mink were released along the River Shingindira in Tadjikistan, with unknown results (Saudskj, 1989). Third, in 1982, 11 European mink were released on Walam Isalnds in Lake Ladoga (Leningrad Region; Tumanov & Rozhnov, 1993), but by 1992 none remained. Modern conservation thinking would have foreseen these failures: the Kuril Islands flood, the Walam Islands are too small for a sustainable population and lack suitable habitat, and no or little thought was given to the genetic or demographic features of the founding populations.

Clearly, the immediate conservation goals must be to identify the causes of the European mink's decline and to remedy them. We are undertaking field-work on the guild of semi-aquatic mustelids in north-eastern Belarus and central Russia, but even when good data become available, reversing the species decline will pose enormous practical problems. The problem of excessive hunting is more difficult to solve. Because the two species cannot be trapped selectively, and indeed many hunters cannot distinguish them at all, any ban on hunting European mink would necessitate a similar veto on hunting American mink (a politically controversial proposal that currently keeps European mink out of the Russian and Belarussian Red Data Books). The American mink is incompatible with the persistence of European mink populations so one might seek selective means of reducing their numbers, but innovations such as virus-vectored immunocontraception are not only far off, but problematic in themselves (Bradley, 1994; Tyndale-Biscoe, 1994). An option is to introduce European mink to island reserves but, to our knowledge, all the suitable islands are already occupied by American mink which would have to be removed first.

Nonetheless, the apparently imminent threat of extinction, and the anticipated goal of providing founding populations for island sanctuaries, both make urgent the need for a captive breeding programme. This was first initiated in Tallinn Zoo in 1984, and re-launched there in 1992 under the auspices of the European mink Conservation & Breeding Committee, thus putting the conservation efforts under international supervision and control (Maran, 1992*b*, 1994*a*). Insufficient founders exist in captivity, the prospect of catching them is low, and the likelihood of housing the target population in European zoos is minimal. However, in practice the black-footed ferret (*Mustela nigripes*) recovery project has been relatively successful with only five founders (Thorne & Russel, 1991), so that are grounds for cautious optimism.

ACKNOWLEDGEMENTS

Our work on the European mink is funded by the Darwin Initiative. We sincerely thank Dr P. Johnson for statistical advice and Miss L. Handoca for preparing the figures. Tallinn Zoo generously accommodates the European Mink Captive Breeding Programme.

References

Barreto, G. R., Macdonald, D. W. & Strachan, R. (1998). The tightrope hypothesis: an explanation for plummetting water vole numbers in the Thames catchment. In *United Kingdom floodplains*: 311–327. (Eds Bailey, R., Gose, P. V. & Sherwood, B. R.). Westbury Academic and Scientific Publishing, London.

Bartá, Z. (1956) The European mink (*Mustela lutreola* L.) in Slovenia. *Ziva* 4: 224–225.

Bevanger, K. & Kenriksen, G. (1995). The distributional history and present status of the American mink (*Mustela vison*, Schreber 1777) in Norway. *Annls zool. fenn.* 32: 11–14.

Blas Aritio, L. B. (1970). *Vida y costumbres de los mustelidos Españoles*. Servico de Pesca Continental, Caza y Parques Nationales, Ministerio de Agricultura, Madrid.

Bluzma, P. (1990). [Habitat condition and the stutus of mammals in Lithuania.] In [*Mammals of the cultivated landscape in Lithuania]*: 4–78. (Ed. Bluzma, P.). Mokslas, Vilnius. [In Russian.]

Bradley, M. P. (1994). Experimental strategies for the development of an immunocontraceptive vaccine for the European red fox, *Vulpes vulpes. Reprod. Fert. Dev.* 6: 307–317.

Braun, A.-J. (1990). The European mink in France: past and present. *Mustelid Viverrid Conserv.* 3: 5–8.

Camby, A. (1990). Le vison d'Europe (*Musteola lutreola* Linnaeus, 1761). In *Encyclopédie des carnivores de France* 13: 1–18. Société Française pour l'Etude et Protection des Mammifères, Paris.

Caro, T. M. (1994). *Cheetahs of the Serengeti Plains: group living in an asocial species*. University of Chicago Press, Chicago.

Chanudet, F. & Saint-Girons, M.-C. (1981). Le répartition du vison européen dans le sudouest de la France. *Annls Soc. Sci. nat. Charente-Marit.* 6: 851–858.

Claudius (1866). [Cited in *Zool. Gart., Frankf.* 7: 315.]

Danilov, P. I. & Tumanov, I. L. (1976a). *[Mustelids of the north-western USSR.]* Nauka, Leningrad. [In Russian.]

Danilov, P. I. & Pumanov, I. L. (1976b). [The ecology of the European and American mink in the north-west of the USSR.] In *[Ecology of birds and mammals in the north-west of the USSR]*: 118–143. Akad. Nauk. Karelski Filial, Inst. Biol., Petrozavodsk. [In Russian.]

Dunstone, N. (1993). *The mink.* T. & A. D. Poyser, London.

Gautschi, A. (1983). Nachforschungen über den Iltis (*Mustela putorius* L.). *Schweiz. Z. Forstwes.* 134: 49–60.

Gerall, R. (1967). Dispersal and acclimatisation of the mink (*Mustela vison* Schreber) in Sweden. *Viltrevy* 4: 1–38.

Granqvist, E. (1981). [European mink (*Mustela lutreola*) in Finland and the possible reason for its disappearance.] *Mem. Soc. Fauna Flora fenn.* 57: 41–49. [In Swedish.]

Graphodatsky, A. S., Volobuev, V. T., Ternovskij, D. V. & Radzhabli, S. I. (1976). [G-banding of the chromosomes in seven species of Mustelidae (Carnivora).] *Zool. Zh.* 55: 1704–1709. [In Russian.]

Groombridge, B. (Ed.) (1993). *1994 IUCN red list of threatened animals.* IUCN, Gland, Switzerland, and Cambridge, UK.

Henttonen, H. (1992). [European mink (*Mustela lutreola*)]. In *[Threatened animals of the world 3. Finland]*: 46–48. (Ed.

Elo, U.). Weilin & Göös, Espoo. [In Finnish.]

Henttonen, H. & Tolonen, A. (1983). [American and European mink.] In *[Animals of Finland 1. Mammals]*: 228–233. (Ed. Koivisto, I.). Weilin & Göös, Espoo. [In Finnish.]

Henttonen, H., Maran, T. & Lahtinen, J. (1991). [The last European minks in Finland.] *Luonnon tutk.* 95: 198. [In Finnish.]

Heptner, V. G., Naumov, N. P., Yurgenson, P. B., Sludsky, A. A., Chirkova, A. F. & Bannikov, A. G. (1967). *[Mammals of the USSR].* Part 2, Vol. 1. Vysshaya Shkola, Moscow. [In Russian.]

Hersteinsson, P. & Macdonald, D. W. (1992). Interspecific competition and the geographical distribution of red and Arctic foxes, *Vulpes vulpes* and *Alopex lagopus. Oikos* 64: 505–515.

Kask, E. (1970). [Bog reclamation.] *Eesti Loodus* 11: 657–659. [In Estonian.]

López-Martin, J. M., Ruiz-Olmo, J. & Palazón, S. (1994). Organochlorine residue levels in the European mink (*Mustela lutreola*) in northern Spain. *Ambio* 23: 294–295.

Löwis, O. von (1899). Jagdbilder aus Livland. *Zool. Gart., Frankf.* 40: 24–26.

Lushnikova, T. P., Graphodatsky, A. S., Romashchenko, A. G. & Radjabli, S. I. (1989). [Chromosomal localization and the evolution age of satellite DNAs of Mustelidae.] *Genetika* 24: 2134–2140. [In Russian.]

Macdonald, D. W. & Smith, H. M. (1991). New perspective on agro-ecology: between theory

and practice in the agricultural ecosystem. In *The ecology of temperate cereal fields*: 413–448. (Eds Firbank, L., Carter, N., Derbyshire, J. F. & Potts, G. R.). Blackwell Scientific Publications, Oxford.

Maëmets, A. (1972). [On the freshwater resources and their utilization.] *Eesti Loodus* 8: 449–455. [In Estonian.]

Maran, T. (1989). Einige Aspekte zum gegenseitigen Verhalten de Europäischen *Mustela lutreola* und Amerikanischen Nerzes *Mustela vison* sowie zur ihren Raum- und Zeitnutzung. In *Populationsökologie marderartiger Säugetiere*: 321–332. *Martin-Luther-Univ. Halle-Wittenberg, Wiss. Beitr.* **P39**.

Maran, T. (1991). Distribution of the European mink, *Mustela lutreola*, in Estonia: a historical review. *Folio theriol. eston.* 1: 1–17.

Maran, T. (1992a). The European mink, *Mustela lutreola*, in protected areas in the former Soviet Union. *Small Carniv. Conserv.* No. 7: 10–12.

Maran, T. (1992b). The European mink, *Mustela lutreola*, Conservation and Breeding Committee (EMCC) founded. *Small Carniv. Conserv.* No. 7: 20.

Maran, T. (1994a). *Studbook for the European mink,* Mustela lutreola *Linnaeus 1761.* 1. European Mink Conservation & Breeding Committee, Tallinn Zoo, Tallinn.

Maran, T. (1994b). On the status and the management of the European mink *Mustela lutreola. Counc. Eur. Envir. Encounters Ser.* No. 17: 84–90.

Maran T. & Henttonen, H. (1995). Why is the European mink (Mustela lutreola) disappearing? A review of the process and hypotheses. Annls zool. fenn. 32: 47–54.

Maran, T. & Raudsepp. T. (1994). Hybrids between the European mink and the European polecat in the wild – is it a phenomenon concurring with the European mink decline? In Second North European Symposium on the Ecology of the Small and Medium-sized Carnivores. Abstracts: 42.

Martenson, A. (1905). Übersicht über das jagdbare und nussbare Haarwild Russlands. Riga.

Mickevicius, E. & Barnauskas, K. (1992). Status, abundance and distribution of mustelids in Lithuania. Small Carniv. Conserv. No. 6: 11–14.

Moutou, F. (1994). Otter and mink in France. Small Carniv. Conserv. No. 10: 18.

Nieminaa, J. (1995). Activity patterns and home ranges of the American mink Mustela vison in the Finnish outer archipelago. Annls zool. fenn. 32: 117–121.

Novikov, G. A. (1939). [The European mink.] Izd. Leningradskogo Gos. Univ., Leningrad. [In Russian.]

Novikov, G. A. (1970). [The European mink.] In [Mammals of the Leningrad Region]: 225–232. Leningrad University, Leningrad. [In Russian.]

Ognev, S. I. (1931). [Mammals of eastern Europe and northern Asia. 2. Carnivora, Fissipedia.] Academy of Science of USSR, Moscow & Leningrad. [In Russian.]

Ozolins, J. & Pilats, V. (1995). Distribution and status of small and medium-sized carnivores in Latvia. Annls zool. fenn. 32: 21–29.

Palomares, F. (1991). Situation of the European and American mink populations in the Iberian peninsula. Mustelid Viverrid Conserv. 4: 16.

Pavlov, M. A. & Korsakova, I. B. (1973). American mink (Mustela vison Brisson). In Acclimatization of game animals in Soviet Union: 118–177. (Ed. Kiris, D.). Volgo-Vjatsk Book Publisher, Kirov.

Popov, V. A. (1949). [Data concerning the ecology of the mink (Mustela vison Br.) and the results of its acclimatization in the Tatar ASSR.] Trudy Kazan. Fil. Akad. Nauk. SSSR 2: 1–140.

Romanowski, J. (1990). Minks in Poland. Mustelid Viverrid Conserv. 2: 13.

Ruiz-Olmo, J. & Palazón, S. (1990). Occurrence of the European mink (Mustela lutreola) in Catalonia. Miscelánea zool. 14: 249–253.

Ruiz-Olmo, J. & Palazón, S. (1991). New information on the European and American mink in the Iberian Peninsula. Mustelid Viverrid Conserv. 5: 13.

Ryabov, P., Lavrov, V. & Sokolov, M. (1991). [The European and the American mink.] Okhota Okhot. Khoz. 12: 12–15. [In Russian.]

Saudskj, E. P. (1989). The European mink (Lutreola lutreola) on the mountain rivers of Kuril and Tadzikhistan. In All-Union conference on problems of ecology in mountain regions, 9–13

October 1989, Dushanbe. Abstracts: 48–52.

Schreiber, A. R., Wirth, R., Riffel, M. & van Rompaey, H. (1989). Weasels, civets, mongooses, and their relatives: an action plan for the conservation of mustelids and viverrids. IUCN, Gland, Switzerland.

Schröpfer, R. & Paliocha, E. (1989). Zur historischen und rezenten Beständerung der Nerze Mustela lutreola (L. 1761) und Mustela vison Schreber 1777 in Europa – eine Hypothesendiskussion. In Populationsökologie marderartiger Säugetiere: 433–442. Martin-Luther-Univ. Halle-Wittenberg, Wiss. Beitr. P39.

Schubnikova, O. N. (1982). [On the results of the introduction of the American mink, Mustela vison, to Russia and on the problems of its relation with the original species, Mustela lutreola L.] In [Game animals of Russia: spatial and temporal changes in their range]: 64–90. (Eds Zabrodin, V. A. & Kolosov, A. M.). Central Government of Hunting Industry & Nature Reserves at the Council of Ministers of the RFSR, Moscow. [In Russian.]

Shashkov, E. V. (197). [Changes in the abundance of the European mink, otter and desman in some central regions of the European part of the USSR during the past 25 years.] Byull. Mosk. Obshch. Ispyt. Prir. (Otd. Biol.). 82: 23–38. [In Russian.]

Shvarts, E. A. & Vaisfeld, M. A. (1993). [Problem of saving vanishing species and the islands (discussion of the introduction of the European mink

Mustela lutreola on Kunashir Island).] *Usp. Sovrem. Biol.* 113: 46–59. [In Russian.]

Sidorovich, V. E. (1992). Gegenwärtige situation des Europäischen Nerzes (*Mustela lutreola*) in Belorussland: Hypothese seines Verschwindens. In *Semiaquatische Säugetiere*: 316–328. (Eds Schröpfer, R., Stubbe, M. & Heidecke, D.). *Martin-Luther-Univ. Halle-Wittenberg, Wiss. Beitr.*

Sidorovich, V. E. (1993). Reproductive plasticity of the American mink *Mustela vison* in Belarus. *Acta theriol.* 38: 175–183.

Sidorovich, V. E. (1995). *[Minks, otters, weasels and other mustelids.]* Uradzhai, Minsk. [In Russian.]

Sidorovich V. E. & Kozhulin, A. V. (1994). Preliminary data on the status of the European mink's (*Mustela lutreola*) abundance in the centre of the eastern part of its present range. *Small Carniv. Conserv.* No. 10: 10–11.

Sidorovich, V. E., Savchenko, V. V. & Bundy, V. B. (1995). Some data about the European mink *Mustela lutreola* distribution in the Lovat River Basin in Russia and Belarus: current status and retrospective analysis. *Small Carniv. Conserv.* No. 12: 14–18.

Storch, I., Lindstrom, E. & de Jonge, J. (1990). Diet and habitat selection of the pine marten in relation to competition with the red fox. *Acta theriol.* 35: 311–320.

Stubbe, M. (1993). *Mustela vison.* In *Handbuch der Säugetiere Europas. Bd. 5, Raubsäuger. Teil 2. Mustelidae 2, Viverridae, Herpestidae, Felidae.* (Eds Stubbe, M. & Krapp, F.). Aula, Wiesbaden.

Szunyoghy, J. (1974). Eine weitere Angabe zum Vorkommen des Nerzes in Ungarn, nebst einer Revision der Nerze des Karpatenbeckens. *Vertebr. hung.* 15: 75–82.

Ternovskij, D. V. (1975). [Will the European mink become extinct?] *Priroda, Mosk.* 1975 (11); 54–58. [In Russian.]

Ternovskij, D. V. (1977). *[Biology of mustelids (Mustelidae).].* Nauka, Novosibirsk. [In Russian.]

Ternovskij, D. V. & Ternovskaja, Y. G. (1994). *[Ecology of mustelids.]* Nauka, Novosibirsk. [In Russian.]

Ternovskij, D. V. & Tumanov, I. L. (1973). [Preserve the European mink.] *Okhota Okhot. Khoz.* 2: 20–21. [In Russian.]

Thorne, E. T. & Russel, W. C. (1991). Black-footed ferret, *Mustela nigripes. AAZAPA a. Rep. Conserv. Sci.* 1990–91: 86–88.

Tumanov, I. L. (1992). The numbers of the European mink (*Mustela lutreola* L.) in the eastern area and its relation to the American mink. In *Semiaquatische Säugetiere*: 329–335. (Eds Schröpfer, R., Stubbe, M. & Heidecke, D.). *Martin-Luther-Univ. Halle-Wittenberg, Wiss. Beitr.*

Tumanov, I. L. & Rozhnov, V. V. (1993). Tentative results of release of the European mink into the Valaam Islands. *Lutreola* 2: 25–27.

Tumanov, I. L. & Ternovskij, D. V. (1975). [The European mink under protection.] *Nasha Okhota* 5: 275–279. [In Russian.]

Tumanov, I. L. & Zverjev, E. L. (1986). [Present distribution and number of the European mink (*Mustela lutreola*) in the USSR.] *Zool. Zh.* 65: 426–235. [In Russian.]

Turjanin, I. I. (1986). [Preserving the wild mammals in the Carpathians.] In *Abstracts of the All-Union conference on monitoring the wild animals* 2: 405–406. [In Russian.]

Tyndale-Biscoe, C. H. (1994). Virus-vectored immunocontraception of feral mammals. *Reprod. Fert. Dev.* 6: 281–287.

Youngman, P. M. (1982). Distribution and systematics of the European mink *Mustela lutreola* Linnaeus 1761. *Acta zool. fenn.* No. 166: 1–48.

van Bree, P. J. H. (1961*a*). On the remains of some Carnivora found in a prehistoric site at Vlaardingen, The Netherlands. *Beaufortia* 8: 109–118.

van Bree, P. J. H. (1961*b*). On a subfossil skull of *Mustela lutreola* (L.) (Mammalia, Carnivora), found at Vlaardingen, The Netherlands. *Zool. Anz.* 166: 242–244.

18

Otters and pollution in Spain

J. Ruiz-Olmo, J.M. López-Martín and M. Delibes

Introduction

Throughout the present century, several species of otter have suffered a significant decline in their numbers, frequently due to direct persecution, hunting and the destruction of their habitat (Foster-Turley *et al.*, 1990). However, where populations have declined in the absence of these causes, contamination is thought to be involved and has been extensively demonstrated in *Lutra lutra* (Mason & Macdonald, 1986; Mason, 1989; Macdonald & Mason, 1994).

Those compounds that are accumulated in the organism through the food chain, reaching high levels in predators, have been specially studied. As well as affecting the mustelid directly, such components can also have an impact on prey, leading to a scarcity of food (Mason, 1989).

Amongst the compounds that are biomagnified, special attention has been paid to organochlorines (polychlorinated biphenyls (PCB), tetradichlordietan DDT, cyclodienes), being those on which most studies have been carried out (Mason, 1989; Olsson & Sandegren, 1991a,b; Smit *et al.*, 1994). Amongst the most notable effects, besides death when the compounds are present in high concentrations, are the faults found in reproduction and the immune system, alterations in the nervous system, with changes in behaviour, and malformations (Mason, 1989; McBee & Bickham, 1990; Kihlström *et al.*, 1992). Several of these effects have been found in *Lutra lutra* (Keymer *et al.*, 1988; Mason & O'Sullivan, 1992).

However, despite the fact that studies on the contents of organochlorine compounds (mainly PCB) in water, sediment, fish and otter tissue have proliferated (see references in Smit *et al.*, 1994) just as Kruuk (1995), points out the cause–effect relationship has still not been verified, and, what is more important, the levels that cause deleterious have not been determined (Kruuk, 1995). The closest studies are those carried out on the American mink (*Mustela vison*) in captivity. These show that levels higher than 50–100 $\mu g/g$ lipid weight of PCB in muscle significantly inhibit reproduction (Jensen *et al.*, 1977; Kihlström *et al.*, 1992; Leonards *et al.*, 1994a). Several studies on feeding contaminated fish to mink show this same effect on reproduction, not only with PCB (Aulerich *et al.*, 1973; Aulerich & Ringer, 1977) but also with Heptachlor (Aulerich *et al.*, 1990).

Various studies on pollution in fish tissue (the otter's principal diet) have demonstrated that a relationship exists between levels of PCBs and the presence or absence of this mustelid (Mason, 1989; Ruiz-Olmo & López-Martín, 1994).

Although a negative correlation has also been found between the status of otter populations and average values of contamination in otter tissue (see references in Macdonald & Mason, 1994; Smit *et al.*, 1994; Gutleb, 1995), it remains to be seen why very elevated average levels of PCB, higher than 200 μg/g lipid weight (Kruuk, 1995) are present in several areas with healthy populations of otters. It is possible that other organochlorines, or heavy metals, in particular mercury, are responsible for the decline in certain populations of *Lutra lutra*.

In Spain otter populations are generally healthy, being found in two-thirds of the country, although they are much scarcer in the eastern third (Delibes, 1990; our personal observations). This distribution seems to demonstrate a relationship between levels of contamination and the disappearance of this species in part of Spain (Ruiz-Olmo & López-Martín, 1994; Ruiz-Olmo, 1995). The current study extends existing knowledge on the levels of both organochlorine compounds and heavy metals in otters, as well as the relationships between pollution and otter distribution.

Materials and methods

Samples
Otters
Since 1986, 41 otters (24 males and 17 females) have been collected for the analysis of organochlorine compounds ($n = 41$) and heavy metals ($n = 19$). Samples were allocated to three subpopulations (Fig. 18.1): (i) a south-western population (Mediterranean bioclimate area), (ii) a north-western population (Atlantic) and (iii) a north-eastern population (Mediterranean). The majority of animals were aged from the cementum rings of the first upper premolar. Whenever possible, total length and weight were measured and an index of body condition (K) was calculated (Ruiz-Olmo, 1995), following the method described by Kruuk (1995). The distribution of age categories was: 0+ years, 16; 1+, 6; 2+, 3; 3+, 1; 4+, 3; 6+, 2; 9+, 2; 10+, 1. Bodies were frozen at $-20\,^\circ$C prior to post-mortem analysis. Tissue samples from the leg muscles and liver were removed for the analysis.

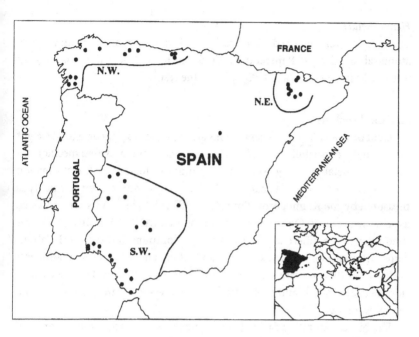

Figure 18.1. Location of the otters analysed in this study and of the three populations compared.

Fish

A total of 44 fish sampling stations were selected throughout the entire river network of Catalonia (north-east Spain). A total of 188 specimens from 12 different species were captured, labelled and stored at −20 °C until analysis (see López-Martín *et al.*, 1995).

Analytical methods
Organochlorine compounds

Samples were analysed by using gas chromatography, following the method described by López-Martín *et al.* (1994). Recoveries for organochlorine compounds ranged from 82–101% ($n = 12$). The limit of detection was 0.1 μg/g of fat. All samples were analysed for Aldrin, Heptachlor, Heptachlor epoxide, hexachloro cyclohexane isomers (α-β-γ-HCH) p,p'-DDE, o,p'-'DDD', p,p'-DD'D, o,p'-DDT, p,p'-DDT and 18 PCB congeners of Aroclor 111260 (IUPAC No: 95, 101, 138, 141, 151, 153, 170, 171 + 202, 174, 180, 182 + 128, 187, 194, 195, 201, 203 + 196). The sum of the five DDT metabolites is shown as tDDT, the sum of the three HCH isomers as HCHs, the sum of PCB congeners as PCBs, and the sum of two heptachlors as Hepo.

Heavy metals
The samples were analysed for mercury, lead, cadmium and chromium. The atomic absorption / ICP method was used. The limit of detection was 0.1 μg/g wet weight for mercury and 0.05 μg/g for the rest.

Statistical analysis
Levels of organochlorine concentrations are expressed as μg/g on a lipid weight basis to reduce variability and allow comparisons (intra- and interspecific) and the heavy metals as μg/g wet weight. Common logarithm transformation ($\log_{10}$) was used in all data analysis to better satisfy conditions for normality as measured by the Kolmogorov–Smirnov coefficient. The results are expressed as arithmetic means. One-way analysis of variance (ANOVA, Fisher PLSD) was used to compare differences between sample locations and species. Relationships between contaminants in fish and otter, and between contaminants within otter tissues, were examined using linear regressions. For the correlations between muscle and liver, data from 29 otters that had examples in both tissues were used.

The bioconcentration factor (BCF), calculated for each group of organochlorine compounds, was considered to be the index between the mean values in fish and otters.

We use the results of the otter survey of 1989–90 in Catalonia (Ruiz-Olmo, 1995) for comparisons of status and distribution.

Results

Organochlorine residues in otters
Residue levels
A broad range of results was found for the whole of the samples, tDDT and PCB being the compounds with the highest levels and present in 100% of the samples (Table 18.1).

The values of HCHs, Hepo and Aldrin varied from non-detection to 9.92 μg/g in muscle. Significant differences were found between the north-western and north-eastern populations for the HCHs in muscle ($F = 6.133$; $p < 0.03$) and in the liver between the south-western and north-eastern populations (Aldrin: $F = 4.978$; $p < 0.05$) and between the north-western and north-eastern populations (HCHs: $F = 8.967$; $p < 0.007$).

With regard to tDDT and its metabolites, the most widespread was p'p-DDE, found in all samples analysed. The values of tDDT were higher in the

Table 18.1. *Average values (arithmetic mean) and ranges of the analysis of organochlorine compounds and heavy metals carried out on otters from Spain*

		Mean	Min.–max.
% Ext. fat	: muscle	1.159	0.10–4.06
	: liver	2.199	0.11–6.60
HCHs	: muscle	1.95	0.03–9.92
	: liver	1.11	n.d.–5.71
HEPO	: muscle	0.26	n.d.–1.53
	: liver	0.27	n.d.–1.64
Aldrin	: muscle	0.24	n.d.–5.84
	: liver	0.35	n.d.–6.40
tDDT	: muscle	14.79	0.19–82.95
	: liver	41.15	0.45–241.76
PCB[†]	: muscle	78.30	1.49–1005.59
	: liver	106.74	2.79–974.82
DDT/PCB	: muscle	0.56	0.01–3.46
	: liver	1.47	0.006–8.64
DDE/DDT	: muscle	0.63	0.07–1.0
	: liver	0.61	0.18–0.97
Hg	: muscle	0.54	n.d.–1.94
	: liver	0.99	n.d.–2.80
Pb	: muscle	0.04	n.d.–0.18
	: liver	0.09	n.d.–0.34
Cd	: muscle	n.d.	–
	: liver	0.04	n.d.–0.22
Cr	: muscle	0.08	n.d.–0.28
	: liver	0.17	n.d.–0.49

The results are expressed in $\mu g/g$ lipid basis for organochlorine compounds and in $\mu g/g$ fresh weight for heavy metals.
[†]Percentage of muscle samples > 50 $\mu g/g$ PCB $= 28.2\%$.
n.d., not detected.

liver than in the muscle ($n = 29$, $F = 8.575$; $p < 0.007$):

$$\log (\text{liver}) = 0.667 \log (\text{muscle}) + 0.585, \ r^2 = 0.426; \ p < 0.0001.$$

For the total population average values of 14.79 $\mu g/g$ lipid weight in muscle and 41.15 $\mu g/g$ in liver were obtained. Differences in both tissues between the

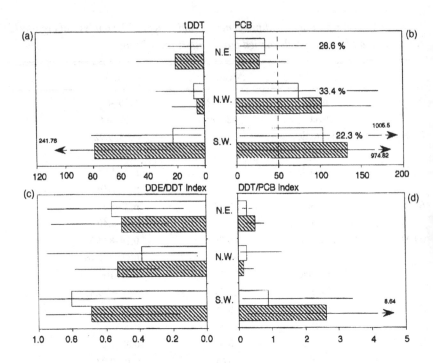

Figure 18.2. Arithmetic mean and ranges of (a) tDDT, (b) PCB, (c) DDE/DDT and (d)DDT/PCB in muscle (open bars) and liver (hatched bars), samples from otters from the north-eastern, north-western and south-western populations.

south-western and north-western populations were found (muscle: $F = 5.962$; $p < 0.03$, and liver: $F = 20.625$; $p < 0.0001$), with tDDT values from the otters in the south-west being higher (Fig. 18.2).

The average values of PCB were higher than those of other compounds, the average in muscle for the whole population being 78.30 μg/g. Of the samples from muscle, 28.2% exceeded the threshold of 50 μg/g. Higher average levels were obtained for the liver (106.74 μg/g) although significant differences were not found between the contents of the two tissues ($n = 29$), which were related according to the following formula:

$$\log (\text{liver}) = 0.758 \log (\text{muscle}) + 0.441, \; r^2 = 0.662; \; p < 0.0001.$$

Significant differences were not found between the three populations analysed for levels of PCB, although the average levels were higher in the south-western population (Fig. 18.2).

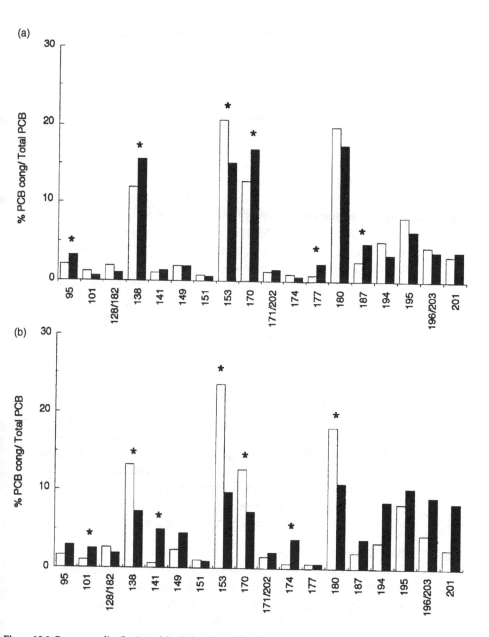

Figure 18.3. Percentage distribution of the different PCB congeners. (a) Distribution in muscle (open bars, $n = 39$) and liver (filled bars, $n = 30$) of otters from Spain. Significant differences among tissues for each PCB are shown with an asterisk when $p < 0.05$ (ANOVA-log 10 transformed). (b) Distribution in otter muscle (open bars, $n = 8$) and in the muscle of fish (filled bars, $n = 35$) taken from stations close to the locations of these otters. Significant differences are shown with an asterisk ($p < 0.05$).

For muscle, a positive correlation was found between the content of PCB and tDDT according to the formula:

$$\log (\text{PCB}) = 0.475 \log (\text{tDDT}) + 1.026, \ r^2 = 0.299; \ p < 0.0003.$$

For none of these organochlorine compounds were any significant differences in levels found between males and females, or between animals of different age or physical condition, either in muscle or in the liver.

PCB congeners
In Fig. 18.3(*a*) the distribution of the average percentages of the different PCB congeners in muscle and liver is shown for the total number of otters analysed. There are significant differences between the two tissues for some PCB congeners (PCB Nos 95, 138, 153, 170, 177 and 187). The maximum values were obtained in both tissues by congeners Nos 138, 153, 170 and 180. Significant differences were not observed for any PCB congener between the three populations. Comparing the distributions of the average percentages of PCB congeners in fish from five stations ($n = 35$ fish) close to where samples were taken of otters from the north-east population ($n = 8$) (Fig. 18.3(*b*)), significant differences were observed in seven PCB congeners: Nos 101, 138, 141, 153, 170, 174 and 180.

Organochlorine residues in fish
For PCB and tDDT the average levels for the total of the samples ($n = 188$) was 17.84 μg/g lipid weight (0.18 μg/g wet weight) and 9.93 μg/g lipid weight (0.08 μg/g wet weight), respectively. The highest values were found in the lowland reaches of rivers, although in some stations below the headwaters abnormally high values were registered (López-Martín *et al.*, 1995).

Comparing the results of otter distribution in Catalonia with concentrations of PCB in fish, it was found that otters did not live permanently at sites where average PCB levels in fish exceed 0.11 μg/g wet weight.

Bioconcentration factors
Comparing the values obtained from six otters of the north-eastern population from three stretches of river with those obtained from fish from the nearest stations ($n = 26$ fish), bioconcentration factors (BCF) were determined for tDDT and PCB. For the total of the samples a BCF of 1.4 (0.9–4.8) was obtained for tDDT, and of 2.9 (2.2–8.2) for PCB (for more details, see López-Martín & Ruiz-Olmo, 1996). Despite the fact that only three locations were compared, the level of contamination indicated a non-linear relationship between the accumulation of PCB in the otter and the levels of these compounds in fish ($r = 0.99$, $p = 0.021$) (Fig. 18.4). Elevated levels of PCB in fish

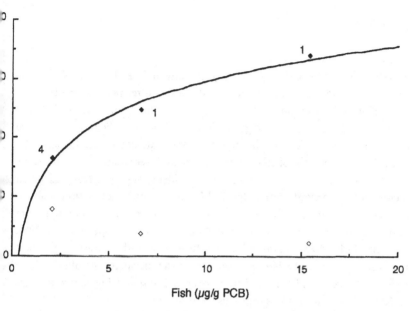

Figure 18.4. Relationship between the PCB concentration in fish and that in otter for the three locations analysed ($r^2 = 0.981$; $p = 0.021$). The BCF values are shown in the same plot for each location. ◇, BCF; ◆, PCB otter.

correspond to levels lower than expected in the otter's musculature (and with low BCF values).

Heavy metals

The average levels of the four heavy metals analysed were low (see Table 18.1), both in muscle (lead (Pb): 87.5% of the samples with less than 0.05 μg/g wet weight: cadmium (Cd): 100%; Chromium (Cr): 62.5%), and in liver (Pb: 41.7%; Cd: 75%; Cr: 100%). Only mercury (Hg) was higher (maximum of 1.94 μg/g in muscle and 2.80 μg/g in liver), although values below the non-detection threshold were more often found in muscle than in liver (62.5% of muscle specimens had values of lower than 0.02 μg/g, as opposed to 50% of liver specimens).

A positive but non-significant relationship was found between the mercury in liver and an index of body condition ($r = 0.561$, $n = 8$, $p = 0.19$) and also between the mercury in liver and age ($r = 0.367$, $n = 11$, $p = 0.26$). There was no relation between the mercury content of the muscle and liver. With the organochlorines, the only significant correlations obtained were between tDDT and mercury in both muscle and liver ($p < 0.05$).

Discussion

Organochlorine compounds

Amongst the compounds analysed in this study, only PCB and DDT were at levels that could be responsible for the decline in otter populations in some areas of Spain. Even though the levels of DDT are high, they are clearly lower than those of PCBs. The average values of 78.30 μg/g of PCB in muscle are exceeded only by those found in otters from the south of Sweden, England, the Czech Republic and the Shetland Islands, Scotland (see summary in Smit *et al.*, 1994). The first three populations are considered highly threatened or in regression. By contrast, in the Shetland Islands the otter population is abundant and, in theory, healthy, which poses an important question on the effect of these compounds on the regression of this species in Europe (Kruuk, 1995), although the localized but intensive oil industry may be the cause of the highest levels in otters from the Shetland Islands. In fact, in Spain, the otter is fairly well distributed and populations are recovering and stabilizing in various areas (J. Ruiz-Olmo & M. Delibes unpublished results).

Average results of PCB in otters from Spain are above the threshold of 50 μg/g where effects on reproduction in the American mink (*Mustela vison*) begin to occur (Jensen *et al.*, 1977; Kihlström *et al.*, 1992; Leonards *et al.*, 1994a). The stabilization–expansion of *Lutra lutra* in Spain does not seem to reflect these high levels. However, although analyses in which populations of otter are compared with average levels of contamination from each country (paying attention to political limits) are suggestive, they disguise certain patterns. Spain is a very heterogeneous country. Thus, even though this study has reported the maximum value for PCB in Europe (1005 μg/g) only 24.4% of the otters analysed exceeded the threshold of 50 μg/g, 9.8% of the samples being greater than 100 μg/g.

If attention is paid to reproductive animals (reproductive females, pregnant, nursing and with embryos, and young) some surprising results are obtained. The female that presented the maximum value of 1005 μg/g PCB had three placental scars. In the young aged 0–5 months ($n = 10$), in which the contaminating lipofiles were incorporated principally *via* the placenta and lactation, 40% exceeded a level of 50 μg/g in muscle (maximum 484.8 μg/g) and 60% in the liver. We do not know whether these high levels may be toxic to cubs, but these results could suggest that 50 μg/g may not be the level at which the otter is affected. In fact, contradictory data exist, with some populations of *Lutra canadensis* and *L. lutra* being more resistant than *M. vison*, and some populations of *M. vison* being more resistant than *L. lutra* (see Smit *et al.*,

1994). In Spain, for example, *M. vison* is spreading rapidly in some places where *L. lutra* has disappeared because of the effects of contamination (Ruiz-Olmo & López-Martín, 1994; Ruiz-Olmo *et al.*, 1997). The differences in diet can be found at the base of these differences, mink being less dependent on fish than otter. In some places in the north-east of Spain where the otter has disappeared due to the effects of high chemical contamination, the average levels of PCB in the muscle of the American mink are low (mean PCB: 13 μg/g PCB; range: 7.82–16.41; $n = 3$: López-Martín *et al.*, 1997). And so, while the effect on mammals of the PCBs in large concentrations is undeniable, it is necessary to carry out studies on the otter rather than to extrapolate from other species, since the threshold level which affects the otter can be higher than that which affects *M. vison*.

It is also necessary to remember that certain populations of otter in Europe, and in particular in Spain, are in recuperation. This may be related to the diminution of the levels of PCB and other organochlorines in nature (Stout, 1986). But we need to explain how the recolonization of certain industrial areas in the south-west of Spain can be reconciled with high levels of contamination in the samples originating from them (average levels of PCB in muscle = 99.06 μg/g). Smit *et al.* (1994) indicate that anomalous cases in Spain and the Shetland Islands do not necessarily imply that PCBs are not responsible for the decline of *L. lutra* in Europe. A decrease in the rate of reproduction in one part of the population would not necessarily cause decline in the total population, especially where other adverse factors are low. Furthermore, even if the average values are high, the majority of the samples have low values. In some areas, otters reproduce sufficiently well to compensate for mortality and lack of reproduction in other areas. This idea was argued in some detail in Mason & Macdonald (1993). The movement of individuals is also important. Ruiz-Olmo (see Chapter 10) shows that even though populations of otters are found below 150 m altitude in the north-east of Spain (contaminated areas), no reproduction has been detected below 310 m. These results demonstrate that non-reproductive populations are maintained by immigrants from areas where reproduction does take place.

Ruiz-Olmo & López-Martín (1994) have shown that otters are distributed in areas with PCB concentrations of less than 0.11 μg/g fresh weight in fish muscle. This does not necessarily indicate that this is the effect threshold for *L. lutra*, since levels at the time of the disappearance of the otter were almost certainly higher, although no data are available from that time. But the result provides supporting evidence of the negative effects of PCB on this mustelid.

The different toxicities of the different PCB congeners prevent assessment of the real significance of the concentrations of the total PCBs. The only comparable data on PCB congener patterns in *L. lutra* come from Broekhuizen & de Ruiter-Dijkman (1988) and from Mason & Ratford (1994). Their patterns are similar to those obtained from otters analysed in this study, in which the PCBs Nos 138, 153, 170 and 180 jointly constituted an average of 65% of the total (Fig. 18.3(*a*)), in both muscle and liver. Larsson *et al.* (1990) put forward the different explanations for the predominance of these PCBs: the proportions in which they occur in the mix of commercial PCBs (these five PCBs constitute 34% of the commercial sample) and their persistence in tissues due to their reduced metabolism. Identical results have been obtained from other species of mustelids (Larsson *et al.*, 1990; Leonards *et al.*, 1994*b*).

The PCB congener patterns differed between otters ($n = 8$) and fish ($n = 35$) for seven PCBs (Fig. 18.3(*b*)). The four PCBs most abundant in otter tissue, represented a total sum of 67.65% of the total PCBs present, whereas in fish this total was 35.33%, a value almost identical to that obtained in the commercial PCBs (Aroclor© 1260, Clophen© A60). A greater proportion of the higher-chlorinated PCBs would be expected in otter tissues, with lower-chlorinated PCBs in fish (Smit *et al.*, 1994); this could not however, be checked, since the PCBs in fish were not analysed in this study.

The highest mean bioconcentration factor (BCF) value was obtained for PCB; the lower bioconcentration of tDDT could be due to greater degradation, metabolization and excretion (Larsson *et al.*, 1990). Among three rivers the BCF results showed four-and fivefold variability for PCB and tDDT, respectively. The BCF in the otter is higher when the levels in fish are lower. In spite of the fact that only three locations were compared, the contamination levels indicate a relationship between the accumulation of PCBs in the otter and the levels of these compounds in fish ($r = 0.99$; $p = 0.021$: see Fig. 18.4). Foley *et al.* (1988), analysing a great number of paired samples of mustelids and their main prey, found a highly correlated relationship between levels in American river otter and mink and those in fish from the same place.

Heavy metals

Kruuk (1995) suggests that mercury may be responsible for the decline in some populations of *L. lutra*. The levels of heavy metals in the tissues of Spanish otters are low, that of mercury being the highest. Gutleb (1995) gives 30 μg/g fresh weight as the concentration which can result in neurological symptoms in mammals. This level is considerably higher than the levels found in the tissues of Spanish otters.

However, these levels could be lower than those of 30–40 years ago, which may have caused the decline of otters in Spain at that time. It is interesting that levels in otters seem to have dropped during the last 20 years, in both America and Europe. No precise data are available, but in any case current levels appear too low to explain the dramatic decline of otters over a large area of Spain (Delibes, 1990).

ACKNOWLEDGEMENTS
The authors thank the governments of Castilla-León, Castilla-La Mancha, Andalucía, Galicia, Extremadura, Asturias, Cantabria and Catalunya for their help during the collection of dead otters. They also thank the people who collaborated in the project. The Catalunya Government financed the organochlorine and heavy metal analysis and the quantification of residues in fish. Emma O'Dowd translated the manuscript, and two anonymous referees reviewed the manuscript and contributed useful suggestions.

References

Aulerich, R. J. & Ringer, R. K. (1977). Current status of PCB toxicity to mink and effect on their reproduction. *Archs Envir. Contam. Toxicol.* **6**: 279–292.

Aulerich, R. J., Ringer, R. K. & Iwamoto, S. (1973). Reproductive failure and mortality in mink fed on Great Lakes fish. *J. Reprod. Fert.* **19**: 365–376.

Aulerich, R. J., Bursian, S. J. & Napolitano, A. C. (1990). Subacute toxicity of dietary heptachlor to mink (*Mustela vison*). *Archs Envir. Contam. Toxicol.* **19**: 913–916.

Broekhuizen, S. & de Ruiter-Dijkman, E. M. (1988). Otters (*Lutra lutra*) met PCB's: de zeehonjes ven het zoete water? *Lutra* **31**: 68–78.

Delibes, M. (1990). *La nutria (Lutra lutra) en España.* Serie Tècnica. ICONA, Madrid.

Foley, R. E., Jackling, S. J., Sloan, R. J. & Brown, M. K. (1988). Organochlorine and mercury residues in wild mink and otter: comparison with fish. *Envir. Toxicol. Chem.* **7**: 363–374.

Foster-Turley, P., Macdonald, S. M. & Mason, C. F. (Eds) (1990). *Otters. An action plan for their conservation.* IUCN/SSC Otter Specialist Group, Gland.

Gutleb, A. C. (1995). *Umweltkontaminanten und Fischotter in Österreich eine Risikoabschätzung für Lutra lutra (L. 1758).* PhD thesis: Wien University.

Jensen, S., Kihlström, J. E., Olsson, M. & Örberg, J. (1977). Effects of PCB and DDT on mink (*Mustela vison*) during the reproductive season. *Ambio* **6**: 239.

Keymer, I. F., Wells, G. A. H., Mason, C. F. & Macdonald, S. M. (1988). Pathological changes and organochlorine residues in tissues of wild otters (*Lutra lutra*). *Vet. Rec.* **122**: 153–155.

Kihlström, J. E., Olsson, M., Jensen, S., Johanson, A., Ahibom, J. & Bergman, A. (1992). Effects of PCB and different fractions of PCB on the reproduction of the mink (*Mustela vison*). *Ambio* **21**: 563–601.

Kruuk, H. (1995). *Wild otters: predation and populations.* Oxford University Press, Oxford.

Larsson, P., Woin, P. & Knulst, J. (1990). Differences in uptake of persistent pollutants for predators feeding in aquatic and terrestrial habitats. *Holarct. Ecol.* **13**: 149–155.

Leonards, P. E. G., Smit, M. D., de Jongh, A. W. J. J. & van Hattum, B. (1994a). *Evaluation of dose-response relationships for the effects of PCBs on the reproduction of mink* (Mustela vison). Instituut voor Milieuvraagstukken, Vrije Universiteit Amsterdam, Amsterdam.

Leonards, P. E. G., van Hattum, B., Cofino, W. P. & Brinkman, U. A. T. (1994a). Occurrence of non-ortho-, mono-ortho- and di-ortho-substituted PCB congeners in different organs and tissues of polecats (*Mustela putorius* L.) from the Netherlands. *Envir. Toxicol. Chem.* **13**: 129–142.

López-Martín, J. M. & Ruiz-Olmo, J. (1996). Organochlorine residue levels and bioconcentration factors in otters (*Lutra lutra* L.) from NE Spain. *Bull. Envir. Contam. Toxicol.* **57**: 532–535.

López-Martín, J. M., Ruiz-Olmo, J. & Palazón, S. (1994). Organochlorine residue levels in the European mink (*Mustela lutreola*) in northern Spain. *Ambio* **23**: 294–295.

López-Martín, J. M., Ruiz-Olmo, J. & Borrell, A. (1995). Organochlorine compounds in freshwater fish from Catalonia, NE Spain. *Chemosphere* **31**: 3523–3535.

López-Martín, J. M., Ruiz-Olmo, J. & Palazón, S. (1997). Organochlorine residue levels in semi-aquatic carnivores from Spain. In *Proceedings of the XIV mustelid Colloquium, Czech. Republic, September 1995*: 56–59. (Eds Toman, A. & Hlavac, V.). Prague.

Macdonald, S. M. & Mason, C. F. (1994). *Status and conservation needs of the otter* (Lutra lutra) *in the western Palearctic.* Council of Europe, Strasbourg.

Mason, C. F. (1989). Water pollution and otter distribution: a review. *Lutra* **32**: 97–131.

Mason, C. F. & Macdonald, S. M. (1986). *Otters: ecology and conservation.* Cambridge University Press, Cambridge.

Mason, C. F. & Macdonald, S. M. (1993). Impact of organochlorine pesticide residues and PCBs on otters (*Lutra lutra*): a study from western Britain. *Sci. total Envir.* **138**: 127–145.

Mason, C. F. & O'Sullivan, W. M. (1992). Organochlorine pesticide residues and PCBs in otters (*Lutra lutra*) from Ireland. *Bull. Envir. Contam. Toxicol.* **48**: 387–393.

Mason, C. F. & Ratford, J. R. (1994). PCB congeners in tissues of European otter (*Lutra lutra*). *Bull. Envir. Contam. Toxicol.* **53**: 548–554.

McBee, K. & Bickham, J. W. (1990). Mammals as bioindicators of environmental toxicity. *Curr. Mammal.* **2**: 37–38.

Olsson, M. & Sandegren, F. (1991*a*). Otter survival and toxic chemicals: implication for otter conservation programmes. *Habitat* **6**: 191–200.

Olsson, M. & Sandegren, F. (1991*b*). Is PCB partly responsible for the decline of the otter in Europe? *Habitat* **6**: 223–228.

Ruiz-Olmo, J. (1995). *Estudio bionómico de la Nutria (*Lutra lutra *L.) en el NE Ibérico.* PhD thesis: Barcelona University.

Ruiz-Olmo, J. & López-Martín, J. M. (1994). Otters and pollution in northeast Spain. In *Seminar on the conservation of the European otter* (Lutra lutra)*: Leeuwarden, The Netherlands, 7–11 June 1994*: 144–146. Directorate of Environment & Local Authorities, Council of Europe, T-PVS (94) 11, Strasbourg.

Ruiz-Olmo, J., Palazón, S., Bueno, F., Bravo, C., Munilla, I. & Romero, R. (1997). Distribution, status and colonization of the American mink *Mustela vison* in Spain. *J. Wildl. Res.* **2(1)**: 30–36.

Smit, M. D., Leonards, P. E. G., van Hattum, B. & de Jongh, A. W. J. J. (1994). *PCBs in European otter* (Lutra lutra) *populations.* Instituut voor Milieuvraagstukken, Vrije Universiteit Amsterdam, Amsterdam.

Stout, V. (1986). What is happening to PCBs? Elements of the environmental monitoring as illustrated by an analysis of PCBs trends in terrestrial and aquatic organisms. In *PCBs and the environment*, **2**: 164–205. (Ed. Waid, J. S.). CRC Press, Boca Raton, FL.

19

The rapid impact of resident American mink on water voles: case studies in lowland England

C. Strachan, D. J. Jefferies, G. R. Barreto, D. W. Macdonald* and R. Strachan

Introduction

The water vole (*Arvicola terrestris*; Arvicolinae) is a 250–300 g herbivorous rodent that inhabits the banks of freshwater-courses in Britain. In contrast, a fossorial form, *Arvicola terrestris scherman*, predominates in central Europe, co-existing with the aquatic form in the rest of Europe. The water vole is adapted to live in burrows, to swim and dive. In Britain it is largely associated with steep riverbanks, with abundant grass and layered vegetation (Lawton & Woodroffe, 1991). Woodall (1993) reported that in English Midland rivers such as the Thames, water vole presence was positively correlated with water depth and with the herbs *Urtica* sp. and *Phragmites* sp. and negatively corre- lated with bank height, bank depth and with herbs of the genera *Polygonum*, *Phalaris*, *Sparganium* and *Juncus*.

Several lines of evidence suggest that water vole populations have been declining in Britain at least since 1900. Jefferies *et al.* (1989) analysed the data contained in County Mammal Reports and literature from the beginning of the century onwards. These were supplemented by a questionnaire survey and by site descriptions from the Waterways Bird Survey organized by the British Trust for Ornithology (BTO). The literature search showed that, formerly, water voles had often been very abundant in all regions of Britain, except South Wales and Northern Scotland. However, by the late 1980s a general perception prevailed that water voles were becoming less numerous. The use of the word 'common' in descriptions of populations decreased from 1900 to 1985. It was also widely held that predation, especially by feral American mink (*Mustela vison* Schreber), was the primary cause of this perceived decline in water voles, followed by habitat disturbance and destruction by drainage, dredging, river works and farming.

To elucidate these perceptions, Strachan & Jefferies (1993) undertook a nationwide survey during 1989–1990. Part of this survey was a systematic

*Author to whom all correspondence should be addressed

search of a pre-selected series of 1926 sites distributed on a grid and part was a search of 1044 known and dated historically occupied sites. The latter retrospective survey confirmed a long-term decline in the water vole population during the twentieth century, and suggested two periods of accelerated loss of occupied sites: the first during the 1940s and 1950s, attributed to water pollution and habitat destruction, and the second during the 1980s, correlated with the spread of the feral mink. There was a significant negative association in site occupation by the two species in each of the 11 National Rivers Authority (NRA; currently Environment Agency) regions. The actual reduction in water vole numbers may be even greater than indicated by occupied site loss as analyses of latrine numbers showed that as the regional occupied site density decreased, so the population size within each occupied site also decreased.

Woodroffe *et al.* (1990) explored the possible impact of mink predation on water vole populations in an area where mink were spreading. They compared vole activity indices at sites in the North Yorkshire Moors National Park before and after the arrival of mink, and in the aftermath of their disappearance from some sites. The various sites were catalogued as 'core sites' (with permanent breeding water vole populations) and 'peripheral sites' (where regular water vole activity was evident but there were no latrines and the animals were never trapped). Mink activity was lowest at core sites, and greatest at peripheral sites or sites without water voles. Across the sites a significant negative correlation between long-term mink activity and water vole activity suggested that water vole numbers were either reduced by mink predation or that, in the presence of mink, water voles dispersed to safer areas (Woodroffe *et al.*, 1990).

These findings for upland habitats have been echoed in a series of surveys in the lowland habitats of the River Thames catchment. There, a combination of historical data and recent surveys indicate that a long-term decline in water voles accelerated rapidly during the 1990s, coinciding with a rapid expansion in the distribution of mink within the catchment (Strachan & Jefferies, 1993; Halliwell & Macdonald, 1996; Barreto *et al.*, 1998). In this chapter, we first present results of a survey that confirms this decline within the 1200 km^2 Thames catchment, and evaluate the widespread disappearance of the water vole within the historical context of the mink's arrival in the Thames valley more than 20 years ago. Second, at a finer scale, and with the aim of answering the question of how quickly can mink exterminate water voles, we present data on the time-course of events on the River Soar in the immediate aftermath of the American mink's arrival there. Finally, we compare these with a control area (the River Amber) where mink were absent. The range in the length of time of the American mink's occupancy enables us to compare water vole

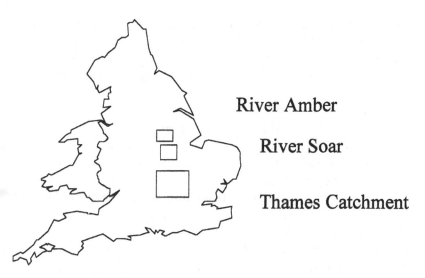

River Amber

River Soar

Thames Catchment

Figure 19.1. Location of areas where this study was carried out: River Amber in Derbyshire, River Soar in Leicestershire and Thames catchment comprising the Thames and tributaries lying upstream of the tidal reach at Teddington Lock.

populations in these three areas 0, 1, 2 and more than 20 years after this alien predator's arrival. Our ultimate goal is to use these results, together with local mink faecal analyses, to tackle the conservation question of whether mink and water voles can co-exist under the conditions prevailing in late twentieth century lowland England; if not, what lies at the root of their incompatibility and can it be remedied?

Methods

The results are gleaned from three study areas (Fig. 19.1). First, the River Thames and tributaries lying upstream of the tidal reach at Teddington Lock (Ordnance Survey national grid reference TQ166715), invaded by mink prior to 1975 (Strachan & Jefferies, 1993). Second, the River Soar in Leicestershire, on one stretch of which (2nd-year site) the mink first bred in 1993, and on a second stretch of which (1st-year site) they first bred in 1994 (C. Strachan, personal observation). The Soar study encompassed a total of 22 km, running from the confluence with the River Trent (SK494308) to Barrow on Soar (SK572174). Third, the River Amber in Derbyshire (SK377524), on which no mink occurred throughout this study (0-year site). The Midlands and southern England are two of the regions of highest density of occupied sites for water voles remaining in Britain (Strachan & Jefferies, 1993). We therefore selected rivers in these regions due to their considerable significance for water vole conservation.

Water voles and mink in the Thames catchment
During the summer of 1995 we conducted a systematic survey of water voles and mink at 130 sites that had been surveyed in 1990 by Strachan & Jefferies (1993). At each site, we searched 600 m along one bank of the water-course for field signs of water voles and mink (footprints, latrines, burrows and feeding remains). A description of the habitat was quantified by using the NRA's River Habitat Survey for future analyses. Data from 1975 were gathered from the Biological Recording Centre Mapping Scheme (BRC Monks Wood, Cambridge), County Museum Records, County Wildlife Trusts and information from local naturalists.

Water vole and mink along the Rivers Soar and Amber
The 22 km Soar study area was divided into 101 contiguous sectors of approximately 200 m each. Within this length of the Soar we identified two 600 m long stretches that were the foci of activity for mink; at the 2nd-year site (SK536221) mink were present almost continuously from early 1993, and bred successfully in both 1993 and 1994; at the 1st-year site (SK498240) mink colonized the stretch only in the autumn of 1993 and remained there, breeding successfully, throughout 1994. Throughout 1993 and 1994 the dispersion of water vole latrines was recorded monthly at each of the 1st-year and 2nd-year stretches on the Soar and at the 0-year site on the Amber. Furthermore, throughout 1994, all 101 sectors of the entire 22 km Soar study area were surveyed monthly along one bank, seeking field signs of water voles and mink; the number of latrines was recorded and mink scats were collected (see below). Habitat information for each of the 101 sectors was also recorded monthly (see below).

Diet of mink on the River Soar
Fresh mink scats were collected monthly along the 22 km Soar study site throughout 1994 by searching the banks, ledges under bridges, beneath overhanging tree roots and the limbs and crowns of pollards along the river. Each scat was labelled, oven-dried and stored for later analysis. Immediately prior to analysis the dried scats were immersed in a warm 1% detergent solution and

left overnight. This had the effect of dissolving any binding mucus, causing the scats to swell and dislodging soil and sand particles stuck to the surface, which fell away leaving the scats clean. Individual scats were then dried in a ventilated cupboard at 24 °C for 48 h prior to weighing (±0.05 g) and analysed by teasing them apart under a 10× binocular microscope. Undigested remains were identified from our own reference collection supplemented by published keys (mammals and birds: Day, 1966 and Teerinck, 1991; fish: Maitland, 1972 and Conroy *et al.*, 1993).

The remains of separate prey items in each scat were segregated and their volume was estimated to the nearest 10% of that of the entire scat. This was then multiplied by the dry weight of the scat to give an estimated dry weight of each item for each scat. The estimated dry weight of each prey item was summed for each month, and for the year, and then expressed as a percentage of the total dry weight of all remains for all prey in the sample (% volume). This method, which is cited by Wise (1978) as suitable for ranking mink prey types in order of importance, has been used in other studies of mink diet (Wise, 1978; Wise *et al.*, 1981; Ward *et al.*, 1986; Dunstone & Birks, 1987) and is therefore useful for comparisons.

Habitat

In order to test whether stretches of the River Soar occupied by mink and/or water voles differed, we recorded habitat features in each sector of the 22 km study area. We recorded (i) bank slope (0–30°, 30–60°, 60–90°); (ii) bank height (< 1 m, 1–2 m, 2–3 m and > 3 m); (iii) vegetation type (grass, reeds, tall ruderal, herbs and emergent); (iv) bank swathe (0, < 1 m, 1–2 m, 2–3 m and > 3 m); (v) tree cover (0, 1–25%, 25–50%, 50–75% and 75–100%); (vi) shrub cover (0, 1–25%, 25–50%, 50–75% and 75–100%); (vii) towpath (present or absent); (viii) piling (present or absent); and (ix) adjacent land use (arable, improved pasture, unimproved pasture, woodland, urban/suburban, and open grazing).

We attempted to identify habitat characteristics associated with particular sections of the Soar. These sections were distinguished according to whether they showed evidence of vole activity, evidence of mink activity, or evidence of both species. For this purpose, Principal Components Analysis (PCA) was used to ordinate the environmental data set, with levels of categorical variables included as dichotomous 'dummy' variables. This analysis allows us to get a description of sites used by water voles and mink but should be taken cautiously as adjacent sites are not truly independent and problems with spatial auto-correlation may arise.

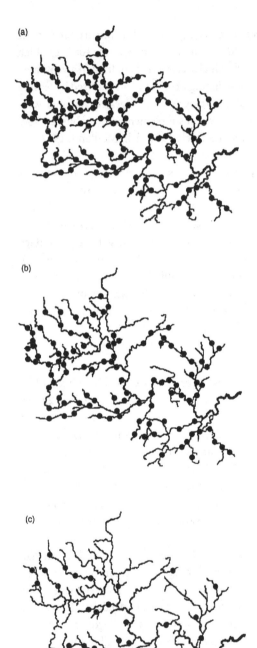

Figure 19.2. Sites supporting water voles in the Thames catchment during (a) 1975, (b) 1990 and (c) 1995. Water voles have disappeared from a significant number of sites in the last 5 years (McNemar test; $\chi^2 = 62.02$; d.f. = 1; $p \ll 0.001$).

Results

The Thames catchment: a long-term perspective
Water voles
The 130 sites known to support water voles in 1975 are shown in Fig. 19.2(a). These sites corresponded to 67 different named rivers, brooks, pools, gravel pits and canals within the Thames catchment. In 1990, 95 of these sites still supported water voles, but by 1995 only 31 did so (Figs. 19.2(b) and 19.2(c)). This indicates a drastic decline in water voles in the last 5 years. The proportion of sites without water voles has changed significantly between 1990 and 1995. (McNemar test; $\chi^2 = 62.02$; d.f. = 1; $p \ll 0.001$). Water voles have disappeared from a significant number of sites. Therefore, a loss of 2.33 sites/annum over the 15 years between 1975 and 1990 accelerated to a loss of 12.8 sites/annum between 1990 and 1995. Healthy populations of water voles are still present in the Rivers Kennet, Pang and Windrush, and in the Kennet and Avon Canal, and sporadically on the Rivers Blackwater, Churn, Coln and Ver, and on the Oxford Canal and Grand Union Canal. We found no evidence of water voles at the sites on the Evenlode, Gade, Leach, Lambourn, Loddon, Mole and Wey. Along the River Thames itself signs of water voles were found at only two sites.

Mink
The spread of American mink is shown in Figs. 19.3(a)–(c). In 1975 only 9 of the 130 sites revealed signs of mink, whereas they had spread to 32 by 1990 and 60 by 1995. This represents 6.92%, 24.62% and 46.15% site occupation, respectively. Site occupancy at 46% is higher than that formerly recorded in 7 out of 11 NRA regions in the 1990 survey. In the 1995 survey no signs of mink were found at any site where water voles persisted, and conversely no water vole sign was found at sites populated by mink.

The River Soar: short-term corollaries of mink arrival
The overall distribution of water voles and mink along the River Soar and within the 1st- and 2nd-year stretches are schematized in Fig. 19.4. Radiating out from the two foci of mink activity, signs of these predators were most abundant between sectors 28 and 40 and between sectors 60 and 101. From January to June 1994, water voles appeared to be distributed evenly between sectors 1 and 65, but were infrequent further upstream in the vicinity of the 2nd-year stretch. To test the statistical validity of the apparent heterogeneity in the dispersion of voles along the riverbank, we aggregated the 101 sectors into blocks of 10 (11 in the final one). For the period February–June, all signs of

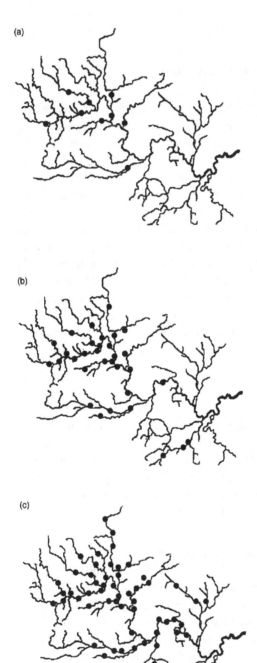

(a)

(b)

(c)

Figure 19.3. The spread of American mink in the Thames catchment since 1975. (a) 1975, (b) 1990 and (c) 1995. In the 1995 survey no signs of mink were found at any site where water voles persisted, and conversely no water vole sign was found at sites populated by mink.

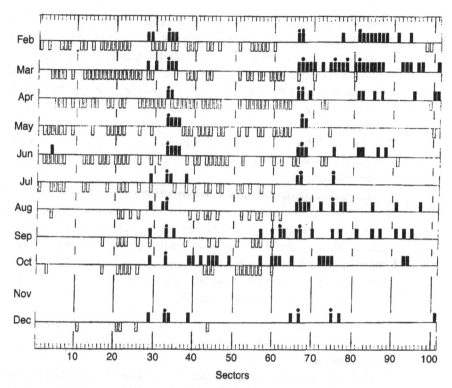

Figure 19.4. Distribution of mink and water voles along a 22 km stretch of the River Soar (Leicestershire). Open bars, water vole sites; filled bars, mink sites; filled circles, mink den locations. Sectors 1–65 comprise the 1st-year site, and the 2nd-year site is located from sector 65 upstream. Sectors 33 and 67 held breeding dens.

vole activity in each block were summed to give a maximum vole presence score of 50 (5 months × 10 sectors). A Kolmogorov–Smirnov test rejected the null hypothesis that vole signs were evenly distributed between the blocks ($D = 0.26$, $p < 0.01$).

Male mink are particularly mobile during the spring rut and Fig. 19.4 reveals that mink signs were prolific in February and March 1994. Thereafter, with the birth of kits in May, mink signs were less widely dispersed until late summer when they began to radiate further afield from their breeding dens, with concomitant reduction in water vole signs between sectors 35 and 65. By October 1994, following two breeding seasons of mink in the 2nd-year site, and one in the 1st-year site, the distribution of water voles appeared to be largely confined to an enclave downstream of the 1st-year site and another sand- wiched between the two mink territories, and by December the upstream enclave had gone. Indeed, whereas the water vole signs had been recorded in 37 sectors during February, in December they occurred in only 5. Figure 19.4 suggests that the elimination of water voles by December from sectors 40–60 follows an expansion of mink into these areas during September and October. To quantify differences between months, we again used the 10-sector blocks to derive vole-presence scores (for each block the proportion of the 10 sectors in that block that yielded vole sign). These scores declined steadily with each month of mink occupancy. The maximum mean vole activity for the 10 blocks (0.32), was in February, when the 1994 mink were newcomers, whereas the minimum (0.04) was in December, by which time they and their kits had eaten into the vole population.

While the distribution of mink scats gave an impression of the extent of the two zones of mink influence on the Soar, it did not precisely delineate those sectors in which mink foraged. On the basis of our unpublished radio-tracking studies of mink on the Thames we assumed that mink were unlikely to rove further than 1 km beyond the most peripheral concentration of their scats. On this basis we erected two categories of mink influence and allocated 200 m sectors of the Soar to these extremes as follows:

1   Low mink presence ($n = 7$): > 1 km from the nearest concentration of mink sign. We assume that only transient mink foraged occasionally in these sectors.

2   High mink presence ($n = 10$): sectors encompassed within concentrations of mink sign. We assume that resident mink foraged within these sectors frequently and in at least two (sectors 33 and 67) we found successful breeding dens.

Comparing these extremes of mink presence, the mean monthly occurrence of water vole latrines was significantly lower within the minks' zone of influence than beyond it (Fig. 19.5) (see analyses below). Indeed, from October onwards no water vole latrines remained in the zones of high mink presence.

To quantify further the impact of mink we redefined high and low mink influence in such a way that each 10-sector block could be assigned to one or other category. Thus, for the whole survey, each mink sign was summed within each 10-sector block, and blocks scoring > 20 were designated high mink influence (this giving roughly equal numbers of high and low influence blocks). A GLM model for the monthly means (± SE) of vole presence scores for these two categories of mink influence suggests that in each block vole scores are influenced by mink ($F_{1,8} = 4.7$, $p = 0.06$). Although there was a highly significant difference in vole-presence scores over the 10 months (repeated measures GLM: $F_{9,72} = 5.8$, $p < 0.01$), there was no evidence for any difference in the temporal pattern of vole activity scores between the two types

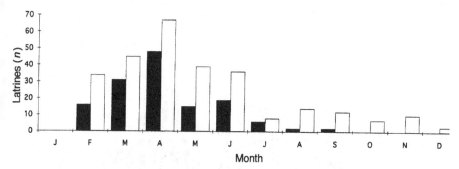

Figure 19.5. Occurrence of water vole latrines along sectors catalogued as having high mink presence or low mink presence in the River Soar (Leicestershire). The number of latrines was significantly lower within the mink zone of influence than beyond it ($F_{1,8} = 4.7$, $p = 0.0611$; see the text for explanation). ■, high mink presence; □, low mink presence.

of mink block (MINK × MONTH, $F_{9,72} = 0.5$, $p = 0.85$). In other words, vole presence differed between months (and generally declined throughout the study), and although it was, on average, lower during every month in those sectors most heavily used by mink, the temporal pattern of vole-presence nevertheless ran in parallel in the blocks heavily and lightly used by mink. A Tukey's mean separation test was used to discriminate months.

The seasonal patterns of water vole latrine abundance in each of the three 600 m sites designated 0-year, 1st-year and 2nd-year of mink occupancy are presented in Fig. 19.6. On the Amber, in the absence of mink, latrines were abundant in summer and less so in winter during both 1993 and 1994. In contrast, vole latrines were rare in both the 2nd and 1st-year sites by the late spring of the first year in which mink bred there. Indeed, in the 2nd-year site (Fig. 19.6(c)) high spring counts in 1993 crashed to only two latrines in August and September, and no latrines at all by October. During the following year of mink occupancy, a single latrine with only two droppings was found in March.

Analyses of scats

A total of 863 scats was collected throughout the study period along the 22 km length of the River Soar (including a large number from the two known breeding dens). Undigested prey remains included eight species of mammals, four of birds (plus one unidentified passerine), one species of amphibian (*Rana temporaria*), and representatives of six families of fish. Amongst invertebrates the orders Coleoptera, Odonata and Hemiptera were frequently recorded.

More than one-third of the volume of faecal remains consisted of mammalian prey (Fig. 19.7(a)) of which the water vole accounted for the greatest

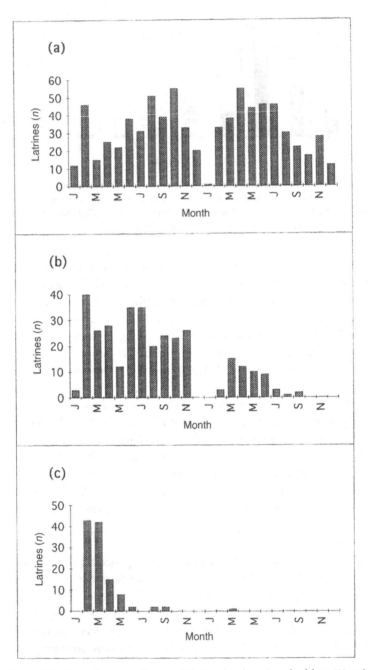

Figure 19.6. Seasonal patterns of water vole latrine abundance in each of three 600 m sites designated (a) 0 year, (b) 1st year, and (c) 2nd year of mink occupancy. Latrines were abundant in the absence of mink but disappeared soon after the arrival of mink.

(a)

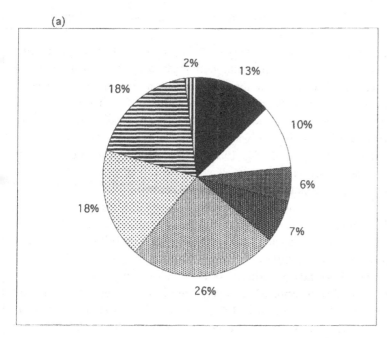

(b)

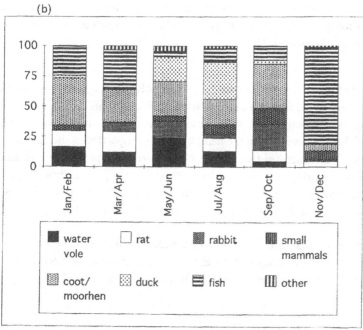

Figure 19.7. Diet of mink in the river Soar expressed as % dry weight from scat analyses: (a) overall figures, and (b) seasonal pattern.

bulk (13.2%). The undigested remains of coots, moorhens and ducks were almost 45% of the faecal volume.

The diet varied seasonally (Fig. 19.7(b)), with small mammal remains increasing from January to June and declining by late autumn. Duck remains occurred most frequently in the summer (it was not always possible to distinguish adult from fledgling remains). Fish remains in scats were at a maximum in the winter and spring. The proportion of water vole was, however, consistently high in winter months (16.3%) and very high in the May/June sample (24.2%) (Fig. 19.7(b)). The varying proportion of water voles in the scats through the year follows a similar trend to that deduced from the water vole distribution survey and latrine counts.

To test the consequences of the mink's impact upon water voles on the mink's subsequent diet, we compared the diet of mink in the 1st- and 2nd-year stretches. We did this by analysing a subsample of scats collected at an active mink den in sector 33 (1st-year stretch of river, with abundant water voles at the start of 1994) and comparing these to a sample of mink scats collected at an active mink den in sector 67 (2nd-year site of mink occupancy, with scant evidence of voles at the start of 1994). The mink scats were analysed in six bi-monthly time blocks. We used a one-tailed t-test on these monthly pairs of samples, to reveal that there was a significantly greater bulk of water voles (% dry weight) in the diet of mink in the 1st-year stretch of the River Soar ($t = 2.08$, $p < 0.05$, d.f. = 5).

Habitat

Figure 19.8 illustrates the location of different categories of sector in August with respect to the first two axes extracted by PCA. In this month, no sector had evidence of both mink and water voles. The first two axes accounted for 32% of the total variation in the habitat data set. Inspection of the pattern of factor loadings suggested that high first-axis scores were positively associated with high tree cover, with high shrub cover and with the presence of a towpath, and negatively associated with grass cover. High second-axis factor loadings were those for bank slope, the presence of reeds and the presence of bank piling.

Sections in which vole activity was recorded tended to be those with grassy banks, with low tree and shrub cover and absence of towpaths. Sections with only mink evidence tended to have low bank slope, no reeds and no piling.

Discussion

The arrival of American mink in both the Thames catchment and the Soar are associated with catastrophic declines in water vole numbers. At the level of the

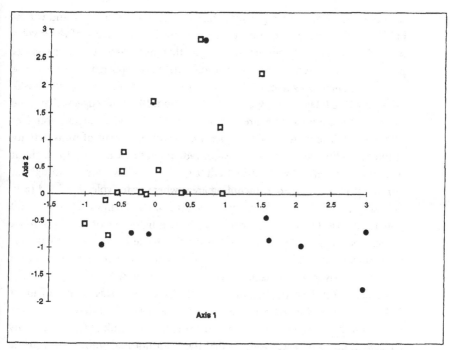

Figure 19.8. Location of different categories of sector in August with respect to the first two axes by Principal Components Analysis. In this month no sector had evidence of both mink and water voles. ●, mink; ☐, water vole.

catchment, our data from the Thames valley indicated that the vole decline had continued for over 20 years since the mink first arrived in the 1970s. However, as recently as 1990, mink were uncommon in the catchment and water voles were still found in three-quarters of the sites they had occupied earlier in this century. Our 1995 survey revealed a rapid increase in mink numbers and a catastrophic decline in sites at which water voles persisted. Closer inspection strongly suggested a causal basis for this correlation in the fortunes of mink and water voles, insofar as no site was found at which the two occurred together. Furthermore, in 1991 Halliwell & Macdonald (1996) live-trapped American mink along 20 km stretches of four rivers in the Thames catchment and found a negative correlation between the numbers of mink caught and the numbers of water vole signs. Our studies in the Thames, therefore, led us to conclude that, at least when mink numbers increase, their impact on water voles could be very rapid, leading to widespread losses within only 5 years. The question arises, within that time-scale, exactly how rapidly, and according to what mechanism, do mink exterminate voles?

Our results from the River Soar provide an answer: within one breeding season, resident mink may greatly reduce evidence of water voles, and within only 2 years eradicate them locally. This, and the high incidence of water voles in the diet of our study population, suggest that mink favoured water voles as prey. This might seem at odds with studies that have reported these voles as a rather infrequent prey item (Day & Linn, 1972; Chanin & Linn, 1980; Birks & Dunstone, 1984; Dunstone & Birks, 1987; Ireland, 1988); we suggest that these studies were conducted in areas where water voles were uncommon (see Chapter 11). The pattern we have observed involves a focus of mink activity around breeding dens, which thereafter radiates out as kits move throughout their maternal territory. Transient adult mink doubtless moved throughout our study area on the Soar, including areas where water voles persisted (and indeed this is also probable in many sections of the Thames catchment still populated by water voles). Furthermore, the mating system of American mink is such that males occupy ranges that overlap those of several females (our radio-tracking of mink on the Thames reveals some males roving over as much as 12 km of river, N. Yamaguchi, D. W. Macdonald & R. Strachan, unpublished data). Yet the extinction of water voles on the Soar was in units of about 2–4 km. We therefore suspect that the main agents of their destruction were female mink and their young. Furthermore, female mink may be especially effective hunters of water voles in that they, but not male mink, can squeeze into the vole's burrows. By combining these aquatic and fossorial skills, female mink comprise, from the water vole's standpoint, a devastating blend of otter (*Lutra lutra*) and stoat (*Mustela erminea*).

Our diet study reveals that the water vole was an important component of the diet of mink colonizing the River Soar, but as the mink became established the decline in water vole numbers was, unsurprisingly, associated with a lesser representation of water voles in mink diet. In a wider study of mink diet on 11 rivers in Derbyshire, Leicestershire, Staffordshire and Nottinghamshire over 1993–94, Strachan & Jefferies (1996) showed that water vole was the single most important species in the diet of colonizing mink. Indeed, in the May–June sample up to 32.2% of the volume of undigested prey remains consisted of water voles.

Principal Components Analysis reveals that mink on the Soar initially colonized sites that had distinct characteristics. We therefore envisage a mechanism whereby voles are initially eradicated from foci of mink activity and thereby confined to habitat less favoured by the incoming predators. It remains to be seen whether these vole sanctuaries are also less favoured by voles. These vole sanctuaries may be subject to predation by transient mink and perhaps resident males, but survive while spared the attentions of breeding females. We envisage that in due course, the colonizing mink will spread, occupying those habitats that were initially refuges for the voles. If breeding females occupy the river in contiguous territories, water voles will be eradicated directly. If 'islands' of sanctuary remain, unoccupied by breeding female mink, water voles may survive there, but fall victim ultimately to the stochastic forces to which fragmented populations are prone. The question arises, then, of what determines the extent to which breeding female mink colonize a river. Halliwell & Macdonald (1996) provide at least one answer to this in that the number of mink trapped in their study in a given stretch of river was correlated with the availability of suitable den sites in the form of pollarded willows.

Thus far we have focused on the seemingly devastating impact of American mink predation on water voles. However, the chronology of the water vole's decline as documented by Strachan & Jefferies (1993) indicates that it began before the arrival of mink. For example, out of 3096 sites occupied in 1900, only 1263 were still occupied in 1980 (Strachan & Jefferies, 1993). Indeed, in our Thames valley study area water voles were already reduced to three-quarters of their former sites by the mid 1970s, at which time mink were relatively uncommon in the catchment. These factors led Barreto *et al.* (1998) to explore factors in addition to mink that might be contributing to the water vole's decline. Amongst several seemingly detrimental features of land-use change this century, including widespread land drainage and river management systems affecting flooding patterns, they identified agricultural encroachment upon riverbank habitats as statistically very strongly linked to the loss of water voles. This led them to formulate the tightrope hypothesis, which suggests that agricultural intensification has reduced riverside habitat to a fragile, and often fragmented line – a tightrope, off which reduced numbers of water voles may be toppled by any or all of a variety of factors. As it happens, American mink have proliferated at a time when, through this linearization, water voles were particularly vulnerable, and the evidence we have presented here supports the view that mink are pushing the voles towards extinction. However, in blaming the mink for the vole's demise there is a risk of confusing the messenger with the message: we suggest that the ultimate cause of the water vole's predicament is not that people introduced an alien predator, but rather that people destroyed their habitat.

The foregoing interpretation raises two linked questions: first, under what conditions, if any, can water voles co-exist with American mink and, second, how does the introduction of the American mink affect the guild of semi-aquatic predators that prey on water voles? As yet, we can offer only speculative answers. First, it is clear that water voles and American mink do co-exist in at least some places and our prediction is that all these places will be typified by expansive stretches of riparian habitat. For example, in Belarus, where some river floodplains are pristine, water voles are the principal mammalian prey of both American and European mink (*Mustela lutreola*), otters (*Lutra lutra*) and European polecats (*Mustela putorius*) (see Chapter 11). There, where reed beds form extensive swathes around waterways, it may be that the arrival of a fourth mustelid predator has reduced water vole numbers, but there is no evidence that it is driving them to rarity. Second, the American mink arrived in the UK at a time when, by chance, other members of its guild were in a parlous state (and water voles in decline nonetheless). Polecats were endangered in the aftermath of the onslaught of late nineteenth century gamekeeping, and otters were rare or absent due to pollution. Instances of character displacement, including examples involving the American mink (Dayan & Simberloff, 1994) support the idea of intra-guild competition, and this may be expressed as spontaneous aggression between carnivores with overlapping niches, as apparently illustrated by aggression between red foxes (*Vulpes* sp.) and both pine marten (*Martes* sp.) (Lindström *et al.*, 1995) and Arctic foxes (Hersteinsson & Macdonald, 1992). Indeed, there is some evidence that American mink spontaneously harass European mink (see Chapter 17). Therefore, had the American mink arrived in Britain at a time when polecats and otters were populous, we might have expected it to face inter-specific competition and, at least in encounters with the otter, to be disadvantaged. This leads us to predict that with the recovery of these native mustelids, American mink numbers may decline. This scenario might have been advantageous for water vole populations if, as Sidorovich *et al.* (Chapter 11) suggest, water voles play a lesser role in the diet of polecat and otter than in that of American mink. However, if Barreto *et al.*'s (1998) tightrope hypothesis is correct, the partial ousting of mink by otters (and possibly a linked recovery of polecats) is unlikely to provide the water voles with respite: they will still be cornered in fragmentary slivers of habitat through which predators, endemic or alien, can sweep with lethal efficiency.

ACKNOWLEDGEMENTS
We gratefully acknowledge others of our mink team, especially Dorcas Walker and Nobuyuki Yamaguchi and the statistical expertise of Dr Paul Johnson. The work on the Soar and Amber formed part of the research programme of the Vincent Wildlife Trust and that on the Thames catchment was generously sponsored by the Environmental Agency (former National Rivers Authority), with additional support from the Peoples' Trust for Endangered Species and Tusk Force. G.R.B. was supported by The British Council.

References

Barreto, G. R., Macdonald, D. W. & Strachan, R. (1998). The tightrope hypothesis: an explanation for plummeting water vole numbers in the Thames catchment. In *United Kingdom floodplains*: 311–327 (Eds Bailey, R., Gose, P. V. & Sherwood, B. R.). Westbury Academic and Scientific Publishing, London.

Birks, J. D. S. & Dunstone, N. (1984). A note on prey remains collected from the dens of feral mink (*Mustela vison*) in a coastal habitat. *J. Zool., Lond.* 203: 279–281.

Chanin, P. R. F. & Linn, I. J. (1980). The diet of the feral mink (*Mustela vison*) in south-west Britain. *J. Zool., Lond.* 192: 205–223.

Conroy, J. W. H., Watt, J., Webb, J. B. & Jones, A. (1993). *A guide to the identification of prey remains in otter spraint.* The Mammal Society, London.

Day, M. G. (1966). Identification of hair and feather remains in the gut and faeces of stoats and weasels. *J. Zool., Lond.* 148: 201–217.

Day, M. G. & Linn, I. J. (1972). Notes on the food of feral mink *Mustela vison* in England and Wales. *J. Zool., Lond.* 167: 463–473.

Dayan, T. & Simberloff, D. (1994). Character displacement, sexual dimorphism and morphological variation among British and Irish mustelids. *Ecology* 75: 1063–1073.

Dunstone, N. & Birks, J. D. S. (1987). The feeding ecology of mink *Mustela vison* in coastal habitat. *J. Zool., Lond.* 212: 69–83.

Halliwell, E. C. & Macdonald, D. W. (1996). American mink *Mustela vison* in the Upper Thames catchment: relationship with selected prey species and den availability. *Conserv. Biol.* 76: 51–56.

Hersteinsson, P. & Macdonald, D. W. (1992). Interspecific competition and the geographical distribution of red and Arctic foxes *Vulpes vulpes* and *Alopex lagopus*. *Oikos* 64: 505–515.

Ireland, M. C. (1988). *The Behaviour and Ecology of the American Mink* Mustela vison *(Schreber) in a Coastal Habitat.* PhD thesis: University of Durham.

Jefferies, D. J., Morris, P. A. & Mulleneux, J. E. (1989). An enquiry into the changing status of the water vole *Arvicola terrestris* in Britain. *Mammal. Rev.* 19: 111–131.

Lawton, J. H. & Woodroffe, G. L. (1991). Habitat and the distribution of water voles: why are there gaps in a species' range? *J. Anim. Ecol.* 60: 79–91.

Lindström, E. R., Brainerd, S. M., Helldin, J. O. & Overskaug, K. (1995). Pine marten–red fox interactions: a case of intraguild predation? *Annls. zool. fenn.* 32: 123–130.

Maitland, P. S. (1972). *A key to the freshwater fishes of the British Isles, with notes on their distribution and ecology.* Freshwater Biological Association, Ambleside.

Strachan, C. & Jefferies, D. J. (1996). An assessment of the diet of feral American mink *Mustela vison* from scats collected in areas where water voles *Arvicola terrestris* occur. *Naturalist* 121: 73–81.

Strachan, R. & Jefferies, D. J. (1993). *The water vole* Arvicola terrestris *in Britain 1989–1990: its distribution and changing status.* The Vincent Wildlife Trust, London.

Teerinck, B. J. (1991). *Hair of west European mammals.* Cambridge University Press, Cambridge.

Ward, D. P., Smal, C. M. & Fairley, J. S. (1986). The food of mink *Mustela vison* in the Irish midlands. *Proc. R. Ir. Acad. (B)* 86: 169–182.

Wise, M. H. (1978). *The Feeding Ecology of Mink and Otters in Devon.* PhD thesis: University of Exeter.

Wise, M. H., Linn, I. J. & Kennedy, C. R. (1981). A comparison of the feeding biology of mink *Mustela vison* and otter *Lutra lutra*. *J. Zool., Lond.* 195: 181–213.

Woodall, P. F. (1993). Dispersion and habitat preference of the water vole (*Arvicola terrestris*) on the river Thames. *Z. Säugetierk.* 58: 160–171.

Woodroffe, G. L., Lawton, J. H. & Davidson, W. L. (1990). The impact of feral mink *Mustela vison* on water voles *Arvicola terrestris* in the North Yorkshire Moors National Park. *Biol. Conserv.* 51: 49–62.

20

Status, habitat use and conservation of giant otter in Peru

C. Schenck and E. Staib

Introduction

Despite being a highly endangered species, the giant otter nevertheless remains largely unstudied (Forster-Turley *et al.*, 1990). The current project was in-itiated in July 1990 by the Frankfurt Zoological Society with the objective of widening the knowledge and understanding of these rare animals and develop-ing a plan for their conservation. After 2.5 years of continuous fieldwork, the project was followed by an annual, 2 month field work period including a monitoring programme, and public relations work during the non-fieldwork period. This chapter provides an overview of the results of the research and conservation project on the giant otter.

Study area

The main part of the study was carried out in south-eastern Peru in the Department of Madre de Dios (Fig. 20.1). The size of the area is approximately 85 000 km², from the Andes in the west to the Brazilian and Bolivian borders. Most rivers are white-water rivers with low visibility, sandy beaches and frequent meanders that eventually erode into oxbow lakes. The Madre de Dios basin is characterized by several major rivers 200 m or more in width and a network of medium-sized and small streams. Water volumes vary considerably between the rainy season and the dry season. The annual average temperature is 24 °C with a maximum temperature of 36 °C and a minimum of 13 °C, influenced by cold Antarctic winds. Two main seasons are distinguished: October–April, with high temperatures and rainfall, and May–September, with lower temperatures and rainfall. Total annual rainfall amounts to approxi-mately 2000 mm in the lower areas. The major part of the area is below 500 m altitude. Evergreen tropical forest with canopies at 40 m above ground domi-nates the river flood plain in the area. Earlier successional stages of vegetation stretch along rivers and oxbow lakes with tracts of seasonally inundate swamp forest, marshes and *Mauritia* palm stands (Terborgh, 1983, 1992; Erwin, 1990).

The biodiversity of the area is extremely high (Terborgh, 1992). With an

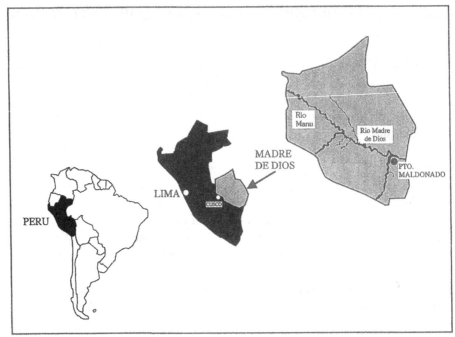

Figure 20.1. The study area: Department Madre de Dios in Peru. PTO., Puerto.

average of 0.8 human inhabitants/km^2, the area is exceptionally sparsely inhabited. Nevertheless regions with settlers and gold miners and clear-cutting of forest and agriculture are found.

Two main study areas were chosen: the Manu National Park and the Tambopata river. Manu, which covers 18 000 km^2, is one of the biggest national parks in the world, and has been protected for more than 20 years. In contrast, the Tambopata river is much more exposed to human influence. The entire Madre de Dios catchment was also extensively investigated.

The status of giant otter in Peru

Local people, natives, hunters, fishermen and park rangers were interviewed to obtain initial information on the status and the past and current distribution of the otter. Information obtained was interpreted carefully: exaggeration and a tendency to report past rather than current events are common problems, as is the confusion of the giant otter with the Neotropical river otter, *Lutra longicaudis*.

Giant otter prefer clear waters without much current (Duplaix, 1980; Laidler, 1984). Given the hydrographic conditions in Peru, the most suitable areas are cochas or oxbow lakes – river segments cut off from the main river. Long-term observation of selected groups showed that otter prefer oxbow lakes (Staib, in prep.). The Otorongo group, for example, which was subject to 403 h of direct observation, was never seen in the respective part of the river, even though an observer was present at the river most of the time. The preference of cochas was also obvious while carrying out the Manu Monitoring Program. Six surveys were conducted between 1990 and 1995, covering 230 river kilometres and 14–18 cochas per year. This census led to a total of 225 otter observations in cochas but only 19 sightings in the river. As a consequence of these initial results, surveys concentrated on cocha habitat. Satellite maps were used to locate the cochas and trails were cut from the main river through dense vegetation. Lake and shoreline surveys were conducted using an inflatable canoe. Travelling slowly along the shore, indirect signs such as tracks, spraint marking places and dens were located. The otters clear these areas of vegetation through repeated trampling. Active scent marks were recognized by a strong odour. Giant otters dig their dens in areas of non-inundated high forest close to the edge of the lake (Duplaix, 1980; Laidler, 1984; Staib, in prep.). All cochas were surveyed for active marking places. After six annual surveys, average activity was calculated for each cocha. If otters were inhabiting a lake, they almost invariably approached the boat, circled it, periscoped and sounded their warning vocalization – a snort. Otters were counted and individual throat patch coloration was recorded using a video camcorder in order to avoid double-countings during the census (Fig. 20.2).

A total of 63 lakes and 8 rivers in Peru were investigated. The occurrence of indirect signs was positively correlated with the number of otters observed during the census ($p < 0.05$; $r = 0.7723$, $n = 23$, Spearman–Rank correlation coefficient) (Fig. 20.3). To compare different river systems, the average number of groups per river kilometre was calculated. The existence of a group was established either by direct observation or by recent indirect signs. The results of surveys in different river systems and the quality of local habitat for otters are shown in Table 20.1. This provides an impression of the current and historical giant otter densities in the Department Madre De Dios (Figs. 20.4 and 20.5). Today the areas best-known for giant otter – Manu National Park is an example – are not areas of highest habitat quality. Areas with high giant otter density no longer exist, due to human influence. The results are confirmed by reports from local people of high otter densities on the lower Madre de Dios river in the past. The results also explain the historically high numbers of exported giant otter skins (23 980 pelts from 1946 to 1973, annual average

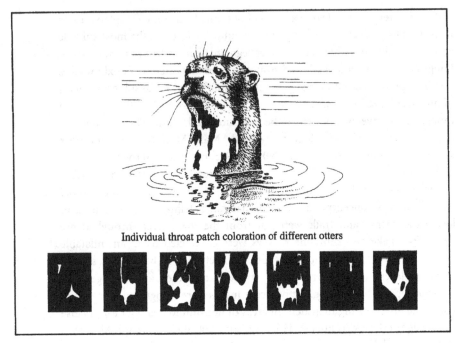

Individual throat patch coloration of different otters

Figure 20.2. A giant otter periscoping and sounding the warning snort. The individual form of the throat patch is visible.

2248 pelts: Mason & Macdonald, 1986). The current distribution of otter in the area, even in the national parks, is far below the densities necessary to account for the quantities of fur harvested in the past.

Nevertheless it is likely that some otters stay in inaccessible swampland and inhabit smaller river systems that are inaccessible. For example, the otter group of Cocha Otrongo spent 65.8% of their hunting time in semi-aquatic vegetation (Staib, in prep.). The annual surveys carried out in Manu National Park, covering the main part of otter habitat in that area, recorded an average of 41 giant otters per year ($n = 6$, max. = 45; min. = 33; $sd = 4.1$). The estimated population of giant otters is below 70.

Survey and long-term observation of selected groups showed that family groups are territorial and stay in the same area for several years (Schenck & Staib, 1994; Staib, in prep.). In Manu National Park four groups have been observed in the same respective lakes over the past 5 years (Staib, in prep.). The Cocha Salvador group consisted of 4–7 otters during the 5 year period from 1990–95 and these lived in the 108.3 ha lake all year round. Other groups used additional swampland and creeks beside the open water bodies of the lakes,

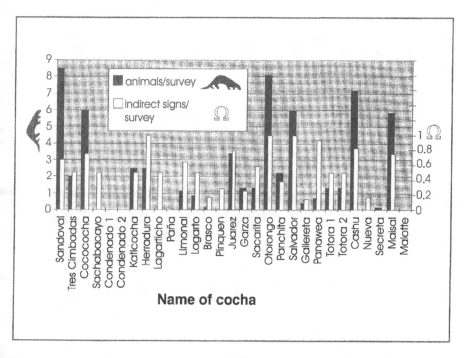

Figure 20.3. Indirect signs and direct observation of giant otters in different lakes of the study area.

Table 20.1. *Habitat characteristics and giant otter densities*

Name of river system	River stretch surveyed (km)	River width (m)	Average cochas/ 100km of river	Giant otter density[†]	Estimated human activity[‡]
Madre de Dios	432	>400	8.6	0.009	3
Manu	227	150–200	7.5	0.044	1–2
Azul	57	50	0.0	0.000	2
Amigos	100	50–80	14.0	0.030	1
Tambopata	130	180–250	4.0	0.030	2–3
Heath	125	100	14.5	0.024	2–3

[†]Density given as average number of groups per kilometre of river.

[‡]1, low; 2, middle; 3, high.

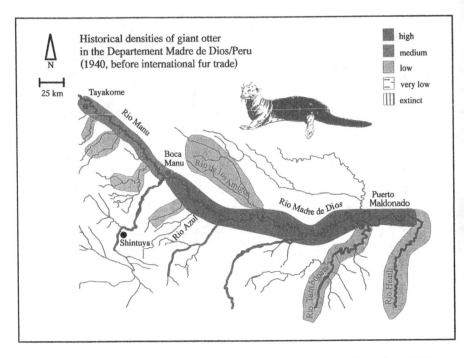

Figure 20.4. Historical densities of giant otter in the Department Madre de Dios/Peru (1940, before international fur trade). The densities are graded by taking into account the available giant otter habitat, in particular the size and number of oxbow lakes, and from reports by local people. Only rivers surveyed during the study are listed.

making it difficult to determine the exact territory size (Schenck, 1996). Using the lake size as an indicator, territory sizes of 55.2 ha for the Cocha Cashu group (4–10 animals), 80.4 ha for the Otorongo group (7–10 animals) and 62.4 ha for the Juarez group (2–6 animals) were determined. A density of 1 animal/15.5–21.7 ha was calculated for the Salvador group. Territories of terrestrial predators such as jaguars (*Panthera onca*) are larger than those of giant otters by a factor of 150 (Rabinowitz & Nottingham, 1986). The occurrence of giant otters in their natural habitat is characterized by low densities over large areas and high local densities in their preferred habitat.

Habitat

The rivers and lakes in tropical rain forest showed conductivity values of 7.5 to 298 μS/cm, with an average of 136.3 μS/cm ($n = 29$; sD = 91.4). In comparison, waters in Central Europe have conductivities of up to 1000 μS/cm (Reichholf,

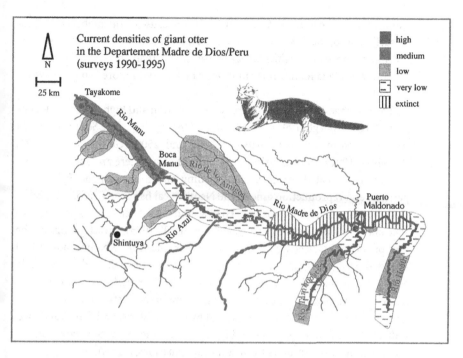

Figure 20.5. Current densities of giant otter in the Department Madre de Dios/Peru (surveys 1990–1995). The densities are graded by taking into account the available giant otter habitat, in particular the size and number of oxbow lakes, and from personal observations. Only rivers surveyed during the study are listed.

1991), and thermal springs on the coast of Peru have conductivities of more than 10 000 μS/cm (Instituto Geologico Minero y Metalurgico; W. Marche, personal communication 1992). Gill net fishing samples showed that rivers and lakes in tropical rain forest are extremely rich in fish (catch per unit effort (CPUE): $\bar{x} = 57.2$ g/m^2/h; max. = 366.4 g/m^2/h; SD = 62; $n = 118$). The average sampling time was 76 min; the established sampling time in most studies exceeded 12 h. For comparison with CPUE values from other studies, the values obtained in our Peruvian study were doubled (Schenck, 1996). The maximum value for CPUE in the study area was three times the value for comparable artificial lakes in the Danube river system near Geisling in Germany (K. Seiffert, personal communication 1993). Values from Lake Malapuzha in south India were exceeded 5 times by the Peruvian average CPUE and 24 times by the Peruvian maximum CPUE (Cowx 1991). The typical food chain does not exist, because of low levels of dissolved minerals in the water. The fish live mainly on organic input from the forest. Falling leaves, fruits and insects represent important sources of food for aquatic animals in tropical forest ecosystems (Goulding, 1980). Because of the higher shoreline-to-surface ratio, the lack of shoreline erosion and the close proximity of trees to the water, the fish densities in lakes are higher than in rivers. This would explain why otters mainly hunt in lakes, which contain more fish and have no current.

The shortage of nutrients enforces specialization and high diversity (Reichholf, 1991; Terborgh, 1992). The number of fish species in the South American tropics is estimated to be between 2000 and 2500 (Geisler, *et al.*, 1971; Reichholf, 1991). In comparison, in Central Europe there are only about 265 species. The great diversity in size, feeding and movement habits, activity patterns and reproduction among the Neotropical fish fauna enforces specialization among fish hunters as well. High abundance of fish enables predators to specialize even in a species-rich system (Reichholf, 1975). Investigations on the diet of the giant otter also showed high specialization. Khanmoradi (1994) analysed about 80 000 scales from giant otter spraints collected in the study area. Only two species (*Satanoperca jurupari* and *Prochilodus caudifasciatus*) made up over 70% of the diet. A significant difference between occurrence of species in the otter spraints, determined by scale analysis, and fish abundance, indicated by gill netting, was found ($p < 0.01$; Kolmogorov–Smirnov-test; 118 gill net samples, 60 spraints from communal marking places).

The high density of prey is one of the basic conditions facilitating the organization of giant otters in territorial groups.

Conservation

Giant otters are subject to severe environmental pressures. The main problem is the conversion of tropical rain forests; in Peru primarily for cattle ranching, agriculture and the exploitation of wood and mineral resources (e.g. oil). The rivers are the most important transport routes, and the colonization and destruction of tropical rain forest starts along big rivers. Since this is their preferred habitat, giant otters are severely affected by such developments. The Giant otter is a bio-indicator for the quality of water systems in the South American rain forest.

Hunting for the international fur trade brought the giant otter close to extinction (Mason & Macdonald, 1986). Although protected by national laws and international treaties for more than 20 years, illegal hunting is still a problem for small and isolated populations. In spite of strict bans otters are killed occasionally for sport, because they are seen as competitors for fish, or more rarely for their pelts. In rare cases adults are killed in order to capture the cubs for pets. Between 1990 and 1992 five skins of recently killed animals were found in the study area. The knowledge that giant otters are endangered animals strictly protected by law is not widely spread in the remote areas of the Peruvian rain forest.

Heavy metal contamination of the environment presents a problem even in remote areas. Gold mining is common in parts of the Amazon and was practised in the Department Madre de Dios for nearly 20 years (Yábar, 1991). It was estimated that the separation of gold dust from river sediments introduces 10–30 tons of mercury per year into the Madre de Dios area (Gutleb *et al.*, 1993, 1997). Of 34 fish samples taken from different rivers, seven samples contained mercury in concentrations of more than 0.5 mg/kg (Gutleb *et al.*, 1997). The allowable daily intake of mercury for humans, based on FAO/WHO recommendations, is 0.03 mg (Mason & Macdonald, 1986). Contaminated fish were found up to 100 km upstream of mineral extraction. Fish may travel great distances and mercury may also be transported by dust in the air.

Domestic animals present another possible threat. It is known that parvovirus and distemper, common diseases of cats and dogs, can have serious effects on wild animal populations (Thorne & Williams, 1988). Giant otter cubs held in captivity died of parvovirus and all mustelids are very susceptible to the canine distemper virus (Wünnemann, 1992; M. Woodford personal communication, 1995). Dogs are common in villages – even inside protected areas. In 1995 blood samples were taken from 16 dogs in and around Manu National Park. Five dogs had parvovirus antibody concentrations showing an infection and one dog had a significant concentration of distemper antibodies. Natives travel with their dogs to the cochas for fishing, thus allowing infection of otters. In addition, solitary otters searching for a new partner travel great distances, covering territories of other otter groups, and thus provide an effective vector for the transmission of parvovirus or distemper.

For specialist piscivores such as the giant otter, overfishing by man results in a lack of food. The use of gill nets, dynamite and other intensive fishing methods leads to a rapid depletion of fish stocks. Only the traditional methods such as bow and arrow or hook are adapted to the abundance of fish in tropical waters and allow for sustainable use.

Tourism is another factor affecting the giant otter – particularly within well-protected areas such as national parks. Until now several thousand tourists per year have been visiting the more remote areas, but tourism in tropical rain forests is increasing quickly and has quadrupled in the Manu reserve over the past 3 years (Pasheco del Castillo, 1995; Dunstone & O'Sullivan, 1996). Giant otters are a special attraction for tourists. The same attributes that made them accessible for fur hunting in the past make them a tourist attraction today. Unlike other otter species, they are active by day, live in groups and are very impressive to observe. Giant otters are already mentioned in advertising materials in tourist agencies and are found on T-shirts and logos. Giant otters are endangered by three different aspects of tourism due to their extreme sensitivity to disturbance:

1   River traffic with large boats equipped with powerful outboard engines may destroy a sub-optimal habitat. Otters were observed to flee into the forest after being surprised by such boats. During the survey, shoreline dens and otter groups were located only in remote areas of the upper Rio Manu, more than 150 km from the Rio Madre de Dios.
2   Canoe trips on cochas are a frequent part of tourist tours. Giant otters first react to boats exactly as they do to big predators such as caimans. They approach the canoe, circle and sound their warning snort. Later they try to chase away the intruder. However, this does not work with tourists who, on the contrary, often paddle closer to the otters to take better photographs. It seems that the otters regard the canoe as a super-predator, leave and sometimes even desert the lake entirely.
3   Such disturbances can have a catastrophic effect when otters raise cubs. It is the experience of the Hagenbeck Zoo in Germany that females will stop lactating when stressed (Wünnemann, 1992). Stress due to tourism was also the probable cause of the loss of litters in 1991 in two different otter groups in Manu National Park. The Lake Salvador otter group – an area in which tourism is most intense – failed to raise cubs for 3 years. Unfortunately, giant otters have cubs at the beginning of the dry season, the same time that the tourist season begins.

Habitat destruction, illegal hunting, heavy metal contamination, diseases, overfishing and tourism – the giant otters is subject to nearly all of the conservation problems that currently exist in the tropics. However, the conflict between giant otter and people is not insoluble. It is possible to counter the main endangering factors through information, education and control. Controlling tourism can serve as an example to reduce the impact of such factors. As a part of the Frankfurt Zoological Society project, the different administrative bodies, beginning with the ministry in Lima (INRENA – Instituto Nacional de Recursos Naturales) were given information. Also local park authorities (e.g. MNP, Manu National Park; ZRTC, Zona Reservada Tambopata Candamo; SNPH, Santuario Nacional Pampas del Heath) were convinced that a conservation management plan in necessary. Alternatives to harmful canoe trips were tested, including the construction of observation towers at the edges of different lakes. This was followed by the production of educational material such as leaflets for tourists and information booklets for schools, park posts and tourist agencies. Training courses for park rangers and seminars for tourist guides were also provided. Bilingual information panels designed by Peruvian artists now outline local guidelines for conduct in the vicinity of otters. As a result of pressure on the agencies, only two lakes in Manu National Park are opened for tourist canoe trips today. The canoe traffic on Lake Otorongo has been halted and a temporary observation tower has been built. The otter group at this lake has been reproducing normally since these measures were implemented. Controlled by a permanent park ranger, a local reserve was established at Lake Salvador.

Educational work show that the giant otter can serve as a flagship species in the conservation of the last remaining tropical rain forest.

ACKNOWLEDGEMENTS

The project was financed by the Frankfurt Zoological Society – Help for Threatened Wildlife and was carried out in co-operation with the Munich Wildlife Society. The Gottfried-Daimler und Karl-Benz-Stiftung supported the project with two fellowships. The participation in the Symposium of the Zoological Society of London was made possible by help from the Stifterverband der Deutschen Wissenschaften. Special thanks are due to the most important collaborator, Jesus Huaman.

References

Cowx, I. G. (Ed.) (1991). *Catch effort sampling strategies and their application in freshwater fisheries management.* Fishing News Books, Oxford.

Dunstone, N. & O'Sullivan, J. N. (1996). The impact of ecotourism development on rainforest mammals. In *The exploitation of mammal populations*: 313–333. (Eds. Taylor, V. J. & Dunstone, N.). Chapman & Hall, London.

Duplaix, N. (1980). Observation on the ecology and behaviour of the giant river otter *Pteronura brasiliensis* in Suriname. *Terre Vie* **34**: 496–620.

Erwin, T. L. (1990). Natural history of the carabid beetles at the BIOLAT Biological Station, Rio Manu, Pakitza, Peru. *Revista peru. Ent.* **33**: 1–85.

Forest-Turley, P., Macdonald, S. M. & Mason, C. M. (Eds) (1990). *Otters: an action plan for their conservation.* IUCN, Gland.

Geisler, R., Knöppel, H. A. & Sioli, H. (1971). Ökologie der Süsswasserfische Amazoniens. Stand und Zukunftsaufgaben der Forschung. *Naturwissenschaften* **58**: 303–311.

Goulding, M. (1980). *The fishes and the forest. Explorations in Amazonian natural history.* University of California Press, Berkeley.

Gutleb, A., Schenck, C. & Staib, E. (1993). Total mercury and methylmercury levels in fish from the Department Madre de Dios, Peru. *IUCN Otter Spec. Group Bull.* No. 8: 16–18.

Gutleb, A., Schenck, C. & Staib, E. (1997). Mercury contamination of fish – a risk for giant otters? *Ambio* **26**: 511–514.

Khanmoradi, H. (1994). *Unter suchungen zur Nahrungsökologie der Riesenotter* (Pteronura brasiliensis) *in Peru.* Diplomarbeit: Universität München.

Laidler, P. E. (1984). *The Behavioural Ecology of the Giant River Otter in Guyana.* PhD thesis: University of Cambridge.

Mason, C. F. & Macdonald, S. M. (1986). *Otters: ecology and conservation.* Cambridge University Press, Cambridge.

Pacheco del Castillo, J. (1995). *Diagnostico de la actividad turistica en la reserva de la biosfera del Manu.* ProNaturaleza, Cusco.

Rabinowitz, A. R. & Nottingham, B. G. Jr (1986). Ecology and behaviour of the jaguar (*Panthera onca*) in Belize, Central America. *J. Zool., Lond. (A)* **210**: 149–159.

Reichholf, J. H. (1975). Biogeographie und Ökologie der Wasservögel im subtropisch-tropischen Südamerika. *Anz. orn. Ges. Bayern* **14**: 1–69.

Reichholf, J. H. (1991). *Der tropische Regenwald.* Deutscher Taschenbuch Verlag, München.

Schenck, C. (1996). *Vorkommen, Habitatnutzung und Schutz des Riesenotters* Pteronura brasiliensis *in Peru.* Doktorarbeit: Universität München.

Schenck, C. & Staib, E. (1994). *Die Wölfe der Flüsse. Riesenotter und ihr Lebensraum Regenwald.* Knesebeck Verlag, München.

Staib, E. (in prep.). *Verhaltensökologie der Riesenotter* (Pteronura brasiliensis) *in Peru.* Doktorarbeit: Universität München.

Terborgh, J. (1983). *Five New World primates: a study in comparative ecology.* Princeton University Press, Princeton.

Terborgh, J. (1992). *Diversity and the tropical rain forest.* Scientific American Library, New York.

Thorne, E. T. & Williams, E. S. (1988). Disease and endangered species: the black-footed ferret as a recent example. *Conserv. Biol.* **2**: 66–74.

Wünnemann, K. (1992). *Das Verhalten von Landraubtieren mit überdurchschnittlicher Cephalisation.* Doktorarbeit: Tierärztliche Hochschule Hannover.

Yábar, R. C. (1991). *Extracción irracional de recursos y propuesta alternativa de explotación via la agrioindustria del platano. El caso de la provincia del Manu.* Tesis Bachiller: Universidad Cusco, Peru.

Index of subjects and authors

Index of animals and plants